DISORDERS OF PERIPHERAL NERVES

EDITION 2

CONTEMPORARY NEUROLOGY SERIES AVAILABLE:

Fred Plum, M.D., *Editor-in-Chief*
Series Editors: Sid Gilman, M.D.
Joseph B. Martin, M.D., Ph.D.
Robert B. Daroff, M.D.
Stephen G. Waxman, M.D., Ph.D.
M-Marsel Mesulam, M.D.

DISORDERS OF PERIPHERAL NERVES

EDITION 2

HERBERT H. SCHAUMBURG, M.D.

Professor and Chairman
Department of Neurology
Albert Einstein College of Medicine and
Montefiore Medical Center
Bronx, New York

ALAN R. BERGER, M.D.

Associate Professor of Neurology
Albert Einstein College of Medicine
Director of Electromyography
Montefiore Medical Center
Bronx, New York

P. K. THOMAS, CBE, M.D., D.Sc., FRCP, FRC PATH

Professor of Neurology
Royal Free Hospital School of Medicine and
    the Institute of Neurology
University of London
London, England

 F. A. DAVIS COMPANY • Philadelphia

Printed in the United States of America

Last digit indicates print number: 10 9 8 7 6 5 4 3 2 1

NOTE: As new scientific information becomes available through basic and clinical research, recommended treatments and drug therapies undergo changes. The author(s) and publisher have done everything possible to make this book accurate, up-to-date, and in accord with accepted standards at the time of publication. The authors, editors and publisher are not responsible for errors or omissions or for consequences from application of the book, and make no warranty, expressed or implied, in regard to the contents of the book. Any practice described in this book should be applied by the reader in accordance with professional standards of care used in regard to the unique circumstances that may apply in each situation. The reader is advised always to check product information (package inserts) for changes and new information regarding dose and contraindications before administering any drug. Caution is especially urged when using new or infrequently ordered drugs.

Library of Congress Cataloging-in-Publication Data

Schaumburg, Herbert H., 1932–
  Disorders of peripheral nerves/Herbert H. Schaumburg, Alan R.
Berger, P.K. Thomas.—Ed. 2
      p. cm.—(Contemporary neurology series   36)
  Includes bibliographical references and index.
  ISBN 0-8036-7734-0 (hardbound:alk. paper)
  1. Nerves, Peripheral—Diseases. I. Berger, Alan R., 1954–
II. Thomas, P. K. (Peter Kynaston), 1926– III. Title.
IV. Series.
  [DNLM: 1. Peripheral Nerve Diseases. W1 CO769N v. 36/WL 500
S313d]
  RC409.S33 1991
  616.8′7—dc20
  DNLM/DLC
  for Library of Congress   91-24890

PREFACE TO THE SECOND EDITION

The favorable reception afforded the first edition and the abundant new information on peripheral nerve diseases encouraged the authors to undertake this second edition. It contains three new chapters: *Diagnosis and Assessment, Rare and Poorly Validated Neuropathies,* and *Rehabilitation in Peripheral Neuropathies.* Two conditions not widely seen in 1983—AIDS and Lyme disease—are discussed in detail in Chapter 6, *Infectious and Granulomatous Neuropathy;* each of the other chapters has been significantly augmented and updated.

New illustrations of histopathology have been added and Dr. Walter Bradley's lucid figures illustrating basic electrodiagnostic principles from *Disorders of Peripheral Nerves* (Blackwell Scientific, 1974) have been further modified and reproduced. We are especially grateful to the late Dr. Webb Haymaker and to Drs. Cedric Raine, Richard Johnson, Donald Price and John Griffin for allowing us to reproduce photomicrographs and diagrams.

We are pleased to acknowledge the following individuals who assisted us in this project: Dr. Rosalind King, Monica Bischoff, Tina Rubano, Patricia Vacchelli, and Kieran Price.

Dr. Sylvia Fields, Ms. Bernice Wissler, and the editorial and publishing staff at F.A. Davis have offered expert guidance and encouragement.

PREFACE TO THE FIRST EDITION

"Disease of the peripheral nervous system stands as one of the most difficult subjects in neurology. Since the structure and function of this system are relatively simple, one might suppose that our knowledge of its diseases would be complete. Such is not the case. At present a suitable explanation cannot be offered in about 40 percent of patients who enter a general hospital with a peripheral nerve disease (usually of chronic progressive type) and the pathologic changes have not been fully determined in any one of them. Moreover, the physiologic basis of many of the neural symptoms continues to elude experts in the field."[*]

The authors do not dispute this unhappy state of affairs, although the figure of 40 percent can perhaps now be improved upon by intensive investigation in special centers.[†] However, it is our conviction, supported by years of undergraduate medical teaching and experience with individuals outside the neurosciences, that a working knowledge of the common peripheral neuropathies can be mastered by any motivated physician.

This volume evolved from the primary author's 1972 syllabus for the undergraduate and graduate courses on peripheral nerve disease at the Albert Einstein College of Medicine. The favorable reception accorded the syllabus, and increasing requests for its enlargement, indicated the need for a concise, elementary monograph on peripheral neuropathy intended for individuals engaged in the practice of general medicine and neurology. Drs. Spencer and Thomas agreed to collaborate in this endeavor; their abundant, scholarly contributions to this volume have justified the decision to challenge the considerable obstacles posed in creating an international, multiauthored text.

---

[*]Adams, RD and Victor, M: Principles of Neurology, ed 2, McGraw-Hill, New York, 1981.
[†]Dyck, PJ, Oviatt, KF and Lambert, EH: Intensive evaluation of unclassified neuropathies yields improved diagnosis. Ann Neurol 10:222, 1981.

We are pleased to acknowledge the following individuals who assisted us in this project: Monica Bischoff, Laurell Edwards, Elaine Garafola, Larry Markowitz, and Patricia Vacchelli. We are especially grateful to Dr. Webb Haymaker for allowing us to reproduce his diagrams of peripheral nerves and their segmental innervation.

Dr. Sylvia Fields, Mr. Richard Heffron and their editorial and publishing staff at F.A. Davis have generously offered expert guidance and encouragement.

CONTENTS

xviii

Part I

**CONCEPTS,
CLASSIFICATION,
AND DIAGNOSIS**

Chapter 1

BASIC CONCEPTS AND GLOSSARY

DEFINITION OF THE PERIPHERAL NERVOUS SYSTEM (PNS)

The PNS may be defined as those portions of motor neurons, autonomic neurons, and primary sensory neurons that extend outside the central nervous system (CNS) and are associated with Schwann cells or ganglionic satellite cells. The concept of a separate peripheral nervous system is obviously artificial, since the cell bodies of many PNS motor neurons lie within the CNS and some peripheral sensory neurons have extensive central projections. The justification for this concept stems from several notions, two of which are especially relevant to this book: one is the predilection for many diseases primarily to affect the PNS, and the other is its ability, in contrast to the CNS, to regenerate.

The PNS, so defined, usually includes the dorsal and ventral spinal roots, spinal and cranial nerves (with the exception of the first and second cranial nerves), dorsal root and other sensory ganglia, sensory and motor terminals, and the bulk of the autonomic nervous system.[2] The connective tissue and vasculature of peripheral nerve have several unique features, including the perineurial and blood–nerve barriers, and play a major role in PNS disorders. Lymphatic vessels are present in the epineurium but not within the fascicles. The salient components of a peripheral nerve are diagramed in Figures 1–1 to 1–4.

RELATIONSHIPS FUNDAMENTAL TO AN UNDERSTANDING OF DISEASE OF MYELINATED PERIPHERAL NERVE FIBERS

Neuron Cell Body and Axon

The axons of peripheral nerves, despite their occasional great length, are simply cytoplasmic extensions of the nerve cell bodies. The volume of cytoplasm in a long myelinated axon is actually far greater than that of the neuron cell body region (perikaryon). PNS axons derive most of the proteins essential for maintenance and function from ribosomes in the nerve cell body. There are no ribosomes in the axon.

Figure 1–3 illustrates prominent cytoskeletal-structural elements (neurotubules and intermediate neurofilaments), which are synthesized by free polyribosomes, aligned, slowly transported down the axon at a rate of 0.2–3 mm per day and disassembled at the terminal.[4,5] Slow transport is exclusively anterograde, its propulsive mechanism unknown (Fig. 1–3). Intermediate neurofilaments convey structural conformity to the axon; they occupy and appear to organize considerable axonal space. Changes in number of neurofilaments directly influence axonal calibre, local accumulations cause axonal swelling (e.g., hexane and CS_2 neuropathies), while depletion results in atrophy (e.g., HMSN and uremic neuropathies). Neurotubules, although

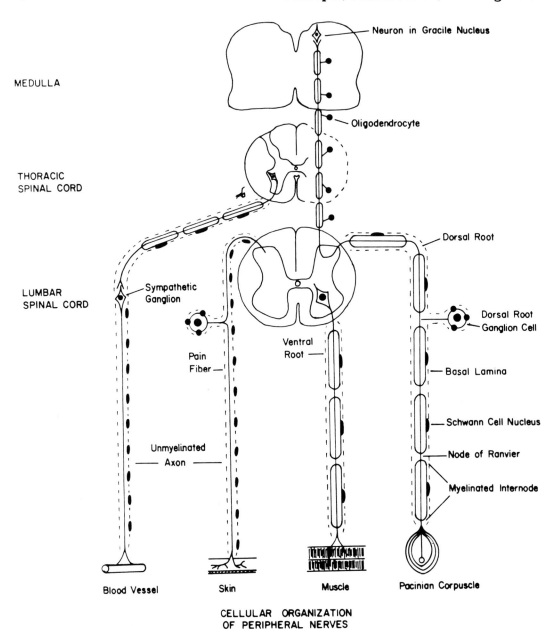

Figure 1–1. The principal components of the peripheral nervous system.

themselves slowly transported, are directly responsible for maintenance of bi-directional *fast* transport (Fig. 1–3). Alterations in neurotubules by agents that depolymerize tubulin (vinca alkaloids) or increase assembly (taxol) disrupt rapid transport.

Small vesicles and particulate organelles, containing proteins essential for membrane maintenance and transmitter function, are synthesized by ribosomes of the rough endoplasmic reticulum and glycosylated by the Golgi apparatus (Fig. 1–3). They are packaged into vesicles and transported anterogradely rapidly along neurotubules at a rate of 400 mm per day. Anterograde vesicular transport is mediated by the

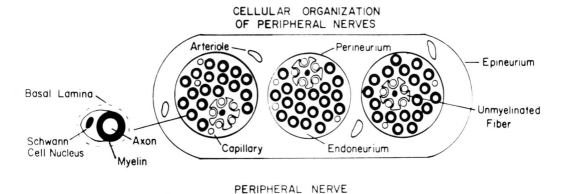

CELLULAR ORGANIZATION
OF PERIPHERAL NERVES

PERIPHERAL NERVE
COMPONENTS

Figure 1–2. A diagram of a peripheral nerve in cross section. The nerve contains three fascicles. The figure on the left represents a high magnification of a myelinated axon in cross section.

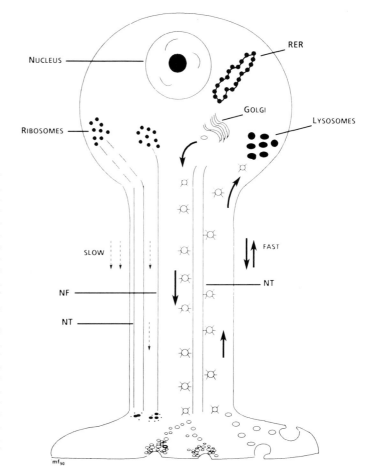

Figure 1–3. Schematic diagram of anterior horn cell and axon illustrating salient features of functional anatomy. Structures depicted on the left, neurofilaments (NF) and neurotubules (NT) are assembled by free polyribosomes and slowly transported and disassembled at the axon terminal. Structures on the right are assembled by rough endoplasmic reticulum (RER) and the Golgi apparatus, then rapidly transported to the terminal for use, recycling, and subsequent retrograde transport. (Adapted from Price, DL and Griffin, JW: Structural substrate for protein synthesis and transport in spinal motor neurons. In Andrews, J and Johnson, R: Amyotrophic Lateral Sclerosis: Recent Research Trends. Academic Press, New York, 1976.)

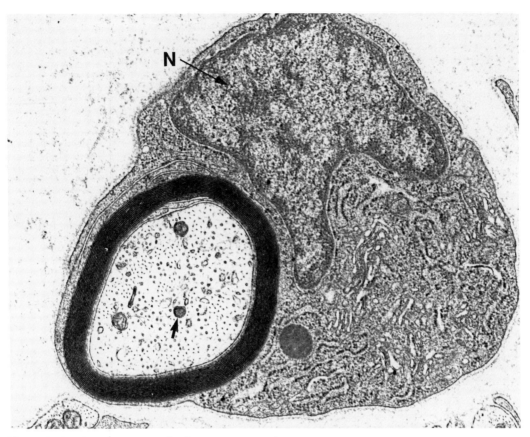

Figure 1—4. Electron micrograph of a cross section of a myelinated axon surrounded by a nucleated (*N*) Schwann cell. The short arrow indicates one of the three axonal mitochondria present in this section. A cluster of five microtubules is at the left of this arrow. Punctate intermediate neurofilaments and filagreed endoplasmic reticulum appear scattered throughout the axon (×35,000). (From Raine, CS: Morphology of myelin and myelination. In Morell, P (ed): Myelin. Plenum Press, New York, 1984.)

propulsive protein, kinesin, depicted in Figure 1–3 as small side arms on vesicles.[10] There is also retrograde rapid transport along neurotubules, at about 200 mm per day, of lysosomes and multivesicular bodies. Retrograde transport returns, for lysosome processing, much of the recycled membrane previously delivered anterogradely, and conveys nonneuronal material (e.g., nerve growth factor, herpes simplex virus, tetanus toxin) from the periphery to the cell body.[4,5] The propulsive mechanism for retrograde transport is undetermined. Some of the toxins described in Chapters 17 and 18 causing distal axonopathy affect bi-directional rapid transport.

In general, injury to the distal portion of the axon does not result in perma-nent damage to the nerve cell body; the latter undergoes transient swelling and breakdown of the endoplasmic reticulum (chromatolysis), but usually survives and supports regeneration of the damaged axon.[6] The converse is not true; severe damage to the nerve cell body or disruption of proximal axonal integrity results in rapid degeneration of the entire distal portion (see later discussion of wallerian degeneration).

Axon, Schwann Cell and Myelin

Schwann cells envelop axons to form unmyelinated and myelinated fibers surrounded by a basal lamina. PNS myelin is derived from the Schwann cell and is dependent both on the Schwann cell itself and the axon for its continued integrity (Fig. 1–4). A single Schwann cell occupies each myelinated internode and almost never associates itself with more than one axon.[1] Death of the axon results in the prompt breakdown of myelin but not the Schwann cell. The opposite is not true; loss of myelin does not usually result in disruption of the axon. This principle is of fundamental importance to an understanding of PNS disorders. An axon denuded of several segments of myelin simply awaits Schwann cell division and remyelination before resuming normal impulse conduction.[7]

Axon and End Organ

The effect of axonal transection on muscle is dramatic. Within weeks or months the muscle undergoes progressive denervation atrophy and will not recover unless reinnervated.[8] The loss of the normal *trophic* effect of nerve on muscle is widely accepted. Less certain are the other alleged trophic functions of nerve for skin, blood vessels, and subcutaneous tissue. Prolonged denervation results in changes of these tissues (red skin, ulcers) sometimes attributed to loss of maintenance function provided by the nerve fiber. Such changes may merely represent the effects of trauma and autonomic dysfunction on these tissues.[9]

Wallerian Degeneration and Axon Regeneration

The morphologic events following a focal crush injury to peripheral nerve are depicted in Figure 2–9 in Chapter 2. Transection of a nerve fiber results in total degeneration of the axon(s) and myelin distal to the site of injury (wallerian degeneration).[3] Within four days the entire distal axon and myelin become fragmented and electrical conduction ceases. The nerve cell body undergoes a chromatolytic reaction, and subsequently, regenerating axonal sprouts emerge from the injured axons at the site of injury. After one week, the distal Schwann cells have divided and are arranged in columns inside their tubes of basal lamina. If regenerating axons reach one of these Schwann cell columns, they can regenerate steadily toward the terminal and be myelinated by the waiting Schwann cells. Injuries that do not disrupt connective tissue continuity of a nerve (closed injuries) often have a good prognosis, since regenerating axons usually arrive at their former peripheral terminations, guided by the pre-existing Schwann-cell columns. Injuries that transect fascicles or the entire nerve, such as knife wounds, are frequently associated with ineffective or aberrant regeneration. Many sprouting axons may never reach the distal stump, but grow in an aberrant fashion (see Figure 2–10, Chapter 2) or a random tangled fashion (traumatic neuroma).

GLOSSARY

Current clinical terminology in peripheral nerve disease appears in Table 1–1.

Table 1–1 GLOSSARY OF CURRENT CLINICAL TERMINOLOGY IN PERIPHERAL NERVE DISEASE

AIDP (acute inflammatory demyelinating polyradiculoneuropathy) A subacute, self-limited, monophasic, immune-mediated, widespread, predominantly motor, demyelinating condition. AIDP is synonymous with the Guillain-Barré syndrome.

Axonal neuropathy (axonopathy) Any PNS condition characterized by the initial appearance of histopathological changes in the axon, followed by myelin degeneration and muscle denervation. Commonly, this process initiates at ends of long, large-diameter fibers (distal axonopathy).

CIDP (chronic inflammatory demyelinating polyradiculoneuropathy) A chronic, progressive, demyelinating, sensorimotor condition. CIDP may pursue either a relapsing or a chronic progressive course.

(continued)

Table 1—1 *(Continued)*

Demyelinating neuropathy (myelinopathy) Any PNS condition characterized by the initial appearance of histopathological changes in myelin or the Schwann cell, followed by demyelination of several or multiple internodal segments and slowed (or blocked) nerve conduction.

Dysesthesias and paresthesias Unpleasant or unusual sensations (burning, tingling, formication) that accompany or herald some PNS disorders. Dysesthesias may be either spontaneous or provoked by stimuli. Previously, dysesthesias referred to evoked sensations only, spontaneous sensations were called paresthesias; the terms are now often used interchangeably.

Electrodiagnostic studies (EDS) Includes electromyography (EMG) and measurements of nerve conduction (NCV), and late responses (F and H responses).

Focal and multifocal neuropathy Indicates local involvement of one or more individual peripheral nerves; it is equivalent to the more cumbersome terms *mononeuropathy* and *multiple mononeuropathy* (mononeuropathy multiplex).

Guillain-Barré syndrome See AIDP.

HMSN I and II A group of slowly progressive, hereditary, distal demyelinating and axonal neuropathies commonly accompanied by skeletal deformity. These heterogeneous disorders are approximately equivalent to Charcot-Marie-Tooth disease.

Hyperesthesia Increased sensitivity to stimuli, usually with an unpleasant quality. The term *hyperpathia* refers to an exaggerated response to a normally painful stimulus, *allodynia* to pain produced by a stimulus such as touch that is normally not painful.

Hypesthesia Decreased sensitivity to stimuli (numbness), synonymous with *hypoesthesia*.

Large-fiber neuropathy PNS disorders characterized by loss of position, vibration, and touch-pressure sensibility, tendon areflexia, and lower motor neuron involvement. Sensory ataxia and pseudoathetosis may be prominent if muscle power is preserved.

Negative symptoms Complaints of loss of function or sensibility (e.g., weakness and numbness).

Neuronopathy Any PNS condition where the initial histopathological changes are manifest in the cell body region (perikaryon, cyton). Neuronopathies may be either sensory (herpes zoster) or motor (poliomyelitis).

Neuropathy (peripheral neuropathy) A broad term including any disorder—infective, toxic, metabolic—affecting the PNS. It replaces the older term *peripheral neuritis*.

Plexopathy (plexitis) Designating disease which is confined to either the lumbar or brachial plexuses.

(continued)

Polyneuropathy (symmetrical polyneuropathy) A generalized process producing widespread and bilaterally symmetrical effects on the peripheral nervous system. It may be motor, sensory, sensorimotor, or autonomic in its effects.

Positive symptoms Complaints of abnormal spontaneous sensations or movement, for example, tingling, fasciculations.

Quantitative sensory testing (QST) The use of biomedical devices that accurately measure thermal or vibratory-touch senses in the distal limbs.

Radiculopathy Disease confined to one (mono) or more (poly) spinal roots.

Small-fiber neuropathies Conditions in which there is prominent disturbance of small myelinated and unmyelinated fibers characterized by diminished pain and temperature sensation, often with spontaneous pain, and autonomic involvement. There is relative preservation of strength, tendon reflexes, and sensory modalities subserved by the larger myelinated fibers (touch-pressure, vibration, joint position).

REFERENCES

1. Berthold, E-H: Morphology of normal peripheral axons. In Waxman, SG (ed): Physiology and Pathobiology of Axons. Raven Press, New York, 1978.

2. Carpenter, MB: Human Neuroanatomy, ed 8. Williams & Wilkins, Baltimore, 1983.

3. Donat, JR and Wisniewski, HM: The spatio-temporal pattern of Wallerian degeneration in mammalian peripheral nerves. Brain Res 53:41, 1973.

4. Griffin, JM and Watson, DF: Axonal transport in neurological disease. Ann Neurol 23:3, 1988.

5. Ochs, S and Worth, RM: Axoplasmic transport in normal and pathological systems. In Waxman, SG (ed): Physiology and Pathobiology of Axons. Raven Press, New York, 1978.

6. Price, DR and Porter, KR: The response of ventral horn neurons to axonal transection. J Cell Biol 53:24, 1975.

7. Raine, CS: Pathology of demyelination. In Waxman, SG (ed): Physiology and

Pathobiology of Axons. Raven Press, New York, 1978.

8. Sunderland, S: Nerves and Nerve Injuries, ed 2. Churchill Livingstone, Edinburgh, 1978.

9. Thomas, PK: Clinical features and differential diagnosis. In Dyck, PJ, Thomas, PK, and Lambert, EH (eds): Peripheral Neuropathy, Vol II. WB Saunders, Philadelphia, 1984, p 1169.

10. Vale, RD, Schnapp, BJ, Reese, TS, et al: Organelle, bead, and microtubule translocations promoted by soluble factors from the squid giant axon. Cell 40:559, 1985.

Chapter 2

ANATOMICAL CLASSIFICATION OF PERIPHERAL NERVOUS SYSTEM DISORDERS

SYMMETRICAL GENERALIZED NEUROPATHIES (POLY-NEUROPATHIES)
FOCAL (MONONEUROPATHY) AND MULTIFOCAL (MULTIPLE MONONEUROPATHY) NEUROPATHIES

The authors endorse an anatomical classification of disorders (Table 2–1) of the PNS based on whether the condition is characterized by generalized symmetrical or focal involvement.[16] This simple classification stresses the site of apparent primary pathologic change and does not suggest the pathophysiologic mechanism.[6] For example, although demyelination is a feature of uremic neuropathy, it is clearly secondary to changes in the axon, and uremic neuropathy is considered as an axonopathy. This classification generally lends itself to clinical-pathologic and electrodiagnostic correlation and is especially useful when initially evaluating a pa-

Table 2–1 ANATOMIC CLASSIFICATION OF PERIPHERAL NEUROPATHY

TWO OVERALL TYPES:
I. Symmetric Generalized
II. Focal and Multifocal

I. Symmetric Generalized Neuropathies (Polyneuropathies)

Distal Axonopathies	Toxic—many drugs, industrial and environmental chemicals
	Metabolic—uremia, diabetes, porphyria, endocrine, deficiency-thiamine, pyridoxine
	Genetic—HMSN II
	Malignancy associated—small-cell carcinoma of lung, multiple myeloma
Myelinopathies	Toxic—diphtheria, buckthorn
	Immunologic—acute inflammatory polyradiculoneuropathy (AIDP) chronic inflammatory demyelinating polyradiculoneuropathy (CIDP)
	Genetic—Refsum's disease, metachromatic leukodystrophy
Neuronopathies	
Somatic motor	Undetermined—amyotrophic lateral sclerosis
	Genetic—hereditary motor neuronopathies
Somatic sensory	Infectious—herpes zoster neuronitis
	Malignancy-associated sensory neuronopathy syndrome
	Toxic—pyridoxine sensory neuronopathy
	Undetermined—subacute sensory neuronopathy syndrome
Autonomic	Genetic—hereditary dysautonomia (HSN IV)

II. Focal (Mononeuropathy) and Multifocal (Multiple Mononeuropathy) Neuropathies
    Ischemia—polyarteritis, diabetes, rheumatoid arthritis
    Infiltration—leukemia, lymphoma, granuloma, schwannoma, amyloid
    Physical injuries—severance, focal crush, compression, stretch and traction, entrapment
    Immunologic—brachial and lumbar plexopathy

---

tient with a peripheral nerve disorder. Exceptions occur: vasculitic and demyelinating neuropathies may eventuate in distal symmetric patterns of dysfunction, and toxic axonopathies and neuropathies may vary in pattern and tempo depending on dose and rate of administration.[4,12,21]

SYMMETRICAL GENERALIZED NEUROPATHIES (POLYNEUROPATHIES)

Axonopathy (Central-Peripheral Distal Axonopathy, Proximal Axonopathy)

This is the most common morphologic reaction of the PNS and CNS to exogenous toxins, and probably also underlies many metabolic and hereditary neuropathies.[17] The biochemical mechanisms and pathophysiology of most axonopathies are poorly understood. Most human axonopathies are distal, but proximal and radicular axonopathies may be encountered, for example, in diabetic and porphyric neuropathy.[16]

HYPOTHETICAL MECHANISMS

In distal axonopathy, a metabolic abnormality initially occurs in the cell body and/or throughout the axon. Eventual failure of axon transport results in degeneration of vulnerable distal regions of axons.[5,13] Long and large diameter fibers are usually first affected, although the reason is unclear. Degeneration appears to advance proximally toward the nerve cell body (dying-back) as long as the metabolic abnormality persists; its reversal allows the axon to regenerate along the distal Schwann cell tube to the appropriate terminal. An identical sequence usually occurs simultaneously in the distal ends of long CNS axons (e.g., dorsal columns, corticospinal and optic tracts), although regeneration is less effective.[14] The distal vulnerability of these axons is also an enigma. Two important determinants are distance from the cell body and fiber diameter.

CARDINAL PATHOLOGIC FEATURES
(see Figure 2–1)

1. Initial distal axonal changes may be generalized or multifocal; the nature of the change may be characteristic of the disorder. Atrophy (dwindling) and focal swelling are especially common.

2. Eventual axonal disintegration resembles wallerian degeneration; the myelin sheath breaks down concomitantly with the axon. Secondary demyelination and remyelination may occur where the axon is still intact. This frequently accompanies axonal atrophy.

3. Distal muscles undergo denervation atrophy.

4. Nerve cell chromatolysis may occur in severe cases.

5. Schwann cells and basal lamina tubes remain in distal nerves and facilitate appropriate peripheral regeneration.

6. Astroglial proliferation triggered by distal axonal degeneration may impede regeneration in CNS.

CLINICOPATHOLOGIC CORRELATIONS

1. Gradual insidious onset: chronic metabolic disease or prolonged, low-level intoxication usually produce prolonged subclinical disease with signs and symptoms gradually appearing later. Biochemical and physiologic axonal abnormalities precede fiber degeneration in some subclinical cases, and likely account for their rapid recovery. High-level intoxications are associated with subacute onset and agents that disrupt fast axoplasmic transport, for example, Vacor, are associated with acute onset.

2. Initial findings frequently in the lower extremities: large and long axons are usually affected early, thus the fibers of sciatic nerve branches are especially vulnerable.

3. Stocking-glove sensory and motor

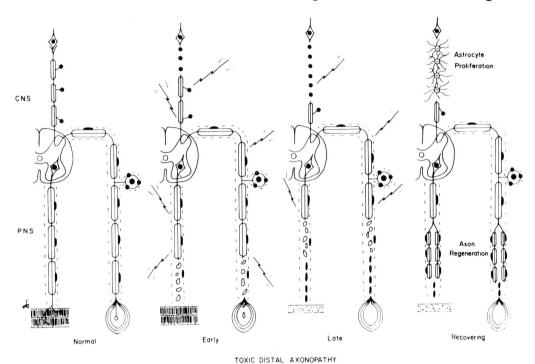

TOXIC DISTAL AXONOPATHY

Figure 2–1. A diagram of the cardinal pathologic features of a toxic distal axonopathy. The jagged lines (lightning bolts) indicate that the toxin is acting at multiple sites along motor and sensory axons in the PNS and CNS. Axon degeneration has moved proximally (dying-back) by the late stage. Recovery in the CNS is impeded by astroglial proliferation.

loss: axonal degeneration commences distally and slowly proceeds toward the neuron cell body resulting in symmetric, distal clinical signs in the legs and arms. The earliest symptoms are usually sensory: toe-tip sensations of tingling or pinprick are common initial complaints. The pattern of sensory loss is depicted in Figure 2–2.

4. Early and symmetric loss of ankle jerks: the axons supplying the calf muscles are of extremely large diameter and are among the first affected in experimental acrylamide and hexacarbon neuropathies.

5. Normal to mildly slowed motor nerve conductions: in contrast to the demyelinating neuropathies, where the motor nerves or roots are diffusely affected. Since some motor fibers remain intact in the axonal neuropathies, motor nerve conduction velocity may remain normal or only slightly slowed despite clinical signs of neuropathy.

Sensory amplitudes are frequently diminished with only mild slowing. Exception: severe impulse slowing may accompany distal axonopathies in which the axon swells and demyelinates focally.

6. Normal CSF protein level: since the pathologic changes are usually distal and the nerve roots spared, most patients with axonal neuropathies have a normal or only slightly elevated CSF protein value.

7. Slow recovery: since axonal regeneration (in contrast to remyelination) is a very slow process, proceeding at a rate of 2 to 3 mm per day, recovery may take many months, several years, or may never completely occur. Function is restored in reverse order to the sequence of loss.

8. Coasting: following withdrawal from toxic exposure, symptoms and signs may intensify for weeks before recovery commences. This does not imply

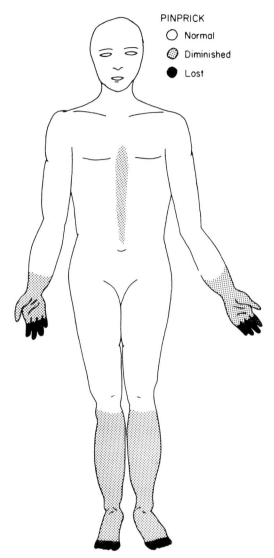

PINPRICK

○ Normal
◍ Diminished
● Lost

Figure 2–2. Stocking-glove pattern of sensory loss of an advanced stage of distal axonopathy. The area of diminished sensation over midthorax (*cuirass distribution*) reflects involvement of distal ends of intercostal nerves.

persistent body burden of toxin but likely reflects continued axonal degeneration and reconstitution.

9. Signs of CNS disease: this has been encountered in individuals recovering from certain toxic neuropathies. Most toxic central-peripheral distal axonopathies are characterized by tract degeneration of the distal extremities of long, large diameter fibers in the CNS pari passu with changes in the PNS. Thus, the clinical signs of degeneration in the corticospinal and spinocerebellar pathways are usually not prominent features early in the illness. However, on recovery from the neuropathy, the patient may manifest hyperreflexia, Babinski's responses, and a stiff-legged, ataxic gait.

Myelinopathy

The term myelinopathy, when applied to the PNS, refers to conditions in which the lesion primarily affects myelin or the myelinating (Schwann) cell. Thus, the moderate segmental demyelination that accompanies some axonal disorders is not evidence of a primary myelinopathy. Stated another way, segmental demyelination (internodal loss of myelin) is not always synonymous with myelinopathy.

AIDP, an immune-mediated inflammatory, demyelinating neuropathy, is the only frequently encountered disease that primarily affects PNS myelin (see Chapter 5). Toxic and infectious myelinopathies (see Chapters 6, 17, and 18) and hereditary disorders of Schwann cell lipid metabolism (see Chapter 14), are rare. The sequences of morphologic change operant in several myelinopathies have been thoroughly studied and are, in general, well understood. The enzymatic abnormalities in several of the hereditary conditions have been elucidated (see Chapter 14).

HYPOTHETICAL MECHANISMS

It is generally held that the segmental demyelination of spinal roots and nerves in the AIDP results from an immune-mediated attack on PNS myelin.[2] The precipitating event or antigen is not known, but the subsequent pathologic events mirror those of experimental allergic neuritis, a condition produced in animals by immunization against peripheral myelin.[20]

The reported segmental demyelination of diphtheritic neuropathy results from toxic inhibition of Schwann cell synthesis of myelin constituents.[9] By contrast, most well-studied myelinopathies appear to be characterized by a primary attack on the myelin itself. The Schwann cell perikaryon survives in most of these conditions and retains the ability to divide and form new myelin.

CARDINAL PATHOLOGIC
FEATURES
(see Figure 2–3)

1. Primary destruction of the myelin sheath occurs, usually leaving the axon intact.

2. Initial attack on myelin mediated by inflammatory cells.

3. Often begins at nodes of Ranvier.

4. Spinal roots are usually heavily involved, but destruction also affects multiple sites in nerve.

5. The Schwann cell divides and remyelinates the axon to form short internodes of thin myelin.

6. Muscle often does not undergo denervation change, but may undergo disuse atrophy if paralysis is prolonged. Axonal loss may occur in primary demyelinating disorders and is occasionally profound. The explanation for this is unclear.

7. Should repeated demyelination occur, Schwann cells divide again, and some of the daughter cells are unable to find a segment of axon to remyelinate. These surplus or supernumerary Schwann cells accumulate around the axons and may form concentric multiple rings—onion bulbs.

CLINICOPATHOLOGIC
CORRELATIONS

1. Onset: in toxic and inflammatory myelinopathies, the process of segmental demyelination occurs over a period of hours, days, or weeks.

2. Initial changes may occur in the lower extremities, but not always distally: the diffuse process may occasion-

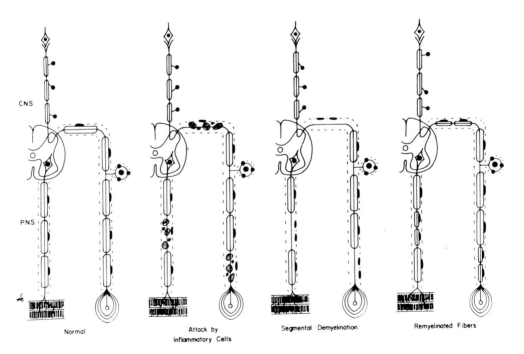

CNS

PNS

Normal   Attack by   Segmental Demyelination   Remyelinated Fibers
            Inflammatory Cells

PRIMARY MYELINOPATHY

Figure 2–3 A diagram of the cardinal pathologic features of an inflammatory PNS myelinopathy. Axons are spared as is CNS myelin. Following the attack, the remaining Schwann cells divide. The denuded segments of axons are remyelinated, leaving them with shortened internodes.

ally become manifest in the short cranial nerves, but more commonly, the nerves to the lower extremities are initially involved. Presumably this occurs because the myelinated axons of the sciatic nerve are longest, contain the most myelin, and are statistically most likely to be involved in a random demyelinating process.

3. Generalized weakness with mild sensory loss: the large-diameter, heavily myelinated motor axons and ventral roots are involved, resulting in diffuse symmetric weakness or paralysis of the extremities and bulbar muscles. Relative sensory sparing may reflect in part the continued function of small-diameter myelinated and unmyelinated fibers. Sensory ataxia may occur from involvement of proprioceptive afferent fibers. The patterns of sensory and motor loss are illustrated in Figure 2–4.

4. Absent tendon reflexes in all extremities: both the afferent and efferent limbs of the monosynaptic stretch reflex are mediated by large-diameter myelinated fibers, especially vulnerable in the toxic and inflammatory myelinopathies. Generalized areflexia is a characteristic of these conditions.

5. Marked slowing of nerve conduction: the widespread demyelination prolongs conduction and may also give rise to conduction block. Conduction velocity in remyelinated fibers with thin myelin sheaths is reduced.

6. Elevated CSF protein: inflammatory and toxic demyelination heavily involve the spinal roots, with leakage of protein into the surrounding subarachnoid space.

7. Rapid recovery: recovery is dependent on remyelination to restore impulse conduction. Effective remyelination of an internode may take only a few weeks and clinical recovery may be dramatic.

8. No signs of CNS disease: most toxic and inflammatory PNS myelinopathies spare the CNS for various reasons. One is that many myelinotoxic agents are unable to cross the blood-brain barrier; another is that many inflammatory conditions are immune-mediated and the response is directed

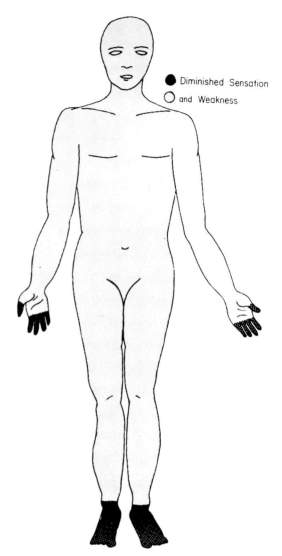

Figure 2–4. Pattern of motor and sensory loss in a severely involved case of AIDP. There is diffuse weakness of limb, intercostal, and facial muscles. Sensory impairment is usually mild and involves only the distal portions of the limbs.

at substances present in peripheral myelin.[2] Some hereditary metabolic diseases of myelin (leukodystrophies) have profound CNS involvement (see Chapter 14).

Neuronopathy

The term *neuronopathy* describes conditions in which the initial morphologic or biochemical changes occur in the neuron cell body.[1] Clinical manifestations of PNS neuronopathies are restricted to the segments innervated by the affected cell bodies. They may be focal, involving one segment, e.g., herpes zoster; multifocal, involving multiple sensory or motor segments, e.g., Sjögren's syndrome and poliomyelitis; or diffuse, e.g., from pyridoxine and doxorubicin. They are a heterogeneous, poorly understood group of conditions and, in the broadest sense, include many disorders of motor, sensory, and autonomic neurons. They may commence prenatally or in infancy, adolescence, or adult life. Infectious neuronopathies include familiar conditions such as poliomyelitis, HIV associated, and herpes zoster ganglionitis. An idiopathic type of diffuse sensory neuronopathy may follow nonspecific infections and some connective tissue diseases, e.g., Sjögren's syndrome, are associated with a multifocal sensory neuronopathy

syndrome. The hereditary motor neuronopathies (spinal muscular atrophies) and some hereditary sensory and autonomic neuropathies (e.g., familial dysautonomia) are generally conceptualized as PNS neuronopathies (see Chapter 15). *Massive doses of intravenous pyridoxine cause a toxic neuronopathy.* Toxic PNS sensory neuronopathies are readily produced in experimental animals by doxorubicin and methyl-mercury.[12] Motor and sensory neuronopathy syndromes have been reported as remote complications of carcinoma.

HYPOTHETICAL MECHANISMS

No single mechanism explains the pathophysiology of these heterogeneous conditions. Indeed, even when the pathologic changes are obvious, as in some of the infectious or hereditary conditions, there is as yet no rationale

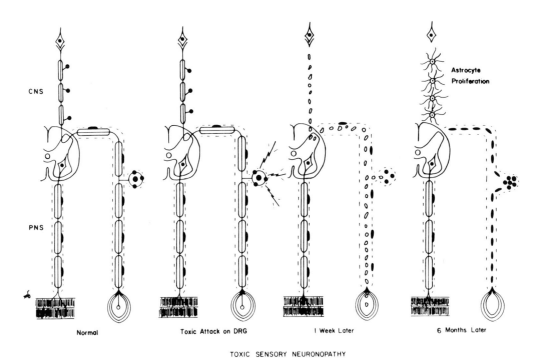

TOXIC SENSORY NEURONOPATHY

Figure 2–5. A diagram of the cardinal features of a rapidly involving toxic sensory neuronopathy. The jagged lines (lightning bolts) indicate that the toxin is directed at neurons in the dorsal root ganglion (DRG). Degeneration of these cells is accompanied by fragmentation and phagocytosis of their peripheral-central processes. The Schwann cells remain; there is no axonal regeneration.

for these events. Studies of experimental megadoses of pyridoxine in PNS diffuse sensory neuronopathy indicate that the pathogenesis and evolution of the changes are best understood as initial disruption of metabolism of sensory nerve cells followed rapidly by degeneration throughout the length of their processes.[7] The dorsal root and gasserian ganglion neurons are believed to be particularly vulnerable to some circulating toxins because of the special permeability of their blood vessels.

CARDINAL PATHOLOGIC FEATURES (EXPERIMENTAL PYRIDOXINE DIFFUSE TOXIC SENSORY NEURONOPATHY) (Fig. 2–5)

1. Circulating pyridoxine leaks through the normally fenestrated blood vessels in dorsal root autonomic ganglia.
2. Pathologic changes appear in the neuronal perikaryon, soon followed by degeneration throughout the length of the axon.
3. Motor cells are not affected and muscle undergoes no change.
4. Regeneration cannot occur and sensory loss is therefore permanent.

CLINICOPATHOLOGIC CORRELATIONS (ACUTE MEGADOSE PYRIDOXINE-INDUCED SYNDROME)[1]

1. Rapid or subacute onset following massive intravenous administration.
2. Initial sensory loss may occur anywhere: characteristic of this disorder is the early appearance of numbness of the face coincident with diffuse sensory loss in the limbs. Presumably this occurs because gasserian ganglion neurons are affected simultaneously with dorsal root ganglion neurons.
3. Diffuse sensory loss and ataxia with preservation of strength: the loss of sensation, sensory ataxia and dysesthesia reflect the disappearance of sensory neurons. In most subacute sensory neuronopathies, large-fiber modalities

are heavily affected so that the proprioceptive deficit is greater than pain or thermal sense loss. Sparing of anterior horn cells accounts for preservation of strength. The pattern of sensory loss is depicted in Figure 2–6.
4. Absent tendon reflexes: one of the characteristics of this condition that reflects the large fiber sensory loss.

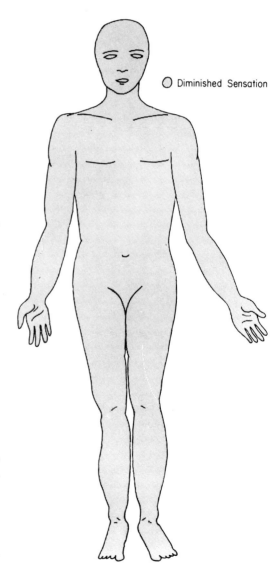

Diminished Sensation

Figure 2–6. The pattern of sensory loss in an advanced stage of the diffuse sensory neuronopathy syndrome. Sensation is diminished, often markedly, throughout. This distribution reflects widespread destruction of sensory ganglion neurons.

5. Normal motor nerve conduction, abnormal or absent sensory conduction studies: this mirrors the pattern of selective nerve cell loss.

6. Variable recovery: this reflects the death of the nerve cell body and consequent permanent loss of axons. Some cells may be only slightly impaired and transiently function poorly, but are able to reconstitute themselves without losing their axons. The phenomenon of collateral sprouting from surviving axons may account for what variable recovery occurs in these conditions.

7. No signs of CNS disease: the pure PNS sensory neuronopathy syndrome is not accompanied by CNS degeneration aside from fiber loss in the central projections of the sensory neurons (dorsal columns). However, some sensory neuronopathy syndromes (for example, carcinomatous sensory neuronopathy, HIV) accompany pathologic processes that involve the CNS as well (see Chapters 6 and 11).

FOCAL (MONONEUROPATHY) AND MULTIFOCAL (MULTIPLE MONONEUROPATHY) NEUROPATHIES

Ischemia

The PNS, unlike the CNS, is usually minimally affected by large-vessel disorders. The principal reason for this resistance is the richly collateralized blood supply of peripheral nerve. In general, ischemia of peripheral nerve is synonymous with widespread small artery or arteriolar disease and is most frequently associated with the necrotizing vasculitides and diabetes mellitus (see Chapter 10).

PATHOGENETIC HYPOTHESIS

There is considerable controversy surrounding the nature and mechanism of vascular injury to peripheral nerve. It is generally held that in the vasculitides the nerve fiber damage results from local ischemia.[3]

CARDINAL PATHOLOGIC FEATURES

1. Compromise of several small arteries at one level in a nerve results in ischemia to an entire segment of nerve (mononeuropathy). Occasionally, multiple levels of several nerves may be simultaneously affected, resulting in a diffuse patchy neuropathy (multiple mononeuropathy). The lesions may summate to produce bilaterally distal symmetric involvement mimicking a distal axonopathy.

2. Axonal degeneration occurs in many fibers and wallerian-like degeneration appears below the level of ischemia. Central fascicular degeneration is often pronounced.

3. Infarct necrosis is rare and connective tissue elements usually are spared.

4. Muscles undergo denervation atrophy.

5. Collateral circulation begins.

6. Regenerative potential is usually good (especially in diabetes mellitus) because of intact connective tissue. The vasculitides may have a poor prognosis because of continuing arteriolar necrosis and involvement of other organs.

CLINICOPATHOLOGIC CORRELATION (DIABETIC MULTIPLE MONONEUROPATHY SYNDROME) (see Chapter 7)

1. Rapid onset is characteristic but not invariable, possibly reflecting occlusion of vessels. Pain frequently accompanies this neuropathy, often local and probably related to ischemia of the nervi nervorum.

2. Initial findings are in the distribution of the ischemic nerves. The distribution of sensory loss in a typical case of multiple mononeuropathy is depicted in Figure 2–7.

3. Weakness is more striking than sensory loss: this may reflect the rela-

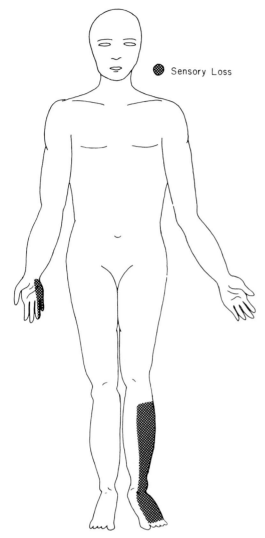

Sensory Loss

Figure 2–7. An illustration of the scattered distribution of sensory loss in ischemic multiple mononeuropathy, with involvement of contralateral ulnar and peroneal nerves.

any diabetic patient and does not reflect ischemia to the distal nerves.

7. Gradual recovery: reflects the slow rate of axonal regeneration and will vary inversely with the locus of the ischemia, that is, the more distal lesion will recover sooner. Mild lesions, featuring predominantly segmental demyelination may recover rapidly, (e.g., as in lesions of the third cranial nerve).

Infiltration

This heterogeneous group includes conditions that disrupt the continuity of nerve fibers and connective tissue and may eventually totally destroy the internal architecture of a nerve. Leprosy, amyloidosis, sarcoidosis, leukemic and lymphomatous infiltrates, perineurial xanthoma, schwannoma, and sensory perineuritis are examples.

MECHANISM

Each condition produces secondary effects on nerve fibers. Most are subacute conditions and randomly destroy fibers. Some, especially the granulomas, give rise to an inflammatory response which, in concert with fibroblast proliferation, totally disrupt axons and Schwann-cell tubes. Eventually, segments of nerve fascicles are converted into bundles of scar tissue through which regenerating fibers cannot pass.

CARDINAL PATHOLOGIC FEATURES (LEPROMATOUS LEPROSY)
(see Chapter 6)

1. Formation of granulomas in vulnerable cutaneous nerves.
2. Axons are disrupted and Schwann cell tubes disorganized at level of granuloma.
3. Wallerian degeneration occurs distal to the level of granuloma, resulting in anesthetic skin.
4. Reactive connective tissue proliferation prevents axonal regeneration.

tive resistance of small myelinated and unmyelinated sensory axons to ischemia. Pain may persist for several weeks.

4. Reflex loss is in distribution of affected nerves: this probably reflects the vulnerability of large-diameter myelinated fibers to ischemia.

5. Motor and sensory nerve potential amplitudes are diminished or abolished. Spontaneous activity reflecting denervation atrophy may be prominent.

6. CSF protein may be elevated in

CLINICOPATHOLOGIC
CORRELATION IN
LEPROMATOUS LEPROSY
(see Chapter 6)

1. Gradual onset: this reflects the indolent granulomatous response to the bacilli.

2. Predominant involvement of superficial cutaneous nerves: the granulomas mainly develop in superficially situated nerves, as *M. leprae* bacilli proliferate more rapidly at lower temperatures. The manifestations are therefore predominantly sensory. Leprosy bacilli initially colonize Schwann cells, especially those associated with unmyelinated and small myelinated ax-

ons, resulting in selective pain and temperature sensory loss and anhidrosis.

3. Permanent anesthesia: the granulomatous lesion totally destroys the architecture of the nerve.

4. Nerve entrapment may occur because of the granulomatous enlargement of nerve trunks.

5. CSF protein normal: the spinal roots are not involved in this disease of nerve.

Physical Injuries

Nerves are very susceptible to the effects of externally applied pressures. In

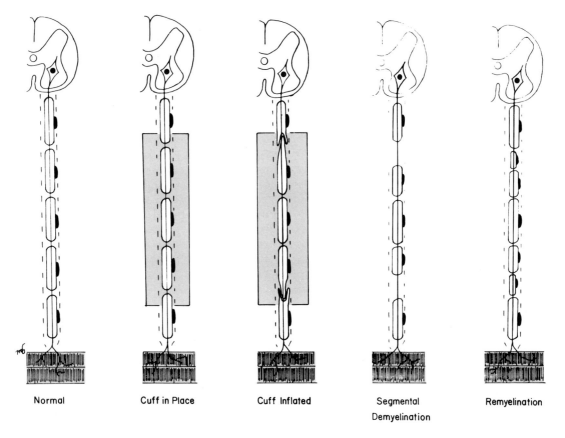

Normal   Cuff in Place   Cuff Inflated   Segmental   Remyelination
                                                Demyelination

CLASS I — ACUTE NERVE INJURY
(e.g. Compression)

Figure 2–8. Class 1 (neurapraxia) nerve injury associated with compression by a cuff. Axon displacement at both edges of the cuff causes intussusception of the attached myelin across the nodes of Ranvier into the adjacent paranode. Affected paranodes demyelinate. Remyelination begins following cuff removal and conduction eventually resumes. Conduction is normal in the nerve above and below the cuff since the axon has not been damaged.

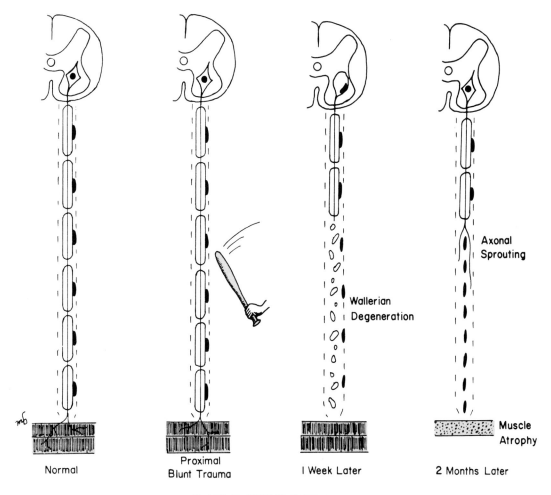

CLASS 2 NERVE INJURY

Figure 2–9. Class 2 nerve injury (axonotmesis) from a crush injury to a limb. Axonal disruption occurs at the site of injury. Wallerian degeneration takes place throughout the axon distal to the injury with loss of axon, myelin, and nerve conduction. Preservation of Schwann cell tubes and other endoneurial connective tissue ensures that regenerating axons have the opportunity to reach their previous terminals and perhaps re-establish functional connections.

general, damage to a nerve fiber appears to increase in proportion to the velocity, force, and duration of the traumatic agent, with the additional factors of traction and friction exaggerating the degree of injury.

There is widespread agreement about the basic three-stage classification of nerve injury, although the pathogenesis of these lesions, especially the mild lesions, remain controversial.[11,18,19] This section outlines and illustrates the salient stages of nerve response to injury. The features of acute and chronic nerve trauma are discussed in Chapter 16.

CLASSIFICATION

This classification is based on three stages (Classes 1, 2, and 3) of seriate vulnerability of components of peripheral nerve to injury; thus slight injury affects myelin, more severe injury, the axon, and the most severe disrupts connective tissue.

Class 1 (Neurapraxia). Conduction block is the hallmark of Class 1 compression injury and may be due either to transient ischemia or to paranodal demyelination. Ischemia results in a rapidly reversible loss of function associated with transient nerve impulse blockade. Paranodal demyelination occurs with more severe compression and is a mild structural nerve injury. Dysfunction persists in the distribution of the affected nerve until paranodal remyelination occurs, usually after a few weeks (Fig. 2–8).[8]

Class 2 (Axonotmesis). Axons are interrupted by a crush lesion, but the Schwann-cell basal lamina and endoneurial tissue remain intact. Wallerian degeneration occurs below the site of injury. Axonal regeneration commences promptly after injury and the growing axons reach proximal targets before distal sites of innervation (Fig. 2–9).

Class 3 (Neurotmesis). The axon is severed and the connective tissue disrupted, ranging from endoneurial and

Schwann-cell tube transection to total nerve severance.[18] Wallerian degeneration is inevitable and axon regeneration is severely limited by distorted connective tissue. Neuroma formation and aberrant regeneration are common (Fig. 2–10).

CLINICOPATHOLOGIC
CORRELATION OF
NERVE INJURIES
(see Chapter 16)

Class 1. This lesion is commonly associated with moderate focal compression of nerve, for example, Saturday night palsy. The motor deficit usually exceeds sympathetic and sensory loss. This reflects the low vulnerability of unmyelinated sympathetic and small myelinated sensory fibers, and the dependence of

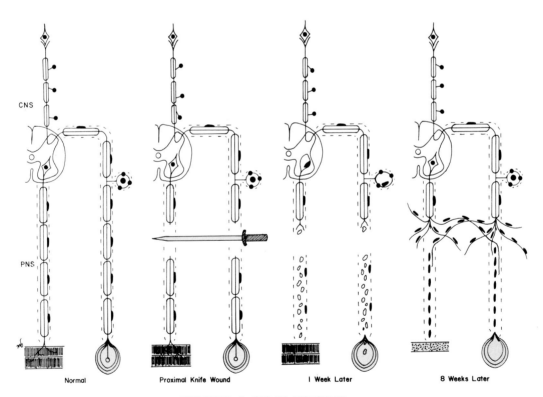

DEGENERATION & ABBERANT REGENERATION
IN (CLASS 3) NERVE INJURY

Figure 2–10. Class 3 nerve injury (neurotmesis) with severance of all neural and connective tissue elements. There is little hope of functional recovery without skilled surgery. Regenerating axons are entering inappropriate Schwann cell tubes (aberrant regeneration).

motor function on larger myelinated axons which undergo focal demyelination. Nerve conduction remains preserved in the intact, still myelinated axons below the injury. The good prognosis and rapid recovery (usually weeks) from Class 1 lesions reflects both the preservation of axonal continuity and the ability of Schwann cells rapidly and effectively to remyelinate the demyelinated segments. Unlike axonal lesions, recovery occurs simultaneously throughout the distribution of the affected nerve.

Class 2. This lesion is commonly associated with severe closed-crush injuries to an extremity. Complete loss of sensory, sympathetic, and motor function may occur from interruption of unmyelinated and myelinated fibers. Nerve conduction fails below the lesion as the axons degenerate; muscle atrophy may ensue. The prognosis is good (especially after distal lesions), since the axons can regenerate within their original Schwann-cell tubes and the pattern of motor and sensory restoration will be appropriate. The course of recovery is slow (usually months) and proximal to distal, reflecting the rate and course of axonal regeneration.

Class 3. These lesions are usually associated with severe traction injuries or open wounds. They have a poor prognosis because connective tissue disruption and proliferation interfere with axonal regeneration. Surgical repair with or without autografts is often required.

REFERENCES

1. Albin, RL, Albers, JW, Greenberg, HS, et al: Acute sensory neuropathy-neuronopathy from pyridoxine overdose. Neurol 37:1729, 1987.
2. Asbury, AK, Arnason, BG, and Adams, RD: The inflammatory lesion in idiopathic polyneuritis: Its role in pathogenesis. Medicine 48:173, 1969.
3. Dyck, PJ, Conn, DL, and Okazaki, H: Necrotizing angiopathic neuropathy: Three-dimensional morphology of fiber degeneration related to sites of occluded vessels. Mayo Clin Proc 47:461, 1972.
4. Gold, BG, Griffin, JW, and Price, DL: Slow axonal transport in acrylamide neuropathy: Different abnormalities produced by single-dose and continuous administration. J Neurosci 5:1755, 1985.
5. Griffin, JW and Watson, DF: Axonal transport in neurological disease. Ann Neurol 23:3, 1988.
6. Jacobsen, J, Sidenius, P, and Braendgaard, H: A proposal for a classification of neuropathies according to their axonal transport abnormalities. J Neurol Neurosurg Psychiat 49:986, 1986.
7. Krinke, G, Schaumburg, HH, Spencer, PS, et al: Pyridoxine megavitaminosis produces degeneration of peripheral sensory neurons (sensory neuronopathy) in the dog. Neurotoxicol 2:13, 1980.
8. Ochoa, J: Nerve fiber pathology in acute and chronic compression. In Omer, GE and Spinner, M (eds): Management of Peripheral Nerve Problems. WB Saunders, Philadelphia, 1980, p 487.
9. Pleasure, DB, Feldman, B, and Prokop, DJ: Diphtheria toxin inhibits the synthesis of myelin proteolipid and basic proteins by peripheral nerve in vitro. J Neurochem 20:81, 1973.
10. Schaumburg, HH and Spencer, PS: Clinical and experimental studies of distal axonopathy: A frequent form of nerve and brain damage produced by environmental chemical hazards. Ann NY Acad Sci 329:14, 1979.
11. Seddon, HJ: Surgical Disorders of the Peripheral Nerves, ed 2, Churchill Livingston, Edinburgh, 1975.
12. Sidenius, P: The effect of doxorubicin on slow and fast components of the axonal transport system in rats. Brain 109:885, 1986.
13. Spencer, PS, Sabri, MI, Schaumburg, HH, et al: Does a defect in energy metabolism in the nerve fiber underlie axon degeneration in polyneuropathies? Ann Neurol 5:501, 1979.
14. Spencer PS and Schaumburg, HH: Ultrastructural studies of the dying-back

process. IV. Differential vulnerability of PNS and CNS fibers in experimental central-peripheral distal axonopathies. J Neuropathol Exp Neurol 36:300, 1977.

15. Spencer, PS and Schaumburg, HH: Ultrastructural studies of the dying-back process. III. The evolution of experimental peripheral giant axonal degeneration. J Neuropathol Exp Neurol 36:276, 1977.

16. Spencer, PS and Schaumburg, HH: Classification of neurotoxic disease: A morphological approach. In Spencer, PS and Schaumburg, HH (eds): Experimental and Clinical Neurotoxicology. Williams & Wilkins, Baltimore, 1980, p 92.

17. Spencer, PS and Schaumburg, HH: An expanded classification of neuro-

toxic responses based on cellular targets of chemical agents. Acta Neurol Scand (Suppl) 7:9, 1984.

18. Sunderland, S: Nerves and Nerve Injuries. ed 2. Churchill Livingston, Edinburgh, 1978.

19. Thomas, PK: Nerve injury. In Bellairs, R and Gray, EG (eds): Essays on the Nervous System. Clarendon Press, Oxford, p 44.

20. Waksman, BH and Adams, RD: Allergic neuritis: an experimental disease of rabbits induced by the injection of peripheral nervous tissue and adjuvants. J Exp Med 102:213, 1955.

21. Xiu, Y, Sladky, JT, and Brown, MJ: Dose-dependent expression of neuronopathy after experimental pyridoxine intoxication. Neurol 39:1077, 1989.

Chapter 3

DIAGNOSIS AND ASSESSMENT

**DIAGNOSTIC ALGORITHM
DESCRIPTION OF THE
  CLINICAL ASSESSMENT
CRITERIA FOR THE DIAGNOSIS
  AND STAGING OF PERIPHERAL
  NEUROPATHY**

The text for Chapter 3 begins on page 26.

DIAGNOSTIC ALGORITHM

The prudent clinician considers the diagnosis of every patient with neuropathy an exercise in inductive reasoning. Central to this process is the recognition of the *key steps* in the diagnostic algorithm and placing the information obtained from the diagnostic techniques in perspective at each decision point. Although the possible causes and manifestations of neuropathy are numerous, data obtained from the history and physical examination can establish the course and distribution pattern of disease, while electrodiagnostic studies (EDS), clinical laboratory, and other specialized studies can determine if the process is axonal, demyelinating, or multifocal. Figure 3–1 and Table 3–1 outline a simple algorithm and relative value scale that we believe is useful in the diagnosis and characterization of neuropathy type. Figure 3–2 on pages 30 and 31 outlines a more comprehensive algorithm with a list of specific disorders.

Step 1: The first major algorithmic decision (Fig. 3–1) distinguishes between true neuropathy and entities mimicking peripheral nerve disease. Pivotal information for this distinction must be obtained from a careful and detailed history and physical examination. EDS can be extremely valuable in the differentiation of radiculopathies and conditions such as distal myopathies or myelopathies that produce a pseudopolyneuritic syndrome with paresthesias in the distal extremities.

Step 2: The second algorithmic decision is to decisively separate the focal mononeuropathies from diffuse disease. The clinical examination supplemented by EDS and other specialized evaluation procedures, such as quantitative sensory testing (QST), can effectively differentiate between dysfunction limited to the distribution of specific nerve segments and that diffusely affecting the peripheral nervous system.

Step 3: Principally utilizes EDS to distinguish multiple mononeuropathies from true polyneuropathies once diffuse neuropathy is established. This distinction is surprisingly difficult to make on clinical grounds alone. Strong clues suggesting multiple mononeuropathies are asymmetric physical findings and a history of focal motor or sensory deficits in the earlier stages of the evolution of the neuropathy, especially if these appear in the setting of a predisposing medical condition. However, some mononeuropathies have an insidious onset of widespread dysfunction often indistinguishable from polyneuropathy, requiring further EDS and sometimes nerve biopsy for diagnosis.

EDS also provide pivotal information for the distinction between axonal and demyelinating polyneuropathy. The axonal polyneuropathies comprise the largest single group. Once the obvious cases of diabetic, alcohol-nutritional, and pharmaceutical-toxic neuropathies are identified, the remaining instances of axonal neuropathy constitute the bulk of the diagnostic problems. Clinical or electrodiagnostic findings rarely help to distinguish one from another and these unclear cases often must undergo considerable laboratory evaluation.

Step 4: This step further divides the demyelinating polyneuropathies into hereditary and axonal and acquired conditions. The hereditary demyelinating conditions usually are readily differentiated since most display profound uniform slowing of nerve conduction, occur in childhood, and frequently affect other family members. Isolated instances of adult HMSN I or II may constitute formidable diagnostic problems. The acquired demyelinating diseases are electrodiagnostically characterized by a pattern of nonuniform slowing of nerve conduction, and are readily subdivided by their evolution into acute and chronic forms. The acute form, the acute inflammatory demyelinating polyradiculoneuropathy (AIDP), is usually identified at initial examination by an experienced clinician. Atypical instances require EDS with a special emphasis on measures of proximal conduction. The subacute or chronic acquired demyelinating disorders may be indistinguishable clinically

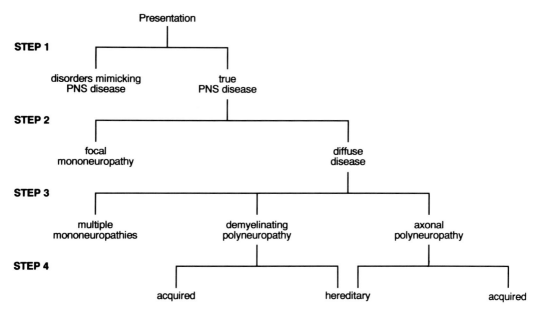

Figure 3–1. Diagnostic algorithm for PNS disease. Figure 3–2 contains a more complete algorithm with specific diseases.

Table 3–1 RELATIVE VALUE OF DIAGNOSTIC TECHNIQUES AT ALGORITHM STEPS

	Pivotal	*Helpful*	*Less Helpful*	*Not Recommended*
Step 1	History, examination	EDS, QST, laboratory investigation		Biopsy
Step 2	History, examination	EDS, QST		Biopsy
Step 3	EDS	History, examination	QST, Biopsy	Laboratory investigation
Step 4	History, examination	EDS, laboratory investigation	Biopsy	QST

EDS = Electrodiagnostic studies; QST = quantitative sensory testing.

Table 3–2 EVALUATION OF ACQUIRED SUBACUTE/CHRONIC SYMMETRIC AXONOPATHIES (FOLLOWING STEP 4)

I. INITIAL WORK-UP

Serum chemistry
Fasting blood glucose, 2-hour glucose tolerance test (Chapter 7)
TSH, T_4 (Ch 8)
ESR, ANA, Rh Latex, CBC with differential (Chapter 10)
SPEP, IPEP, UPEP, urine for Bence-Jones protein (Chapter 11)
Cryoglobulins (Chapter 11)
B_{12} Vit E, cholesterol (Chapter 9)
Chest x-ray, stool heme, urinalysis, breast/prostate/lymph node exam (Chapter 11)
Lyme titer (Chapter 6)
Heavy metal screen (if indicated by history or examination) (Chapter 18)
Baseline QST (vibration, thermal)

II. SUBSEQUENT WORK-UP (WITHIN ABOUT 1 MONTH)

If above are negative:
Clinically examine and perform EDS on family members (Chapter 15)
Anti-Ro, Anti-La, slit-lamp exam, Schirmer test, lip or lacrimal gland biopsy (Chapter 10)
Skeletal survey (Chapter 11)
Schilling test if B_{12} normal but continued strong clinical suspicion
Anti-ganglioside antibody testing if predominantly motor neuropathy
Consider HIV testing (Chapter 6)

Laboratory tests to be repeated about every 4–5 months

ANA, Rh Latex
SPEP, IPEP, UPEP (consider immunofixation)
Lyme-borreliosis titer
Cryoglobulins

III. THIRD EVALUATION (ABOUT 3 TO 6 MONTHS)

Consider repeat EDS to confirm axonal nature
CSF examination (especially cells and protein)
Repeat QST
Consider nerve and muscle biopsy at an appropriate facility (Chapter 4)
Consider evaluation of home and occupational environments for unsuspected toxin exposure (including pharmaceutical agents)

Table 3–3 EVALUATION OF ACQUIRED SUBACUTE/CHRONIC DEMYELINATING NEUROPATHIES (FOLLOWING STEP 4)

I. INITIAL WORK-UP

Serum chemistry, CBC with differential
SPEP, IPEP, UPEP, skeletal survey (Chapter 11)
Anti-ganglioside and anti-MAG antibodies (Chapter 11)
ESR, ANA, Rh latex (Chapter 10)
CSF examination
HIV testing, if clinically appropriate or CSF pleocytosis present (Chapter 6)
Consider Lyme-borreliosis titer (Chapter 6)
If EDS shows uniform demyelination: examine family members clinically and electrophysiologically (Chapter 15)

II. SUBSEQUENT WORK-UP (WITHIN 1–2 WEEKS)

If clinically warranted: lipidosis screen (Chapter 14)
If clinically warranted, treat demyelinating neuropathy (Chapters 5,6,11)
If the above initially negative, repeat SPEP, IPEP, UPEP, and skeletal survey in 4–6 months

Table 3–4 EVALUATION OF MULTIFOCAL AXONAL MONONEUROPATHIES (FOLLOWING STEP 3)

I. INITIAL WORK-UP

CBC with differential
ESR, ANA, cryoglobulins, Rh factor (Chapters 10 and 11)
Lyme-borreliosis titer (Chapter 6)
Chest x-ray
Fasting blood glucose, 2-hour glucose tolerance test (Chapter 7)
HIV titer, if clinically appropriate (Chapter 6)
Consider anti-Ro, anti-La, slit-lamp examination, Schirmer test, lip or lacrimal gland biopsy (Chapter 10)
Leprosy evaluation if patient lives in or visits endemic area (Chapter 6)

II. SUBSEQUENT WORK-UP (DICTATED BY TEMPO OF NEUROPATHY)

Nerve and muscle biopsy (may immediately follow initial evaluation in some instances, especially if progression is rapid) from the distal axonopathies. *In our experience, this is the most common peripheral neurodiagnostic error; it usually results from technically inadequate EDS and can have disastrous consequences if the patient with a chronic inflammatory neuropathy is untreated because of misdiagnosis.* Careful EDS are vital!

Although the outlined diagnostic algorithm is a useful exercise, there are important limitations:

1. Many of the steps in the decision-making process rely on information obtained from EDS. While there is no substitute for EDS individually focused on the patient's problem, the results can be uninformative or, worse, misleading if performed by an inexperienced person. Amplitude measures, often critical for the diagnosis of axonopathies and/or for the identification of subtle neuropathies, are particularly subject to technical error. It is the authors' experience that critical facets of EDS should be repeated in most instances. Occasionally, data obtained from EDS may not be sufficiently decisive to distinguish between a demyelinating and axonal neuropathy.

2. More than one cause of neuropathy may be disclosed by the investigation, especially in older subjects. For example, mild glucose intolerance is frequent in the elderly and may not be relevant. In these cases the particular clinical features of the neuropathy can help.

3. A diagnosis suggested by the algorithm may need to be pursued aggressively, e.g., findings consistent with a hereditary neuropathy require that distant living relatives have EDS, while the suspicion of an environmental toxin frequently dictates a home or workplace analysis.

DESCRIPTION OF THE CLINICAL ASSESSMENT

History and Physical Examination

The history should address both positive and negative symptoms relevant to motor, sensory, and autonomic function, the onset and progression of symptoms and their relationships to systemic illness; medications and environmental factors should also be determined. The family history is critical in any unexplained polyneuropathy, and family relatives should be examined clinically and with EDS whenever possible.

The history usually focuses the physical examination. For instance, an asymmetric onset indicates that a seemingly diffuse distal axonal neuropathy should be meticulously checked for focal nerve involvement since it may actually represent the end stage of a multiple mononeuropathy; likewise, the indolent progression of distal symmetric negative sensory and motor symptoms dictates a careful search for thickened nerves and pes cavus, which may suggest hereditary neuropathy.

Following the physical examination the following analysis is useful:

• *Overall pattern and distribution:* focal or multifocal versus bilaterally symmetric, proximal versus distal, and upper limb versus lower limb predominance.

• *Modality:* motor, sensory, autonomic, mixed.

• *Fiber size selectivity:* large fiber type (weakness, loss of joint position, vibration and touch-pressure sensibility, sensory ataxia, and pseudoathetosis); small fiber type (pain and temperature loss, autonomic dysfunction), or generalized.

Electrodiagnostic Studies

EDS include electromyography and measures of nerve conduction velocity and of compound sensory and motor action potential amplitudes. The electromyographer should be familiar with the nuances of peripheral nerve disease, perform an independent history and examination if the problem has not been formulated clearly by a referring neurologist, and tailor his physiologic studies to the problem.

Unlike electrocardiography and

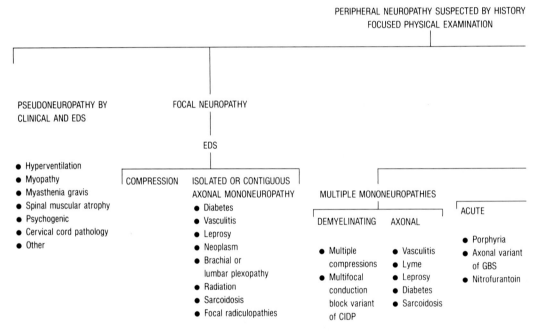

Figure 3–2. Diagnostic algorithm for PNS disease, with specific diseases.

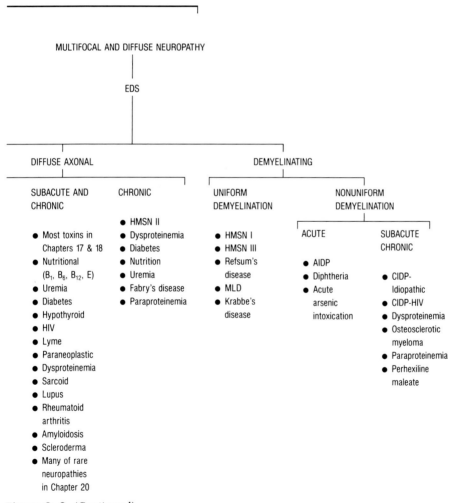

Figure 3–2. (*Continued*).

electroencephalography, there is no routine EMG or nerve conduction study. While electrodiagnosis is most helpful in evaluating nerve disease affecting large myelinated fibers, it is especially useful in addressing three issues: confirmation of neuropathy and elimination of nonneuropathic disorders, localization of focal nerve lesions, and prediction of pathology. The techniques and role of electrodiagnosis are discussed more fully in Chapter 4.

Nerve Biopsy and Quantitative Sensory Testing

Nerve biopsy is usually unnecessary to establish the *presence* of neuropathy, should only be performed where specialized facilities are available, and is most useful in disorders where the supporting tissues are affected (e.g., vasculitis, leprosy, sarcoidosis, amyloidosis) or if an inherited storage disorder is suspected. Biopsy is sometimes justified as a last resort in a puzzling neuropathy that has undergone extensive fruitless evaluation. The role of nerve biopsy is discussed more fully in Chapter 4.

Quantitative sensory testing (QST) devices deliver reliable vibratory or thermal stimuli; they can assess both vibration and thermal senses. They are especially useful in repeated assessments in longitudinal studies of treatment efficacy (e.g., aldose reductase inhibitors in diabetic neuropathy) and in rapid screening of large populations (e.g., prevalence of neuropathy in a population exposed to putative toxic chemicals). QST is discussed more fully in Chapter 4.

CRITERIA FOR THE DIAGNOSIS AND STAGING OF PERIPHERAL NEUROPATHY

This is an uncommon issue in the everyday practice of medicine or neurology since most patients have florid signs and symptoms strongly suggestive of peripheral nerve disease at the initial visit. It is an issue in special centers studying the detection and treatment of metabolic (diabetic, uremic) and toxic (heavy metal, pharmaceutical) neuropathies in large populations. Such studies require a systematic approach, analogous to those employed in clinical oncology where early detection and staging are at a premium. This problem is especially formidable since there is both a premium on detection of subtle, sometimes asymptomatic instances, but nerve biopsy—the absolute determinant of disease—is invasive and inappropriate as a longitudinal measure. It is suggested that four quantitative clinical measures be obtained: a symptom score, neurological examination score, computer-assisted quantitative sensory assessment score, and nerve conduction scores. Abnormalities on any two of these four measures suffice for a diagnosis of neuropathy. This notion is strongly supported by a meticulously controlled study of diabetic neuropathy where individuals with two or more abnormalities displayed nerve biopsy changes, while controls and patients with less than two abnormalities had normal biopsies. Criteria that are appropriate for one type of neuropathy (e.g., cisplatin) will not necessarily be applicable to other neuropathies (e.g., amyloid).

Chapter 4

LABORATORY INVESTIGATION OF PERIPHERAL NERVE DISEASE

ELECTRODIAGNOSTIC STUDIES
INVESTIGATION OF AUTONOMIC
  FUNCTION
QUANTITATIVE SENSORY TESTING
NERVE BIOPSY

ELECTRODIAGNOSTIC STUDIES

Electrodiagnostic studies (EDS) are essential in the early evaluation and differential diagnosis of peripheral nervous system disorders; they should be performed in every instance of peripheral nerve disease. Although commonly referred to as *electromyography* (EMG), these studies are actually composed of two separate complementary procedures, *nerve conduction studies* and *needle EMG.* The procedure should be tailored to the diagnosis; unlike electrocardiography and electroencephalography, there is no routine EMG or nerve conduction study. These evaluations are best performed by specialized electrodiagnostic consultants, or by neurologists with a particular interest in neuromuscular disease. A thorough electrodiagnostic evaluation, especially needle EMG, causes discomfort and is expensive. Nerve conduction studies involve electrical stimulation of peripheral nerves to elicit sensory and motor responses usually recorded by surface electrodes.[5,6,10] Abnormalities of conduction velocity and of the evoked potential reveal the underlying pathophysiology of the neuropathy. Table 4–1 gives representative normal values for

Table 4–1 NORMAL VALUES FOR NERVE CONDUCTION*

Nerve	Stimulus Site	Recording Site	Conduction Velocity (meters per second)	Amplitude
MOTOR			RANGE**	LOWER LIMIT OF NORMAL RANGE (In mV)*
median	elbow/wrist	abductor pollicis brevis	49–70	4.5
ulnar	elbow/wrist	abductor digiti minimi	47–70	4.5
peroneal	knee/ankle	extensor digitorum brevis	39–60	2.5
tibial	knee/ankle	abductor hallucis	39–60	2.5
SENSORY				(In µV)***
sural	lateral malleolus	midcalf	39–70	5
median	second digit	wrist	44–70	6
ulnar	fifth digit	wrist	45–70	6
radial	first digit	wrist	45–70	8

*Approximate values for adults below the age of 60 years. Conduction velocity and sensory-nerve action-potential amplitude decline in later life.

**It is advisable that precise ranges of normal be established for individual laboratories, as some of the values will be altered by variations in technique.

***Orthodromic recording

Table 4−2 COMPARISON OF LABORATORY
INVESTIGATIONS OF PERIPHERAL NERVE DISEASE

Procedure	Value in Differential Diagnosis	Value in Monitoring Progression	Frequency of Performance	Cost	Discomfort
EDS	High	High	Every case	High	High
Autonomic	Very limited	High for a few select diseases	Rare	Variable	Low
QST	Little	High	Common	Low	None
Biopsy	Specific for a few select diseases	Not done	Variable among centers	High	High

nerve conduction. Needle EMG assesses the electrical activity of a muscle from a recording needle electrode inserted directly into the muscle being examined. The nature of any spontaneous electrical activity, and the assessment of motor unit parameters and recruitment, indicate if axonal degeneration or reinnervation has occurred. Table 4−2 is a comparison of the application of the four types of laboratory investigation discussed in this chapter.

Indications and Usefulness

Electrodiagnostic studies may support, expand, or contradict the clinical diagnosis. Depending on the patient's complaints, electrodiagnostic testing may be needed to:
1. Localize a problem to the peripheral rather than central nervous system
2. Differentiate peripheral nerve disease from dysfunction of neuromuscular transmission or muscle
3. Localize the pathology to individual nerves, roots, plexus, or motor neurons
4. Characterize dysfunction as axonal, demyelinating, or both (see Chapter 2)

In addition, electrodiagnostic testing may indicate whether the neuropathy is generalized or multifocal—information often impossible to discern by clinical examination. Electrodiagnostic studies may also indicate severity and chronicity of neuropathy by documenting the degree of axonal loss and motor unit reinnervation.

Each electrodiagnostic examination should be tailored to the individual problem; *there is no such thing as a routine study.* Electrodiagnostic evaluations can be misleading, either because they are inadequately planned, technically unsound, or wrongly interpreted. In the authors' experience, misleading electrodiagnostic evaluations are common (usually performed by inadequately trained physicians) and result in erroneous therapies.

Techniques

SENSORY CONDUCTION
STUDIES

Measurements of the conduction velocity and amplitude of sensory and mixed nerve action potentials directly assess sensory and sensorimotor nerve integrity (Fig. 4−1). Recordings may be either orthodromic, when the nerve is stimulated distally with potentials recorded at one or two proximal sites, or antidromic, when the opposite stimulating and recording positions are used. As with all electrophysiologic parameters, normal values need to be established by the performing laboratory. Antidromic potentials generally have larger amplitudes than their orthodromic counterparts because of a closer proximity between the recording electrode and the nerve. Supramaximal stimulation must be used to ensure activation of all available sensory axons. Strict attention to limb temperature and to proper positioning of recording and stimulating electrodes is critical.

Sensory or mixed nerve action potential recordings provide the most sensitive electrodiagnostic indicator of sensory dysfunction and as such are fre-

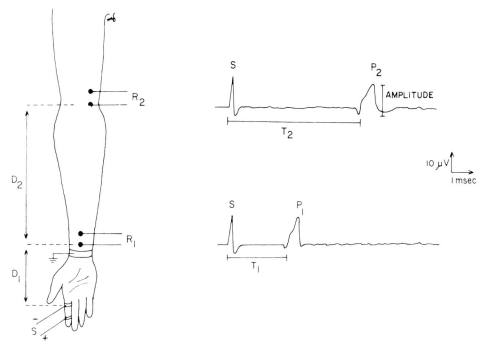

Figure 4–1. The technique for orthodromic recording of sensory nerve action potentials is illustrated, using the median nerve as an example. The digital nerves are stimulated (S) with the cathode at the proximal interphalangeal joint and potentials (P_1, P_2) are recorded at two proximal sites (R_1 and R_2). Latencies (T_1 and T_2) and distances (D_1 and D_2) are measured. Conduction velocity between the two recording sites is calculated as $D_2/T_2 - T_1$. The amplitude is measured from baseline to peak (as shown) or peak to peak. (Adapted from Bradley, W: Disorders of Peripheral Nerve. Blackwell Scientific Publications, Cambridge, 1974, p 88, with permission.)

quently abnormal early in the course of most peripheral neuropathies. Sensory potentials may be difficult to record, especially in elderly patients or those with peripheral nerve disease, and often require computer averaging to increase the signal-to-noise ratio. Since abnormalities of distal latency, conduction velocity, and—especially—evoked potential amplitude represent pathology affecting large-diameter, fast-conducting fibers, selective small-fiber involvement may not produce any detectable abnormalities. Reduced action potential amplitude directly reflects loss of sensory axons, either through fiber degeneration or conduction block/temporal dispersion due to segmental demyelination. Distal latencies and conduction velocities may remain normal, however, even in severe axonal disease, if a few large-diameter fibers remain intact. Amplitude measurements, rather than conduction velocity, are a more sensitive indicator of axonal peripheral neuropathy. Distal latencies and velocity measurements are sensitive in demyelinating neuropathies because large-fiber dysfunction is common. In addition, acquired demyelinating neuropathies are characterized by nonuniform slowing of conduction velocity within the many axons of the nerve, resulting in temporal dispersion of the evoked potential (manifested as low amplitude, increased duration, multiphasic, serrated potentials). Lesions proximal to the dorsal root ganglia do not affect the sensory potential. Therefore, normal sensory conduction studies help to distinguish radicular from peripheral nerve disease.

MOTOR CONDUCTION STUDIES

Motor conduction studies involve the supramaximal stimulation of mixed or motor nerves and surface recordings over the muscle's end plate region (so as to avoid contamination by muscle fiber conduction). Unlike sensory conduction studies, in which velocities may be obtained with single or multiple sites of stimulation, motor conduction determination always requires stimulation at both a distal and proximal site to eliminate the delay related to neuromuscular transmission (Fig. 4–2). Pertinent measurements include distal latency, conduction velocity, and compound muscle action potential (CMAP) amplitude, configuration, and duration. Since the CMAP represents the surface recorded summation of multiple muscle fiber potentials, its amplitude parallels the number of available motor axons, but is again influenced by temporal dispersion of conduction in the motor axons. CMAPs are usually easier to obtain than sensory potentials

and do not require averaging techniques. Motor conduction velocities can be determined for different nerve segments by varying the site of stimulation. As with sensory conduction, distal latencies and conduction velocities reflect function of the largest-diameter, fastest-conducting fibers.

Motor potential amplitudes are diminished by dysfunction of the motor perikaryon unit anywhere from the motor neuron to the nerve terminals. In most axonal sensorimotor neuropathies, changes in motor potential amplitudes lag behind sensory potential changes. In axonal neuropathies, motor conduction velocities are usually unchanged until there is substantial loss of fast-conducting fibers. Demyelinating neuropathies are characterized by prolonged distal motor latencies, slowed motor conduction velocities and tempo-

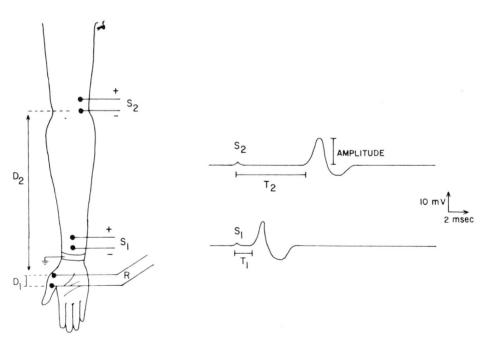

Figure 4–2. The examination of motor nerve conduction is illustrated for the median nerve and recordings from the abductor pollicis brevis muscle (R). The median nerve is stimulated supramaximally at the two sites (S_1 and S_2) and compound muscle action potentials are recorded. Latencies (T_1 and T_2) and distances (D_1 and D_2) are measured and conduction velocity between the stimulation sites is calculated as $D_2/T_2–T_1$. The amplitude is measured from the baseline to the peak of negative (upward) deflection, as shown. (Adapted from Bradley, W: Disorders of Peripheral Nerve. Blackwell Scientific Publications, Cambridge, 1974, p 86, with permission.)

ral dispersion of CMAPs, especially with proximal stimulation. Occasionally, conduction block is prominent. Low amplitude CMAPs with spared sensory potentials, in disorders in which there is sensory involvement, indicate disease at or proximal to the root. Focal slowing of motor conduction or conduction block across a possible entrapment site may indicate a compressive lesion.

ASSESSMENT OF CONDUCTION IN PROXIMAL NERVE SEGMENTS AND SPINAL ROOTS

Motor conduction in the proximal portion of upper-limb nerves can be directly assessed by stimulating motor roots with either a high voltage stimulator or needle electrode and recording from proximal or distal limb muscles.[3] Both in the lower and upper limbs, proximal nerve segments can be evaluated indirectly by late-responses studies (F responses, H reflexes). F responses are obtained with supramaximal stimulation of distal motor nerve segments that cause antidromic activation of a few motor neurons and subsequent orthodromic conduction back to the recording muscle (Fig. 4–3). F response amplitude is considerably smaller than that evoked by direct stimulation. The positions of the recording electrodes are identical to those for distal motor studies. F responses are best recorded by utilizing nerves whose recording muscles lie at a distance from the spinal cord (e.g., intrinsic hand or foot muscles). F responses are sensitive but not specific indicators of peripheral nerve dysfunction; abnormal minimal latencies may be seen in peripheral neuropathies, radiculopathies, and plexopathies. Prolonged F response latencies or failure to elicit a response (impersistence), in the presence of normal distal motor conduction, suggests proximal nerve segment disease, usually at the root or plexus level. In early acute demyelinating neuropathies, marked weakness may be present despite normal distal nerve conduction. In these cases, F responses often cannot be obtained,

indicating conduction block in proximal nerve segments. With the recovery of strength, F responses reappear.

Proximal conduction in sensory fibers can be assessed by examining somatosensory evoked potentials.[5] In the upper limb these can be recorded over the brachial plexus (N9 potential) and over the spinal cord (N13–N14 potential) following stimulation of the median or ulnar nerves at the wrist. For the lower limbs, it is possible to record a potential over the lumbosacral cord equivalent to the cervical N13–N14 potential, on stimulation of the tibial nerve. It is also possible to record dermatomal potentials following stimulation of the lumbosacral dermatomes. This may prove useful in the diagnosis of root compression syndromes,[9] although the technique has not yet been fully evaluated.

These techniques are useful also for assessment of dysfunction in conditions where the centrally-directed axons of the primary sensory neurons are affected, for example, central axonopathies and distal axonopathies.[14] Unfortunately, the SER is unobtainable in some instances because of destruction of sensory axons in the peripheral nerves.

The H reflex from the calf muscles (Fig. 4–3) is the electrical equivalent of the Achilles reflex; both latency and amplitude are measured. Although present in multiple muscles in infancy, the adult H reflex is practically limited to the gastrocnemius or soleus muscles following tibial nerve stimulation. H reflex abnormalities occur with generalized peripheral neuropathies, sciatic neuropathies, and S1 radiculopathies; they are sensitive but not specific indicators of nerve dysfunction. Unlike the F response, the H reflex is obtained with submaximal stimulation which selectively activates tibial nerve IA afferent fibers destined for the monosynaptic reflex arc. Supramaximal stimulation abolishes the H reflex. The tibial H reflex is affected early in demyelinating neuropathies and may be abnormal in axonal neuropathies before distal conduction studies show changes.

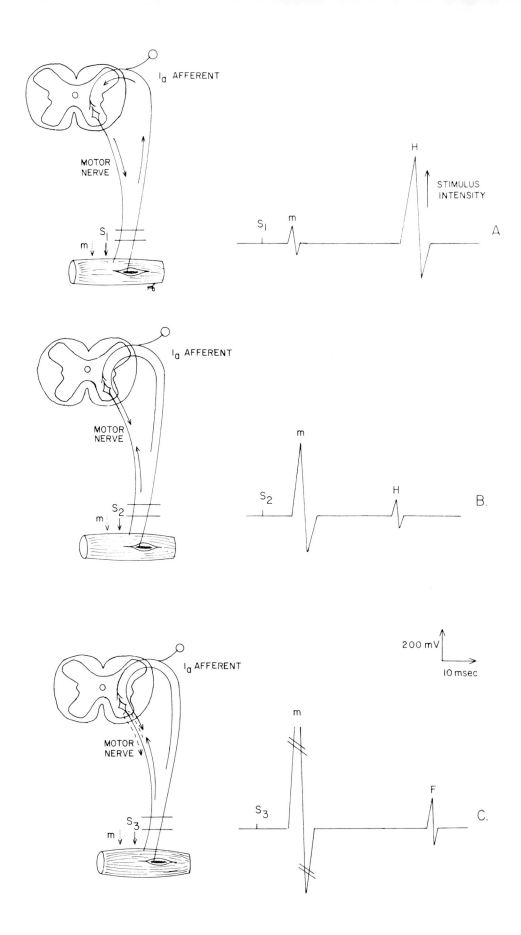

I$_a$ AFFERENT

MOTOR
NERVE

S$_1$

m

H

S$_1$ m

STIMULUS
INTENSITY

A.

I$_a$ AFFERENT

MOTOR
NERVE

m

S$_2$

m

S$_2$

H

B.

200 mV

10 msec

I$_a$ AFFERENT

MOTOR
NERVE

m

S$_3$

m

S$_3$

F

C.

BLINK REFLEX STUDIES

The blink reflex records the reflex contraction of the orbicularis oculi after supraorbital nerve stimulation. It measures conduction in the trigeminal and facial nerves. The blink reflex is the electrical counterpart of the corneal reflex and is composed of an oligosynaptic pontine reflex (R1 potential or early component) and a polysynaptic pathway from the pons to lateral medulla (R2 or late component). Different patterns of blink reflex abnormalities are associated with dysfunction at selective sites along the reflex pathway. When combined with direct facial nerve stimulation at the stylomastoid foramen, blink reflex studies can be used to provide important information regarding conduction along the entire facial nerve pathway.

RECORDING OF CONDUCTION IN SMALL FIBERS

Specialized procedures are available for recording from small myelinated and unmyelinated fibers, either employing collision techniques, which are applicable to motor and sensory fibers, or by recording with near-nerve needle electrodes and averaging very large numbers of responses. These techniques are not usually employed for ordinary clinical studies but may be informative in selective small fiber neuropathies. For research purposes, in vitro recordings can be made of the evoked compound action potential of excised portions of nerve.

NEEDLE EMG

Direct needle electrode recordings of the electrical activity present in normal and injured muscle provides useful information regarding motor unit integrity. Insertion of a recording electrode into a resting normal muscle detects no electrical activity aside from brief insertion discharges. Spontaneous activity (fibrillation or positive sharp wave potentials with the needle at rest) may be present in conditions where muscle fibers are denervated; it is a useful indicator of motor axon degeneration, but does not develop until 10 to 14 days after denervation and is not detectable in all denervated muscles. Spontaneous activity may occur in some myopathies. Fasciculation potentials represent the involuntary random firing of single motor units; they occur with anterior horn cell disease, radiculopathies, peripheral neuropathies, or a number of benign conditions.

In addition to detecting spontaneous activity, needle recordings allow quantitative analysis of voluntary motor units (amplitude, duration, degree of polyphasia). Chronic motor unit reinnervation is characterized by large-amplitude, long-duration polyphasic motor units; this is in contrast to most myopathies, where muscle fiber loss produces short duration, small polyphasic motor unit potentials. The manner and completeness of motor unit recruitment can be assessed. Neuropathic conditions are generally characterized by impaired voluntary recruitment, while myopathies display normal motor unit recruitment patterns despite clinical

---

Figure 4–3. (A) The technique for recording late responses is illustrated. A mixed nerve is stimulated with a small current causing excitation of the low-threshold 1a fibers with minimal stimulation of motor fibers. A long-latency potential (H reflex) is indirectly evoked following motor neuron synaptic stimulation of 1a afferents, at a time when the short-latency directly evoked compound muscle action potential is of minimal amplitude. (B) With larger stimulus currents, more motor axons are excited; the larger antidromic potential collides with and reduces the reflexly-generated H response, while the orthodromically-induced M response is larger. (C) With supramaximal stimulation of the motor nerve, the H reflex is completely blocked while the M response is maximal. Under these circumstances, the antidromic volley in the motor axons leads to retrograde firing of anterior horn cells, which generates descending impulses that give rise to low-amplitude long-latency muscle action potentials (F waves).

weakness. Needle EMG analysis can be performed on multiple proximal and distal muscles, so that the distribution and severity of nerve dysfunction can be ascertained.

Case History: Axonal Polyneuropathy

M.K., a 54-year-old salesman, experienced 9 months of numbness and burning paresthesias in his feet. Within the last month numbness ascended to his mid shin, his fingers have become numb and his gait is slightly unsteady. There were no systemic complaints. Examination revealed a stocking-glove distribution of impaired appreciation of pinprick and cold to the midshin and over all fingers. Vibration appreciation was markedly diminished up to the ankles and slightly in the fingers; joint position sense was mildly impaired. Minimal weakness of foot muscles was present. Tendon reflexes were absent at the ankles, trace at the knees, and normal elsewhere. Gait was steady but there was marked swaying without falling on Romberg testing. Coordination was normal.

Electrophysiologic studies demonstrated low amplitude peroneal (0.9 mV, NL = >2.5 mV) and tibial motor potentials (1.2 mV, NL = >2.5 mV), recording from the extensor digitorum brevis and abductor hallucis, respectively. Motor potential amplitudes were essentially unchanged when stimulation was at proximal sites and conduction velocities were within normal limits. F responses were impersistent and delayed in minimal latency in peroneal and tibial nerves. The sural and superficial peroneal sensory potentials were absent as were tibial H reflexes bilaterally. Radial and median sensory potentials were mildly reduced in amplitude (3 μv median; 6 μv radial; antidromic recordings) but displayed normal conduction velocities. Upper limb motor conduction, including F responses, was normal. Needle EMG demonstrated a combination of active denervation (fibrillation, positive sharp waves) and chronic motor unit reinnervation in intrinsic foot and distal leg muscles. Voluntary motor unit recruitment was reduced in distal leg muscles.

Routine laboratory investigations were re-markable for the presence of Bence-Jones protein and a high total protein to albumin ratio. Bone marrow examination was diagnostic for multiple myeloma.

Comment. This patient demonstrated electrophysiologic findings characteristic of axonal polyneuropathy. Sensory potentials were absent or of low amplitude early in the course of the disease. Tibial H reflexes were prolonged in minimal latency or absent. Motor potential amplitudes were subsequently reduced and F wave minimal latency mildly prolonged. Both motor and sensory conduction velocities remained essentially normal until late, when loss of fast-conducting axons caused mild slowing. Hands were affected after the legs and, initially, less severely. The earliest needle EMG changes included decreased voluntary motor unit recruitment and active/chronic denervation in intrinsic foot muscles. As the disease progressed, needle EMG abnormalities appeared in proximal muscles.

Case History: Demyelinating Polyneuropathy

R.B., a 25-year-old student, developed mild numbness and weakness in the right leg over a day, which was followed by similar complaints in the left leg. Within 2 days there was diffuse weakness of the arms and legs but only mild sensory complaints. He was unable to arise from bed. Bowel and bladder functions were normal. Physical examination demonstrated marked diffuse weakness in the legs and moderate distal weakness in the arms. Vibration and joint position sense was diminished in the legs more than the arms; appreciation of pinprick and cold was normal. Tendon reflexes were absent in the legs and arms, except for the right biceps.

Electrophysiologic studies, performed after 1 week, revealed normal motor potential amplitudes from the peroneal and tibial innervated muscles but prolonged minimal latencies (8.1 msec and 10.3 msec, respectively; NL<7 msec). Motor conduction velocities were 23.8 m/sec for peroneal and 31 m/sec for tibial nerves. Motor potentials evoked by distal and proximal stimulation were temporally dispersed and complex in configuration. Late responses were absent from the peroneal nerve and impersistent from the tibial nerves, with prolonged minimal latencies and increased maximum-minimal range. The sural sensory potential was normal in amplitude and conduction velocity. Tibial H reflexes were absent. The median sensory potential was absent. Median distal latency was prolonged at 5.1 msec and motor conduction was slowed to 33 msec, despite normal amplitude. Median F responses were impersistent and prolonged in minimal latency. Ulnar motor and sensory conduction studies were normal except for F responses which were both impersistent and prolonged in minimal latency. Needle EMG demonstrated reduced voluntary motor unit recruitment in all leg and distal arm muscles without evidence of denervation.

Comment. This patient's electrophysiology clearly indicated a demyelinating disorder characterized by slowed conduction velocities, prolonged motor and sensory distal latencies, early involvement of late responses, and temporal dispersion of evoked potentials, especially following proximal stimulation. Impersistence of late responses, despite relatively normal distal conduction, indicates demyelination and conduction block at proximal nerve segments; it is one of the earliest electrophysiologic findings in AIDP. Involvement of median rather than sural nerve is characteristic of AIDP, reflecting the patchy nature of the disorder. Needle EMG demonstrated reduced motor unit recruitment; evidence of denervation appears if subsequent axonal degeneration has occurred.

Interpretation of Electrodiagnostic Results

AXONAL POLYNEUROPATHIES

In a similar pattern to the clinical examination, the severity and distribution of electrophysiologic findings in distal axonopathies should obey a length-dependent relationship. Physiologic abnormalities are first present in distal segments of the peroneal and tibial nerves and only later affect the arms and face. The electrophysiologic hall-mark of axonal neuropathy is a reduction in sensory and motor potential amplitudes, reflecting axonal degeneration. In most sensorimotor neuropathies, sensory potential amplitudes (e.g., sural) are diminished before motor. Distal latencies and conduction velocities are preserved as long as a few fast-conducting fibers remain. F responses can be normal, mildly prolonged in minimal latency, or unobtainable. The tibial H reflex is usually absent early, paralleling disappearance of the Achilles reflex. The initial abnormality on needle EMG is reduced voluntary motor unit recruitment, reflecting diminished numbers of motor units. Active denervation is detected in distal leg muscles, eventually appearing in more proximal leg muscles and intrinsic hand muscles as neuropathy progresses. Changes in motor unit configuration, indicating motor unit reinnervation, are common in chronic neuropathies.

DEMYELINATING POLYNEUROPATHIES

Since acquired demyelinating neuropathies generally affect conduction in the largest, fastest-conducting fibers, changes in distal latencies and conduction velocities are early electrophysiologic findings. Latency and conduction velocity changes may occur without reduced evoked potential amplitudes. Motor and sensory potentials show temporal dispersion due to nonuniform slowing of conduction among fibers of different conduction velocities. Late responses are typically either prolonged in minimal latency or unobtainable. Needle EMG characteristically shows reduced motor unit recruitment consistent with motor axon loss or conduction block. Spontaneous fibrillation or positive sharp waves may be present on EMG if axonal degeneration has occurred.

Although the distribution of electrophysiologic abnormalities occasionally mimics that of distal axonopathies, acquired demyelinating neuropathies commonly have a multifocal, patchy pattern; i.e., nerves in the arms may be affected earlier or to a greater degree than in the legs, or be asymmetric either within the same limb or compared with the opposite side. The facial and trigeminal nerves, which are often spared in distal axonopathies, may show early changes on blink reflex testing.

FOCAL COMPRESSIVE NEUROPATHIES

Entrapment or compressive neuropathies are characterized by motor and/or sensory changes within the distribution of the affected nerve. Focal slowing of sensory, mixed, or motor conduction frequently occurs across the compressed segment; in other cases, focal slowing of conduction is not present but needle EMG demonstrates active or chronic denervation in the appropriate distribution. Electrodiagnostic studies are useful not only in localizing the site of compression but in determining severity and course.

MULTIFOCAL NEUROPATHIES

Electrodiagnostic studies are especially helpful in detecting asymmetry of nerve involvement unapparent on clinical examination.

INVESTIGATION OF AUTONOMIC FUNCTION

Autonomic dysfunction is present in many PNS disorders; except for primary dysautonomias, diabetes, amyloidosis, porphyria, and AIDP, it usually is subtle compared to motor or sensory impairment and its precise documentation is generally less useful in differential diagnosis than EDS or biopsy. Physiologic tests are available to evaluate, quantitatively and qualitatively, autonomic nervous system function. The methods of testing vary from the simple and inexpensive, which can be practically em-

ployed by most electrophysiologic laboratories, to the more sophisticated, which are best reserved for specialized centers. The following summary is a brief description of methods commonly employed to evaluate autonomic function.

Tests of Sudomotor Function

THERMOREGULATORY SWEAT TEST

This test involves dusting the patient with an indicator which changes color when contacted by sweat. Sweating is induced by heating the patient until central temperature has risen 1°C. This procedure delineates the distribution of sweating abnormality but does not define the locus of pathology to pre- or postganglionic fibers.

QUANTITATIVE SUDOMOTOR AXON REFLEX TEST (Q-SART)

This test measures the sweat output quantitatively over a defined skin area and is a measure of postganglionic sudomotor function.[11] Sweat glands are stimulated by acetylcholine, electrophoresed by a known current for a specified duration. Sweat output is measured by quantitating the change in relative humidity of dry nitrogen gas passed over the designated skin site, usually a 1 cm site over the forearm or foot. Q-SART can be used to evaluate the severity and progression of postganglionic autonomic disorders.

SYMPATHETIC SKIN RESPONSE

Recordings of skin resistance and potentials can be used to detect sympathetic sudomotor deficits, especially when due to peripheral neuropathies.[13] Recording electrodes are placed on the dorsal and ventral surfaces of the hand or foot. The stimulus ranges from the patient coughing to a low level electrical shock. The sympathetic skin response is recorded on an ordinary EMG machine; this is thus a simple and available measure of sympathetic sudomotor function. Unlike the Q-SART, sympathetic skin responses are highly variable, easily habituate, and are not easily quantitated.

Tests of Parasympathetic Cardiac Autonomic Function

The normal variation in beat-to-beat cardiac rate (RR interval variation) on respiration (sinus arrhythmia) is mediated reflexly by the parasympathetic autonomic nervous system through the vagus nerve.[13] Changes in normal heart rate fluctuation can be measured by monitoring the electrocardiogram either on cassette or FM tape with computer-assisted analysis or, more simply, on a conventional EMG machine by triggering on a target QRS complex and quantitating the fluctuation of subsequent QRS complexes relative to the fixed potential (Fig. 4–4). Measurements are performed during rest, the Valsalva maneuver, and on deep breathing. Reductions in cardiac variability

(loss of R–R variation) usually represent impaired vagal function. Reduction or loss of the increase in heart rate that occurs on standing or with a sustained handgrip are also useful measures.

Infrared Thermography

Infrared thermography displays skin temperature and has been advocated as a reliable procedure for detecting radiculopathy, entrapment mononeuropathies, peripheral nerve trauma, and reflex sympathetic dystrophy. While this technique appears sensitive in detecting thermal abnormalities associated with reflex sympathetic dystrophy, its broader diagnostic use is unclear.[8]

QUANTITATIVE SENSORY TESTING

Indications

Quantitative sensory testing (QST) is the use of precisely measured and repeatable sensory stimuli to determine

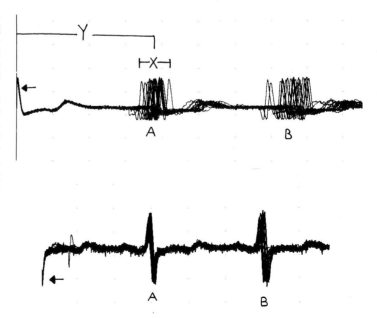

Figure 4–4. R-R interval variation testing. The top tracing represents normal R-R interval variation during normal breathing. R-R variation can be roughly measured in the electromyography laboratory. Twenty consecutive triggered QRS complexes (*arrow*) are superimposed and simultaneously recorded with two following untriggered QRS complexes (*A* and *B*). Percent R-R interval variation is calculated by mean R-R variation (X msec) divided by the R-R interval measured from the midpoints of the triggered and untriggered potential complexes (Y msec) multiplied by 100. The bottom tracing is from a patient with chronic dysautonomia and illustrates reduced R-R interval variation in QRS complexes labeled *A* and *B*.

the absolute threshold of sensation within specific somatosensory modalities. Precise delineation of sensory perception is clearly useful in early detection and for monitoring progression or recovery. It is of less help in differential diagnosis and currently available only in specialized centers. The authors advocate the further deployment of these inexpensive, painless procedures to neuromuscular specialists, diabetologists, and to centers evaluating toxic neuropathies, e.g., to oncologists and infectious disease specialists. Vibration, thermal, and pain perception have been well studied and several commercially available QST devices exist for each. When properly performed, QST enhances the neurologic examination and can increase the sensitivity and reliability of the clinical assessment of sensory function.[1,4,12] The principal strengths of QST include simplicity, permitting testing by nonprofessional personnel, and the availability of mean and standard deviation values that are accurate and age-controlled. QST is especially recommended for rapid screening of large populations (e.g., workers at risk for toxic neuropathy) and/or for the longitudinal evaluation of patients at risk for subtle sensory dysfunction (e.g., cumulative trauma disorders). These procedures are also useful for parametric evaluation of sensory loss in clinical studies involving iatrogenic and metabolic polyneuropathies.

Technical Considerations

Adequate QST requires the use of equipment that allows: (1) specification of the stimulation and calibration parameters; (2) isolation of the test stimuli from alternative cues (e.g., the use of a constant ramp intensity function can yield temporal cues that can distort threshold); (3) a sufficiently wide range of intensity to allow testing of normal and impaired individuals; and (4) the capacity to utilize sensitive psychophysical procedures such as two alternative forced-choice methods.

NERVE BIOPSY

Indications

Nerve biopsy is a *diagnostic* test; it is of little help in early detection or in monitoring progression of PNS disease, and is useful in only a small number of patients with peripheral neuropathy. Nerve biopsy and tissue evaluation is expensive; the procedure may be accompanied by considerable discomfort and currently is over-utilized as an early diagnostic procedure. It is performed best in centers with special expertise and facilities for full examination of the specimen.

The indications for nerve biopsy are well defined.[2,7,15] Biopsy is most helpful in identifying systemic illnesses that produce multiple mononeuropathy syndromes, for example, amyloidosis, sarcoidosis, and leprosy; simultaneous muscle biopsy may be informative in vasculitis and sarcoidosis. Inherited metabolic illness such as metachromatic leukodystrophy, Krabbe's disease, adrenoleukodystrophy, and Fabry's disease are associated with specific changes in peripheral nerve, but are now more readily identified by biochemical analysis of peripheral blood samples. Tomacula suggest an inherited tendency to pressure palsies. Demyelinating neuropathies may sometimes be identified on biopsy if the specimen is examined by epoxy resin and teased fiber techniques. Biopsy may be helpful in distinguishing acquired chronic inflammatory demyelinating polyneuropathy from inherited demyelinating polyneuropathies, such as HMSN I. Immunoglobulin deposition on myelin can be demonstrated immunohistochemically in many cases of demyelinating polyneuropathy associated with IgM paraproteinemia. Nerve biopsy is valuable in the diagnosis of amyloidosis and immunohistochemical studies can differentiate between AL (immuno) amyloid and inherited transthyretin-derived amyloid (Chapter 12). Biopsy appears justified in cases of diffuse cryptogenic neuropathy whose investigation has failed to suggest an etiology. Conditions masquerading atypically as distal axonopathies (chronic inflammatory demyelinating neuropathy, vasculitis, and sarcoidosis) are occasionally revealed only at biopsy.

Individuals whose diagnosis is relatively secure on clinical grounds (diabetes, alcoholic-malnutrition, porphyria, uremia, AIDP, and those metabolic-toxic disorders whose etiology is clearly established) do not require biopsy. Distal symmetric axonal neuropathies, of the type associated with most metabolic or toxic conditions, show similar nonspecific morphologic features, and biopsy is rarely as informative as meticulous medical evaluation.

Technical Considerations

SURGERY

The sural nerve at the ankle is favored for nerve biopsy; however, the superficial peroneal nerve or the radial nerve at the wrist are sometimes used. All are sensory nerves and a suitable length may be excised under local anesthesia. Following biopsy at these sites, cutaneous sensory loss appears in the distribution of the nerve, and may be accompanied by dysesthesias for several weeks. Removal of the whole nerve is customary in many centers; it is indicated in cases in which vasculitis, amyloidosis, or granulomatous neuropathy is suspected, as the lesions may be scattered. Fascicular biopsy is usually adequate for diagnosis in most cases and is followed by less sensory loss. It is helpful if a single individual in an institution becomes thoroughly familiar with the technique and performs all diagnostic biopsies. Although this seems a simple and trivial procedure, even an experienced surgeon can easily biopsy a vein or cause histologic artifacts by rough handling a specimen of nerve.

HISTOLOGIC TECHNIQUE

Three preparations should be available for every nerve biopsy: conventional paraffin sections, teased fibers, and epoxy-embedded sections for light and electron microscopy. Each has advantages and each may be done on a separate segment of the fixed tissue obtained at biopsy. A sample should also be taken and kept for frozen sections if required. If vasculitis is suspected, a quick answer can sometimes be obtained from frozen sections stained by hematoxylin and eosin.

Conventional Paraffin-Embedded Tissue. Paraffin sections, following routine staining, are useful for assessing cellular infiltrations, blood vessel changes, and granulomatous and neoplastic infiltrations. Special stains for amyloid and *M. leprae* bacilli are sometimes indicated, although leprosy bacilli are more readily identified by electron microscopy. Subtle changes in axons, myelin, and Schwann cells are not well appreciated in conventional stained tissue. Immunolabeling for identification of lymphocyte subsets may be helpful in the evaluation of inflammatory infiltrates.

Single Teased Fibers. This technique readily allows the rapid identification of axonal degeneration and segmental demyelination in long lengths of nerve fibers. Quantitative studies of internodal length and diameter are easily performed but are laborious; the relationship between length of internodes and myelinated fiber diameter can be expressed graphically or statistically. Normally there is little variation between internodal lengths in a single fiber. Teased-fiber studies are not required on every biopsy specimen but are helpful in confirming a demyelinating neuropathy and are mandatory if tomaculous neuropathy is suspected.

Epoxy Resin–Embedded Tissue. Light microscope examination of such sections is an especially useful technique for assessing changes in large numbers of axons, Schwann cells, and myelin since all are stained and well preserved by this technique. Both loss of myelinated fibers and the size of fiber affected can be appreciated by casual examination. The exact change in the population and caliber spectrum can be further determined, if necessary, by quantitative morphometric assessment of fiber numbers and diameters. Ultrathin sections may be cut from the same epoxy blocks and processed for electron microscopy. Electron microscopy[15] is useful for determining ultrastructural features in axons, myelin, and Schwann cells and identifying both subtle changes and specific pathologic features characteristic of some neuropathies, such as cellular inclusions in inherited storage disorders. As stated previously, it is an effective method for detecting leprosy bacilli.

REFERENCES

1. Arezzo, JC and Schaumburg, HH: Office and field diagnosis of neurotoxic disease. J Amer Coll Toxicol 8:2, 311–319, 1989.
2. Asbury, AK and Johnson, PC: Pathology of Peripheral Nerve. WB Saunders, Philadelphia, 1978.
3. Berger, AR, Busis, NA, Logigian, EL, Wierzbicka, M, and Shahani, BT: Cervical root stimulation in patients with radiculopathy. Neurology 37:329–332, 1987.
4. Bove, F, Litwak, MS, Arezzo, JC, and Baker, E: Quantitative sensory testing in occupational medicine. In Baker, E (ed): Seminars in Occupational Medicine. Thieme Medical Publishers, New York, 1986, p 185.
5. Brown, W: The Physiological and Technical Bases of Electromyography. Butterworth, Boston, 1984.
6. Brown, W and Bolton, C: Clinical Electromyography. Butterworth, Boston, 1987.
7. Dyck, PJ, Karnes, J, Lais, A, et al: Pathologic alterations of the peripheral nervous system in humans. In Dyck, PJ, Thomas, PK, Lambert, EH, and Bunge, R (eds): Peripheral Neuropathy, ed 2, Vol 1. WB Saunders, Philadelphia, 1984, p 760.
8. Editorial Assessment: Thermography in neurologic practice: Report of Academy of Neurology Therapeutics and Technology Assessment Subcommittee. Neurology 40:523, 1990.
9. Katifi, KA and Sedgwick, EM: Evaluation of the dermatomal somatosensory evoked potential in the diagnosis of lumbo-sacral root compression. J Neurol Neurosurg Psychiat 50:1204, 1987.
10. Kimura, J. Electrodiagnosis in Diseases of Nerve and Muscle: Principles and Practice, ed 2. FA Davis, Philadelphia, 1989.
11. Low, PA, Caskey, PE, Tuck, RR, et al: Quantitative sudomotor axon reflex test in normal and neuropathic subjects. Ann Neurol 14:573, 1983.
12. Moody, L, Arezzo, JC, and Otto, D: Screening occupational populations for asymptomatic or early neuropathy. J Occupational Medicine 28: 1–12, 1986.
13. Shahani, BT, Day, TJ, Cros, D, Khalel, N, and Kneebone, C: RR interval variation and the sympathetic skin response in the assessment of autonomic function in peripheral neuropathy. Arch Neurol 47:659, 1990.
14. Thomas, PK, Jefferys, JGR, Smith, IS, and Loulakakis, D: Spinal somatosensory evoked potentials in hereditary spastic paraplegia. J Neurol Neurosurg Psychiatry 44:243, 1981.
15. Vital, C, Vallat, J-M: Ultrastructural Study of the Human Diseased Nerve, ed 2. Elsevier, New York, 1987.

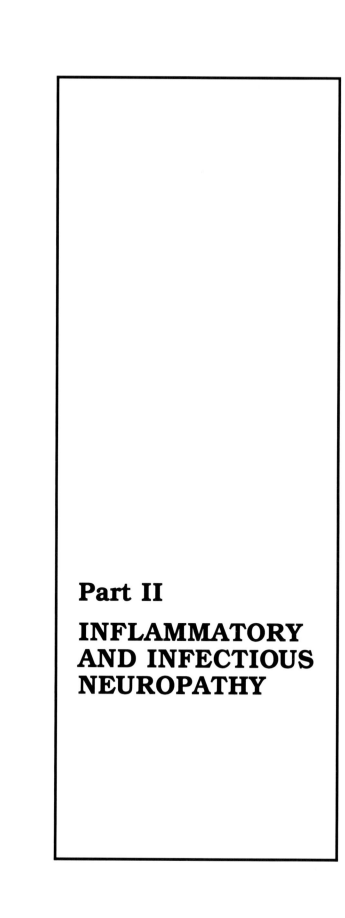

Part II

INFLAMMATORY AND INFECTIOUS NEUROPATHY

Chapter 5

INFLAMMATORY AND DEMYELINATING NEUROPATHY

ACUTE INFLAMMATORY DEMYELINATING POLYRADICULONEUROPATHY (GUILLAIN-BARRÉ SYNDROME, POSTINFECTIVE POLYNEUROPATHY)
CHRONIC INFLAMMATORY DEMYELINATING POLYRADICULONEUROPATHY (CIDP)

ACUTE INFLAMMATORY DEMYELINATING POLYRADICULONEUROPATHY (GUILLAIN-BARRÉ SYNDROME)

Definition

Acute inflammatory demyelinating polyradiculoneuropathy (AIDP) is a rapidly evolving, paralytic illness of unknown etiology. The salient pathologic feature of AIDP is widespread inflammatory PNS demyelination, presumed to be secondary to a hypersensitivity reaction.[5] Although the etiology is unknown, many cases follow nonspecific viral infections, and some are clearly associated with other events such as immunization, surgery, and infection with mycoplasma pneumoniae and Campylobacter jejuni.[3,38]

Pathology, Animal Models, and Pathogenesis

PATHOLOGY

Endoneurial infiltration by mononuclear inflammatory cells is the hallmark of this condition. This is followed by widespread, multifocal segmental demyelination with relative sparing of axons (see Figure 2–3, Chapter 2). The endoneurial infiltrates are mostly lymphocytes and monocytes, with occasional plasma cells. They are frequently perivascular. Polymorphonuclear leukocytes are present only in the most fulminant cases. The blood vessels appear normal.

Both segmental demyelination and inflammatory infiltrates are usually most pronounced in the ventral roots, limb-girdle plexuses, and proximal nerve trunks; changes also occur in dorsal roots, autonomic ganglia, and distal peripheral nerves.[36] Axonal degeneration is an occasional finding and is widely held to result from axonal interruption in zones of especially intense inflammation ("bystander" effect). Chromatolytic change may then occur in the anterior horn and dorsal root ganglion cells. Loss of these cells, or proximal axonal transection, probably accounts for poor clinical recovery in some cases. The primary lesions usually are confined to the PNS. Occasional perivascular inflammatory changes may occur in the CNS; CNS demyelination is not described.[5]

Within 2 weeks of the onset of acute demyelination, Schwann-cell proliferation occurs—a prelude to remyelination of the denuded internodes. In later stages of the disease, remyelination is the predominant feature and the inflammation subsides. A few inflammatory cells may persist for years within the endoneurium of otherwise normal-appearing nerves, and possibly are significant in the pathogenesis of recurrent cases.[5]

The morphologic events of the acute demyelinating process have been extensively investigated[37,59] and in combination with animal studies have strongly contributed to an understanding of the pathogenesis of AIDP. Ultrastructural study of the acute demyelination in this syndrome reveals phagocytes disrupting the Schwann cell, stripping away and phagocytizing normal-appearing myelin. The axon usually is unaffected. This pattern of myelin loss does not occur in the other demyelinating neuropathies (diphtheria and others) and appears unique to AIDP and related conditions among human disorders.

ANIMAL MODELS

PNS inflammatory demyelination also occurs in three animal diseases: experimental allergic neuritis,[6] coonhound paralysis,[10] and Marek's disease.[46] Experimental allergic neuritis (EAN) is a disorder induced in any mammal by injection of peripheral nerve. EAN can also be produced by sensitization with galactocerebroside,[44] which is discussed later in this chapter under Chronic Inflammatory Demyelinating Polyradiculoneuropathy. Peripheral nerve-induced EAN has long been considered especially relevant to AIDP, since it mimics the clinical and pathologic features of the human illness, although this notion has recently been questioned.[6] Its salient features are: (1) generalized ascending paralysis begins about two weeks after injection with mammalian peripheral myelin or myelin P1 or P2 basic proteins; (2) paralysis is accompanied by widespread PNS inflammatory demyelination and recovery follows disappearance of the inflammation and remyelination; (3) the disease can be transferred passively to another animal by lymphoid cells,[2] that is, this appears to be a disease of delayed hypersensitivity; and (4) the disease can be prevented by immunosuppression of the animal.[3]

Coonhound paralysis is an acute illness occurring in susceptible dogs after a raccoon bite[10] or injection with raccoon saliva,[22] and Marek's disease is a disorder of susceptible fowl associated with infection by an oncogenic group B herpes virus.[46] Neither of these naturally occurring conditions has been as well studied as EAN, and their relevance to AIDP remains sub judice.

PATHOGENESIS

It is generally held that AIDP is an immune mediated disorder; both lymphocyte-mediated delayed hypersensitivity and circulating antibody-mediated demyelination have been suggested as mechanisms. The delayed hypersensitivity is held as analogous to that of experimental allergic neuritis, but this is by no means fully established. The demyelination in EAN allegedly is controlled by lymphocytes that have become transformed in response to the injected antigen (peripheral myelin or P2 basic protein). The transformed lymphocytes attract macrophages to peripheral nerve and presumably direct an attack on myelin. The myelin sheath is removed, impulse conduction is disrupted, and weakness or sensory loss ensue as a result of conduction block or, if axons also degenerate, as a consequence of denervation. Since the process is directed against peripheral myelin, all nerves and roots are potential targets, and the involvement is frequently diffuse, albeit with a proximal predilection. Recovery is associated with remyelination, which usually begins within a few weeks of the onset of the inflammatory event and may be rapid.[3,5]

Two cardinal problems remain if EAN is to be accepted as the model of AIDP. First, several disease patterns occur in EAN; and second, AIDP must somehow be related to the diverse conditions that precede it.[3] Also, it may develop in situations where cell-mediated immunity is depressed,[12] as in Hodgkin's disease and HIV infection. Theories abound, but none is wholly satisfactory.[31,51,58]

It is suggested that circulating com-

plement-fixing anti-PNS myelin antibodies have a role in the pathogenesis
of the disease.[25] One study reports that
these antibodies were present in all
AIDP patients tested; the titers were
highest on admission and decreased as
patients improved.[2,25] A number of circulating antibodies have been identified, both to a variety of gangliosides[48]
and to a neutral glycolipid of human
peripheral nerve myelin. The glycolipid
reacting antibody cross reacts with
Forssmann hapten.[26] Removal of antibodies may underlie the successful
early intervention with plasma exchange in some instances.

The uncertainty surrounding the exact immune mechanism operant in
AIDP may reflect the fact that AIDP may
not be a single entity, and there may be
multiple triggering events. Humoral immunity to several antigens, cell mediated immunity, or immune complexes
may have roles in different cases.

Clinical Features

INCIDENCE

AIDP is a worldwide illness. It is the
most frequent cause of acute paralytic
illness in young adults (1.2 cases per
100,000) in developed countries.[1] The
age distribution is bimodal: the majority of cases occur among young adults,
and there is a second, lesser peak in the
group aged 45 to 64 years. Males are
slightly more susceptible. Cases occur
throughout the year, with a slightly
higher incidence in the colder months.

PREDISPOSING FACTORS

Table 5–1 lists antecedent events
that have been implicated in AIDP.
Studies of HLA antigens among Guillain-Barré syndrome patients have revealed no consistent pattern that would
indicate a predisposition.[47] Sixty percent of cases have an antecedent upper
respiratory or gastrointestinal illness
within one month of onset.[57] Gastrointestinal infection with *Campylobacter*

**Table 5–1 ANTECEDENT EVENTS
FOR AIDP**

Viral Infections	Vaccines
Measles	Rabies
Mumps	Swine Flu
Rubella	
Influenza A	Surgery
Influenza B	
Varicella-zoster	Fever Therapy
Cytomegalovirus	
Epstein-Barr virus	Malignancy
Infectious mononucleosis	Hodgkin's disease
Vaccinia	Carcinoma
Variola	Lymphoma
Infectious hepatitis	
Coxsackie	Pregnancy
ECHO	
HIV	Wasp sting

Other Infections
  Mycoplasma pneumoniae
  Salmonella typhosa
  Listeriosis
  Brucellosis
  Tularemia
  Ornithosis
  Campylobacter jejuni

jejuni is an especially common preceding event.[38] In cases that have followed
immunization against rabies or with
A/New Jersey influenza (swine flu) vaccine, symptoms have usually occurred
within 10 days to 3 weeks of the injection.[23] Infectious mononucleosis may
be accompanied by an inflammatory demyelinating polyneuropathy indistinguishable from AIDP.[3] Usually a menin-
goencephalitic illness is associated with
the neuropathy of infectious mononucleosis, and signs of central nervous
system involvement may dominate the
clinical picture. AIDP may also develop
in patients with HIV infection (see
Chapter 6).

SIGNS AND SYMPTOMS

There is now broad agreement on the
semiology of AIDP.[3,4,56] It may be described most simply as a rapidly
progressive, largely reversible, motor
neuropathy. Progressive and usually
symmetric weakness, combined with
hyporeflexia, are the cardinal features.
In one half of the cases, weakness be-

gins in the distal lower limbs and may appear to spread upward (ascending paralysis) over the entire body. Difficulty in walking is a common early complaint. It may take several forms, including bilateral foot-drop or a waddling, wide-based, unsteady gait. In other cases the weakness may begin or predominate in the upper limbs. The eventual degree of paralysis is variable, encompassing a broad spectrum that includes the occasional individual who never progresses beyond a mild foot-drop, to others with extreme weakness of all four extremities and facial muscles. Severe involvement may eventuate, with flaccid quadriplegia and inability to breathe, swallow, or speak. Some patients become totally "locked in," losing even eye movement. Limb weakness is generally symmetrical, and early muscle atrophy uncommon.

Facial weakness, present in many cases, is a striking feature of this illness, and helps to distinguish AIDP from most other neuropathies, apart from those related to sarcoidosis. The tendon reflexes are usually absent, even in cases where weakness is confined to the distal extremities. On occasion, feeble tendon reflexes may be elicited in strong proximal muscle groups. Preservation of tendon reflexes in severely weakened muscles seriously challenges the diagnosis.

Sensory symptoms, usually distal paresthesias, are present in most cases, rarely persist or progress (in contrast to the weakness), and generally are not accompanied by signs of a profound loss of sensation. Mild impairment of distal position and vibration sensation, and slight loss of pinprick over the toes, are commonly present. Marked sensory loss accompanied by pain or severe dysesthesias is very unusual, but lumbar and proximal limb pain may usher in the disease.

Autonomic dysfunction occurs in many cases, probably reflecting involvement of the myelinated preganglionic fibers and the ganglia. Orthostatic hypotension and hypertension are frequent, are very difficult to treat, and

can seriously complicate the management of patients with respiratory compromise. In individuals who appear otherwise clinically stable, fatalities may follow sudden, unexplained fluctuation in blood pressure or cardiac dysrhythmias. Tachycardia occurs in one half of the cases, and transient bladder paralysis has been documented in several individuals. Hyponatremia from the syndrome of inappropriate antidiuretic hormone (SIADH) may rarely accompany AIDP. Persistent, severe bladder or bowel dysfunction has not been described.

The "core" syndrome of rapidly progressive diffuse weakness constitutes the majority of cases and is easily recognized. The inflammatory demyelinating process is random and about 15% of cases develop unusual patterns of weakness. Several unusual, regional variant patterns have been identified and are listed in Table 5–2.[39]

Signs suggesting central nervous system involvement may rarely appear in otherwise typical cases. Features indicating this include a sharply defined abdominal sensory level, unequivocally extensor plantar responses, and vertical nystagmus. In contrast to CIDP where "overlap syndromes" clearly occur,[50] there is no evidence of CNS demyelination in AIDP from histopathologic or imaging studies. The significance of these "central signs" in AIDP is unclear.

Increased intracranial pressure and papilledema may rarely occur late in the illness. Raised intracranial pressure may or may not be associated with an increase in the CSF protein level. The pathogenesis of the raised intracranial pressure remains obscure.

Most individuals do not appear systemically ill while becoming weak, and constitutional signs such as fever, chills, and weight loss are unusual.

COURSE

Rapid progression of weakness is characteristic of AIDP; paralysis is maximal by one week in more than half, by 3 weeks in 80%, and 1 month in 90%.

Table 5–2 REGIONAL VARIANT AIDP SYNDROMES

1. **Ptosis Without Ophthalmoplegia**
   The most common of the regional variants. Ptosis may initially improve following neostigmine, leading to erroneous diagnosis of myasthenia. Eventually, typical diffuse AIDP develops in most, without progression to ophthalmoplegia.
2. **Pharyngeal-Cervical-Brachial Paralysis**
   Strength and reflexes in the lower limbs are spared and sensation is normal throughout in this striking condition. Botulism and diphtheria are diagnostic considerations.
3. **Paraparesis Alone**
   Strength and tendon reflexes in the arms are spared; cranial nerves and respiration remain normal. Spinal cord lesions and carcinomatous meningitis are diagnostic considerations.
4. **Limb Ataxia-Ophthalmoplegia-Areflexia**
   (Fisher Syndrome)
   The constellation of complete ophthalmoplegia, diffuse areflexia, and limb ataxia is generally agreed to be a variant of AIDP and not a CNS disorder[40]. Brain-stem disorders may be difficult to distinguish from this condition.
5. **Diffuse Axonal Radiculoneuropathy**
   This rapidly evolving postinfective disorder may cause flaccid quadriplegia and diffuse motor cranial neuropathy with poor recovery.[19]
6. **Ophthalmoplegia Alone**
7. **Purely Sensory AIDP**
   It is suggested that postinfective demyelination may be confined to the dorsal roots, resulting in a purely sensory AIDP.[11] Such instances are extremely rare and may represent overlap with the acute sensory neuronopathy syndrome (Chapter 19).

The remaining 10% will usually progress for variable intervals up to 8 weeks. AIDP should not be diagnosed with confidence in individuals who are maximally weak within 2 days. Those progressing for more than 2 months are considered instances of CIDP.

Recovery usually begins within 2 to 4 weeks after progression ceases. The pattern of recovery is extremely variable, normally proceeds at a steady pace, and within 6 months 85% of cases are ambulatory. Occasional individuals experience a dramatic improvement and are able to return to work within 2 months following a quadriparetic episode. Rare cases show little or no improvement. Recurrences occur in approximately 3% of cases, and the clinical profile is an acute monophasic illness resembling the original episode. Recovery may be slow. Several recurrences may occur separated by asymptomatic intervals of 10 to 15 years, and may be antedated by immunizations, surgery, or nondescript illnesses.[55] It is possible that some recurrent cases represent atypical instances of CIDP.

PROGNOSIS

Although in time most individuals have an excellent functional recovery, this is not a benign illness. The overall mortality is 5% and more than 50% of all individuals retain evidence of damage to the peripheral nervous system. Sixteen percent remain significantly handicapped by weakness and 5% are severely disabled. There may be persistent abnormalities of nerve conduction studies despite adequate functional recovery.

Some features of the initial clinical illness are of help in predicting the eventual outcome.[32] In general, individuals who experience only mild distal extremity weakness and have an onset of improvement within weeks of the first sign have excellent recovery. Four factors are consistently associated with poor outcomes in AIDP: a mean amplitude of distally stimulated compound muscle action potentials of 20% of normal or less, older age, rapid onset of neurological deficit, and need for ventilatory support. Curiously, objective sensory loss, papilledema, and CSF pleocytosis are not significant as prognostic signs.

Laboratory Studies

ROUTINE CLINICAL LABORATORY FINDINGS

A moderate increase in circulating polymorphonuclear leukocytes is present early in the illness, and lymphocytosis occurs later. Serum immunoglobulins IgG, IgA, and IgM may be elevated.

SPECIAL LABORATORY TESTS

Circulating activated, protein-synthe-sizing lymphocytes are present in early cases, and their numbers allegedly correlate with the severity of the illness.[3] Complement-dependent antimyelin antibodies are present in the serum of many acute cases (see Pathogenesis). Serologic evidence of infection with *Campylobacter jejuni*, cytomegalovirus or Epstein-Barr virus often is described in individuals with AIDP. Antibodies to HIV may be detected at the onset of illness, but the fully developed acquired immune deficiency syndrome does not usually appear until after recovery (Chapter 6).[9]

CEREBROSPINAL FLUID

The CSF protein is usually normal during the first three days of illness; it then steadily rises and may reach extraordinarily high levels. The CSF protein may remain elevated for several months, even after recovery is underway.[3]

Mononuclear cells, usually less than 10 per mm³, are present in up to one half of the cases. Exceptionally, they may number up to 50 cells per mm³.[3] CSF pleocytosis is especially common in HIV associated cases.

ELECTRODIAGNOSTIC STUDIES

Early in the illness, distal motor nerve conduction may be normal. Presumably, in such cases the disease process is confined to spinal roots and proximal nerves. Analysis of the F response, a measurement of proximal motor conduction, is of great value in the early cases. Conduction in proximal motor axons can also be studied by high voltage electrical stimulation of spinal roots with recordings from upper-limb muscles. Prolongation of latencies and evidence of conduction block may be demonstrable. Should the demyelination be present in distal nerves as well, more profound slowing of motor conduction, characteristic of

segmental demyelination, will be evident and conduction block may be demonstrable. Marked reduction in the amplitude of evoked compound muscle action potentials on distal stimulation (to a level less than 20% of normal) within the initial 2 weeks correlates with prolonged recovery. The profoundly reduced amplitudes in these cases probably reflect axonal damage and not preterminal demyelinative conduction block.[8]

Sensory nerve conduction is often normal and electromyography usually does not reveal widespread denervation changes, unless axons have been damaged in the inflammatory process or nerves have become compressed in a chronically bedridden individual.

NERVE BIOPSY

Nerve biopsy is rarely indicated in AIDP. The diagnosis can usually be established on clinical grounds with the aid of less invasive techniques. The distal cutaneous or subcutaneous nerves chosen for conventional biopsy are frequently not involved in the inflammatory demyelinating process and the specimen may be normal. Should occasional fibers display wallerian degeneration, the biopsy results may be misleading.

Treatment

PLASMA EXCHANGE

Plasmapheresis is clearly the treatment of choice for individuals acutely ill with AIDP; it should be instituted early. Plasmapheresis shortens the time on a respirator, the time to achieve independent walking, and the proportion of patients improved at 1 month and 6 months. Most studies suggest administration of continuous flow plasmapheresis to a total of 200–250 ml/kg over a 14-day period.[32] It is uncertain that all GBS patients need plasmapheresis; clearly those who plateau with mild weakness need not undergo this expensive, uncomfortable procedure. It is difficult to identify such mild cases with certainty at the onset of weakness. Our practice is to follow closely the mildly affected patients for the initial 2 weeks of hospitalization. Should they experience progressive inability to walk, reduction in respiratory capacity or bulbar insufficiency, pheresis is instituted. In our experience plasmapheresis appears less useful if administered after 2 weeks.

Therapeutic roles of corticosteroids, immunosuppression, and gamma globulin in AIDP have not been established.

ACUTE SUPPORTIVE TREATMENT

Respiratory. The patient suspected of having AIDP must be admitted to the hospital even if the involvement is minimal, since the neuropathy often progresses rapidly and unpredictably. We admit such patients directly to a respiratory care unit where they remain until the condition has stabilized or is clearly improving. The tidal volume, vital capacity, blood pressure, and ability to cough and swallow should be closely monitored, as these can change without warning.

If mechanical ventilation is necessary (as determined by the degree of respiratory effort, the vital capacity, and the blood gases), it should be instituted early rather than waiting until the last moment. Endotracheal intubation is usually performed initially, and replaced by a cuffed tracheostomy after 5 days if there is insufficient improvement. It is dangerous to rely on arterial blood gas measurements in monitoring patients with respiratory muscle weakness. Significant hypoxia and hypercarbia may not be present until just before respiratory collapse, owing to the compensatory increase in respiratory rate. If only blood gas measurements are available, then rely on the PO_2 rather PCO_2 for evidence of respiratory compromise. Simple bedside measurements of respiratory function, such as expiratory vital capacity or peak inspiratory force, are more valuable and sensitive parameters of respiratory status. Maintenance of a good mouth-bellow seal, at times assisted by the examiner, is essential. The patient should be encouraged to give a best effort on the initial try, since subsequent attempts consistently underestimate true respiratory function. Intubation should be considered when respiratory parameters show a progressive decline or if the expiratory vital capacity falls below 12–15 ml/kg. Earlier intubation is best for individuals when significant bulbar weakness causes inaccurate assessment of respiratory parameters, or when there is difficulty clearing secretions.

Once intubated, large tidal volumes (15 ml/kg) and 5–10 cm H_2O of end-expiratory pressure should be employed to prevent atelectasis. The use of intermittent mandatory ventilation, preferably delivered via a volume-cycled ventilator, will allow the patient to regulate the respiratory rate as dictated by metabolic needs, thereby maintaining central ventilatory responsiveness. It also avoids ill-timed ventilatory assistance that superimposes breaths on end-inspiration. Since positive pressure ventilation decreases venous return to the heart, care must be taken in individuals with circulatory embarrassment (i.e., congestive failure, reduced blood volume). Connective tissue within the lung and thoracic wall will shorten if not regularly opposed by stretch. Passive breathing exercises are useful in maintaining the distensibility of the lungs and thoracic wall and in preventing the frozen-chest syndrome. Once intubated, intermittent positive pressure breathing will promote expansion of respiratory connective tissue.

Impaired expiratory power causes ineffective cough and accumulation of bronchial secretions. Coughing machines are often useful; these devices use positive pressure to inflate the lungs followed by negative pressure to force expiration of air and bronchial secretions. Postural drainage may facilitate expectoration of accumulated secretions.

Autonomic. Loss of autonomic reflexes may eventuate, with peripheral pooling of blood, poor venous return, and low cardiac output. Positive-pressure respiration may further impede venous return and aggravate this situation. Rapid decrease in blood pressure may occur, and should be treated initially by fluid replacement. Pharmacologic manipulation of blood pressure in AIDP patients is extremely perilous and should be avoided unless absolutely necessary. Attempts to prevent cardiac dysrhythmias by measures such as the combined administration of atropine and a beta adrenergic blocking drug have not yet been adequately evaluated.

Some patients will be unable to swallow or gag. Feeding should be done through a small nasogastric tube. The patient should be sitting when food is given, and kept sitting for 30 to 60 minutes afterwards to minimize the risk of aspiration.

The patient may be unable to urinate because of abdominal muscle weakness or autonomic involvement. Abdominal compression may be helpful and catheterization is occasionally necessary. Fecal impaction should be avoided by the use of stool softeners, laxatives, and enemas as required. The paralyzed patient must be turned frequently to avoid pressure sores. A sheepskin or water bed may be used. Low-dose subcutaneous heparin has been employed in an attempt to prevent deep venous thrombosis in the legs and lessen the consequent risk of pulmonary embolism. If the eyes do not close completely, corneal damage must be prevented with protective solutions and the eyes taped shut at night; a temporary lateral tarsorrhaphy may be required. Finally, patients may experience severe psychologic stress because of paralysis and the intensive-care environment, and mild sedation occasionally may be necessary. The patient may be unable to communicate at all, but hospital personnel must not forget that the patient can see, hear, and understand. As stated earlier, AIDP is one cause of the "locked-in" syndrome.

If the patient with AIDP can be carried through the acute stage of progressive paralysis (usually two or three weeks), strength will gradually return. Since most patients make a gratifying recovery after months of weakness, the importance of extremely fastidious supportive care in the acute stage cannot be overstressed.

CHRONIC SUPPORTIVE CARE

Maintaining Joint Mobility. Preservation of joint mobility is of crucial importance during the period of severe weakness. Failure to do so may result in fixed deformities due to painfully contracted joints that resist movement, regardless of the degree of nerve regeneration. Specific attention should be directed to: (1) encouraging patient activity; (2) proper positioning when bed-bound; (3) range-of-motion exercises; and (4) proper splinting.

Encouraging Patient Activity. There is no basis for the widely held belief that activity accelerates neurologic deterioration. In many cases, weakness or severe sensory loss dictates a period of bed rest until adequate recovery begins. Patients with mild neuropathies usually do well out of bed, and continued performance of daily activities probably is beneficial in maintaining good joint mobility and coordinated motor function. Bed rest is particularly liable to aggravate gait ataxia due to proprioceptive loss.

Proper Positioning. Proper positioning of bed-bound patients is critical; this minimizes dependent edema and prevents contractures, decubitus ulcers, and pressure on susceptible peripheral nerves. Patients should be turned regularly, about every 2 hours, and skin areas should be examined for evidence of tissue damage or breakdown.

The distal portion of an immobilized extremity commonly develops edema when left in the recumbent position. Increased local tissue pressure compromises circulation, which promotes skin breakdown and decubitus ulcer formation. A simple, effective measure to relieve edema is elevation of the distal limb segment about 35 to 40 degrees, thereby enhancing venous drainage. Intermittent compression by a pneumatic sleeve or stocking, using pressure of 60 to 100 mm Hg for 60 to 90 minutes, twice daily may relieve swelling, as will snug wrapping and rewrapping of the swollen area. Elastic stockings as sole therapy are not helpful for significant edema but may assist in preventing fluid reaccumulation.

Passive range-of-motion (ROM) exercises of all joints, especially the shoulders, hips, and knees, should be initiated during the period of marked weakness or as soon as the patient is pain free. Three to five repetitions, performed twice daily, are sufficient to ensure adequate joint mobility. Gradual gentle stretches, maintained for longer periods, are better than stronger stretches rapidly performed. Passive ROM exercises should be continued as long as active ROM is limited either by weakness or soft tissue contractures. Between sessions, the limbs are positioned so that gravity produces a constant mild stretch to counteract contracture formation. Established contractures frequently require prolonged, gentle passive stretch. Mild heating using ultrasound helps during these stretching sessions, since elongation of collagen fibers is facilitated by warmer temperatures. Dynamic splints, as described below, can provide constant stretch between therapy sessions.

Proper Splinting. Should contractures develop, or intermittent ROM exercises fail to prevent deformity, splinting of the affected joint by dynamic or static splints is indicated. Static splints are nonmovable, maintain joints in fixed positions, and support the extremity or joint by substituting for weak muscles. Static splints prevent overstretch of paralyzed muscles and often support the extremity in the position of greatest function; they may also be used progressively to stretch a joint, sustaining and augmenting the benefits derived from passive ROM exercises. In acute peripheral neuropathies, static

splints have three uses: (1) to position the thumb in abduction, thereby avoiding web space adduction contractures; (2) as cockup splints to avoid overstretch of weak wrist extensor muscles; and (3) to prevent plantar flexion contractures of the foot when the muscles of dorsiflexion are weak.

Severe contractures may require dynamic splints; these are static splints to which an outrigger has been applied on either the volar or dorsal surface. Extending from this outrigger are rubber bands or springs that extend to the segment that is to be supported or stretched. The amount of tension applied is critical and may be regulated by changing the length and thickness of the rubber bands. As joint ROM increases, the tension is adjusted to provide appropriate stretching forces. Passive ROM exercises may be performed during the day, whereas a dynamic splint is applied at night.

Pain Relief. Patients bed-bound from peripheral neuropathies suffer from a variety of different pains. Apart from acute lumbar pain at the time of onset, three types of pain have been associated with AIDP.[41] One is a deep ache or cramping, predominantly in the lower extremities, occurring especially at night and unrelieved by changing position. Quinine occasionally lessens this discomfort. A second variety is a sharp discomfort affecting large joints, worsened by movement. During recovery, patients also experience uncomfortable burning dysesthesias, lasting for several weeks. These sensations gradually resolve as recovery continues. Parenteral steroids are occasionally effective for this latter pain, whereas quinine, tricyclic antidepressants, anticonvulsants, and nonsteroidal anti-inflammatory agents are of little use. Epidural morphine is suggested in cases of intense pain, refractory to conventional therapies.[21]

Emotional Rehabilitation. Rapid progressive loss of functional independence leaves patients confused, depressed, and uncertain of the future.

Before effective rehabilitation, they must both accept their condition and be willing to interact with the therapist. Patients must be warned that despite rehabilitative efforts, muscle wasting may continue to progress for some time.

Emotional adjustment may take considerable time, depending on the degree of impairment. Most commonly, the patients realistically accept their current physical and functional limitations and eagerly enter the rehabilitative program aiming for specific goals. Many individuals experience a period of grief and distress before adjusting to their condition. Even if unduly prolonged, this despondency passes once the patient achieves some recovery. Patient support groups, such as the Guillain-Barré Society, exist in most large North American and British cities; they are a valuable source of information about local resources. We recommend that physicians and patients read *No Laughing Matter* by Joseph Heller and Speed Vogel (Avon Books, New York, NY, 1987), an eloquent account of the first author's debilitation and recovery from AIDP.

Differential Diagnosis

With the decline of poliomyelitis and diphtheria in Europe and North America, a typical instance of AIDP is not usually a difficult diagnosis for a neurologist or experienced internist. However, even the core cases should be confirmed by electrodiagnosis to avoid mistakes.

HYPOKALEMIA AND TICK PARALYSIS

In addition to being an embarrassing mistake, considerable risk may accrue to the patient with severe hypokalemia if not promptly treated. Therefore, serum electrolytes are immediately obtained in all cases of acute paralysis, and if there are indications of its likeli-

hood, a careful search for a tick, *dermacentor andersoni,* is undertaken.

ACUTE MYELITIS

Cervical spine fracture and epidural abscess have, on occasion, been confused with AIDP. A sharp sensory level and sphincter paralysis at the onset of the illness should raise suspicion of spinal cord involvement.

BOTULISM

Early prominence of extraocular muscle and lower cranial nerve muscle involvement, pupillary abnormalities, the absence of sensory complaints, and normal CSF protein all raise the possibility of botulism. On occasion, clinical distinction between these entities may be very vexing, especially if there is only mild pupillary involvement in the botulism cases. Evidence of ingestion of tainted canned or packaged food can be helpful in diagnosis.

POLIOMYELITIS

Poliomyelitis is generally a febrile illness, in the past frequently occurring in epidemics and is associated with nuchal rigidity and headache resulting in asymmetric paralysis and usually accompanied by more than 50 mononuclear cells per mm^3 in the CSF.

PORPHYRIA

This may closely mimic AIDP. The limb weakness is often proximal in both conditions, but greater involvement of the upper extremities is more characteristic of porphyria. Urine measurement of porphobilinogen and delta aminolevulinic acid should be performed in suspected cases of acute intermittent porphyria. Electrophysiology shows an axonal rather than a demyelinating process.

TOXIC NEUROPATHIES

Dapsone, nitrofurantoin, and thallium all produce neuropathies that may have a subacute onset and are occasionally confused with AIDP. Adolescents who abuse paint thinner or glue (sniffers, huffers) may have rapid onset of a predominantly motor neuropathy and respiratory paralysis. These individuals rarely have cranial nerve palsies, muscle atrophy is profound, and, in general, sensory complaints are a persistent problem. A careful history that specifically inquires about medication or drug use will usually disclose those problems.

LYME BORRELIOSIS

Infrequently a patient infected with *Borrelia burgdorferi* following a tick bite will develop a fulminant multifocal inflammatory polyradiculopathy mimicking AIDP. The presence of a cerebrospinal fluid pleocytosis coupled with appropriate serum antibody titres indicate Borrelia infection.

CHRONIC INFLAMMATORY DEMYELINATING POLYRADICULONEUROPATHY (CIDP)

Definition

Affected individuals have an initial illness clinically similar to AIDP, although usually with a more gradual onset, but subsequently undergo either a chronic relapsing or a chronic progressive course.[15,49] The salient histologic features of CIDP are remarkably similar to those of AIDP. It is generally held that CIDP has a common pathogenetic mechanism with AIDP and may be conceptualized as a recurrent or progressive multiple sclerosis-like disorder of the PNS.[13] The nosologic limits of CIDP remain poorly delineated; for the present, we still distinguish between CIDP and the chronic multifocal or diffuse demyelinating neuropathies that accompany osteosclerotic myeloma, benign IgM gammopathy, and lymphoma. Some of these conditions may be immune mediated but none is clearly associated with early,

prominent endoneurial accumulations of lymphocytes and macrophages may occur in CIDP.

Pathology and Pathogenesis

Thinly myelinated or demyelinated axons, varying degrees of onion-bulb formation, and loss of myelinated fibers are the light microscope hallmarks of CIDP. Endoneurial edema and mononuclear cell infiltrates are occasionally present. Figure 5–1, from a nerve biopsy in a patient with CIDP, shows active stripping of myelin by macrophages. It is alleged that onion-bulb formation is a more prominent feature of the relapsing cases, and lymphocytic infiltration more common in the chronic progressive form. These changes are distributed throughout proximal nerves and spinal roots. Multifocal or localized pathology occurs but is rare, and may give rise to striking hypertrophy of proximal nerve trunks, for example, the brachial plexus. Several recent clinical reports utilizing neuroimaging have described accompanying multifocal CNS white matter lesions in CIDP.[33,50] There is no histologic study of the CNS lesions.

It is generally held that the pathogenesis of CIDP is similar to AIDP, that is, it is an autoimmune disorder of delayed hypersensitivity analogous to EAN. This notion is reinforced by two factors: the continued presence of inflammatory cells in the nerves of individuals with AIDP long after clinical recovery from the monophasic illness[5] and, more importantly, some animals with EAN (usually an acute monophasic illness) pursue a chronic progressive or relapsing course.[6] It is not certain which antigen is responsible for chronic relapsing EAN, but it is assumed that it is the same as for acute EAN, the different clinical behavior being related to age and species-related factors involved in disease suppression.[6,35] The histologic features are identical with acute EAN except for particulars attributable to chronicity. Foci of CNS demyelination accompany the PNS disease of EAN in

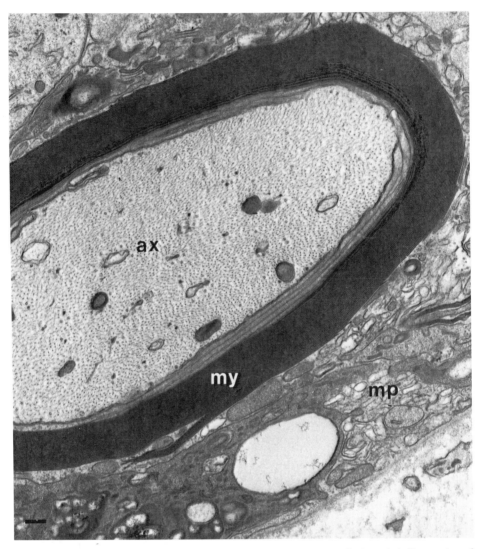

Figure 5–1. Electron micrograph from a nerve biopsy from a patient with chronic inflammatory demyelinating polyneuropathy. An intact axon (*ax*) is surrounded by a myelin sheath (*my*) that is being stripped off by a macrophage (*mp*) that contains myelin debris. Bar = 0.2 μm.

some species; this may be relevant to reports of multifocal CNS white matter lesions in humans with CIDP.

The neuropathy induced in rabbits by sensitization with galactocerebroside is also suggested as a model of human CIDP.[44] The galactocerebroside EAN animal displays a long latent period to disease, a subacute onset, and a chronic course. Another contrast with the EAN model is that the im-

munologic process in the galactocerebroside-immunized animal appears not to be cell-mediated; lymphocytes have not been observed during the demyelination. A recent study failed to detect anti-galactocerebroside antibodies in individuals with CIDP or AIDP and does not support the notion of a humoral response to this lipid in the human disease.[43] There is conflicting evidence for humoral demyelinative factors in CIDP;

some studies find evidence of antimyelin antibodies in human serum; others do not. Similarly, some describe animal studies wherein successful passive transfer of demyelinating factors from affected humans has been achieved, others have failed.

Clinical Features

INCIDENCE

CIDP is a relatively common neuromuscular disorder. At the Albert Einstein/Montefiore Medical Center its incidence approximates that of amyotrophic lateral sclerosis. It occurs throughout the world and in all ages; its peak incidence is in the fifth and sixth decades, and males are more frequently affected.[30] The mean age of onset is younger in the chronic relapsing than in the chronic progressive cases.[30]

PREDISPOSING FACTORS

About one-third of patients with CIDP experience an antecedent, nonspecific illness but the association is not as strong as with AIDP. One study has suggested an association with pregnancy and reports of HLA antigen testing describe disease susceptibility genes.[20,29] CIDP clearly may be associated with HIV infection, but the onset of neuropathy usually antedates development of the fully developed acquired immune deficiency syndrome (AIDS).[9] In endemic areas it is prudent to test all individuals with CIDP and AIDP for HIV. There is one report of an association with persistent hepatitis B infection, with relapses that paralleled liver dysfunction and immune deposits in the vasa nervorum.[24]

SIGNS AND SYMPTOMS

Although widely believed to be a predominantly motor neuropathy similar to AIDP, many actually present with a sensorimotor neuropathy with only slight motor predominance. Only 22% have a striking motor predominance, 6% are purely sensory.[30]

Weakness is present in both proximal and distal muscles, while paresthesias are usually distal. Weakness of lower cranial nerves and intercostal muscles may occur. An asymmetric onset is frequent and may initially be mistaken for an entrapment syndrome when electrodiagnostic studies display local conduction block. Positive sensory complaints are especially common and help to differentiate CIDP from the hereditary neuropathies. Hyporeflexia or areflexia develop in almost all. Action and intention tremors of uncertain mechanism occur in 3%, papilledema is present in 7%, and 11% have nerves enlarged to palpation.[15]

In our experience, many cases deviate from the core syndrome; some have slow progression over decades, others may present as ataxia, still others resemble entrapments or motor neuron disease. CIDP is the "great imitator" of PNS disease, a modern-day syphilis. Atypical cases present formidable diagnostic problems, often are mislabeled even by experienced clinicians, and are not correctly identified until meticulous electrodiagnostic testing is performed. This applies particularly to the localized forms.[10a]

COURSE AND PROGNOSIS

The development and course of the illness are considered the salient features that distinguish between recurrent examples of AIDP and CIDP. Recurrent AIDP is characterized by an acute onset and a maximal neurologic deficit within 4 weeks of the initial complaint in each episode. CIDP generally has a more protracted, often stuttering onset and an indolent progression. Two broad clinical varieties are recognized, relapsing and progressive, with considerable overlap.[15,30,49] For example, many relapsing cases do not recover completely and could be considered instances of stepwise progression.

The course of the relapsing group displays a considerable range in the interval between relapses, the severity of an episode, and rate and degree of recovery. Subsequent attacks are generally similar to the initial episode and disability varies considerably. With treatment, improvement is generally good between episodes. Life-threatening episodes with respiratory insufficiency are more common early in the illness. Persistent disability following repeated attacks is variable, and in many instances, attacks cease after a few years. Patients with a relapsing course are held to have a more favorable prognosis than those with the progressive form, but one study found the differences in disability were not significant at prolonged follow-up.[30]

The course of the nonrelapsing form is either stepwise or gradual. If untreated, this condition may become disabling or fatal. With treatment, the prognosis significantly improves. One series with a mean 6.5-year follow-up disclosed that 34% had minimal or no disability, 31% had mild motor and sensory signs, 24% were moderately disabled, 3% required assistance in daily activities, and 7% died.[30]

A syndrome of multifocal demyelinating neuropathy with conduction block has been recently defined and suggested as a variant of progressive CIDP.[27,34,53] This condition has a strikingly asymmetric onset and initially may mimic entrapment syndromes, if sensory findings predominate,[27] or motor neuron disease if only weakness appears.[34] Correct diagnosis depends on sophisticated electrodiagnostic testing and it is likely that this disorder goes widely unrecognized. In our experience, most patients with this condition respond to treatment; the response in the others is variable.

Occasional cases have clinical evidence of a concomitant multifocal relapsing CNS disorder that resembles multiple sclerosis.[33,50] The temporal conjunction of the CNS and PNS manifestations suggest that they both stem from the same process and are not the chance coexistence of two independent conditions.

Laboratory Studies

CLINICAL LABORATORY FINDINGS

Mild elevations of plasma gamma globulin may occur, and antibodies to HIV are detectable in infected individuals with CIDP.

CEREBROSPINAL FLUID

The CSF protein is elevated at some stage of the illness in almost every case of CIDP, but may fluctuate to normal levels. Mononuclear cells, usually less than 10 per mm^3, are occasionally present. In CIDP associated with HIV disease, a moderate pleocytosis is usual.

ELECTRODIAGNOSTIC STUDIES

With the possible exception of nerve injury, there is no condition in which nerve conduction studies are more important than in CIDP; properly performed, they are the key to correct diagnosis. There is a characteristic pattern of nerve conduction changes with differential slowing of velocity in proximal and distal limb segments of the same nerve. This is in striking contrast to the uniform slowing along the entire length of the nerve in the hereditary demyelinating neuropathies.[28] The electrophysiological findings in CIDP are as follows:

1. Multifocal nerve involvement in which some nerves may be relatively spared compared with others;

2. Slowing of conduction velocity, temporal dispersion of evoked nerve and muscle compound action potentials, and, frequently, conduction block; the latter two items distinguish CIDP from hereditary demyelinating neuropathies, in which all fibers are usually affected to the same degree; there is no dispersion or conduction block;

3. As with all demyelinating neuropathies, there is a poor correlation between conduction slowing and weakness;

4. Often significant axonal loss—probably more common than with AIDP.

NERVE BIOPSY

Nerve biopsy may be extremely helpful in the diagnosis of these conditions if one is fortunate enough to select a diseased area. The histologic picture is characteristic and consists of:

1. Reduction in the number of myelinated fibers;
2. Axons with abnormally thin myelin, or that are demyelinated;
3. Varying degrees of onion-bulb formation. Endoneurial edema and scattered mononuclear cell infiltrates are present in some instances. Cell labeling studies and/or electron microscopy may be required to detect the presence of inflammatory cells.

Treatment

It is generally held that plasma exchange (PE), corticosteroids, and human immune globulin (HIG) therapy, are the treatments of choice for CIDP. Controlled trials of therapies have demonstrated that both the progressive and the relapsing variants can be ameliorated.[16,17,30,52] There are no firm guidelines about which therapy will be most efficacious and treatment patterns vary from center to center. Mildly affected individuals who have stabilized or older individuals with slowly progressing mild disease are initially best followed closely with supportive therapies. The benefits of PE, corticosteroids, or HIG in these cases sometimes fail to balance the risks. When treatment is indicated, we at the Albert Einstein Hospital/Montefiore Medical Center initiate therapy in young persons with either HIG or alternate day PE.

Using PE, patients with the relapsing disorder often need only a small number of exchanges in a brief period of time. Those with the progressive disorder frequently require exchanges at regular intervals over long periods of time

to maintain benefit. About one-half of our patients have clearly benefited from plasma exchange. Improvement has sometimes been striking in the recurrent cases, commencing before the sixth exchange. Improvement following PE in the progressive cases is rarely dramatic and generally requires at least four weeks of alternate day pheresis. Should no improvement appear after 12 weeks, PE is abandoned and corticosteroid therapy is instituted. Plasma exchange is not the initial treatment in older individuals because of the risks of volume depletion and hypotension.

Recent reports have claimed striking benefit from administration of high doses of pooled human immunoglobulin (HIG).[7,18,54] This therapy, although expensive, offers distinct advantages over PE and corticosteroids. HIG therapy is simple, safe, can be given at home following an initial in-hospital challenge, and has few side effects. Improvement often begins within 2 weeks and can be dramatic; relapses occur, requiring repeat administration of HIG.[54]

Corticosteroid therapeutic regimens suggested for CIDP usually emphasize initial high doses slowly declining over a 4- to 6-month period. We initiate 60 mg of prednisone daily for 1 month and gradually shift to an alternate day regimen for 6 months. If improvement commences, dose levels are lowered steadily; should weakness arise, the dose must be increased. As with PE, improvement is often more apparent in the relapsing cases and appears sooner. The response in the progressive form is often only modest. CIDP in children is reported consistently to respond favorably to corticosteroid therapy.[45]

Immunosuppressive therapy with azathioprine, cyclophosphamide, cyclosporin, or lymphoid irradiation[42] is advocated for individuals unresponsive to PE, HIG, and corticosteroids. Isolated case reports describe improvement but the only controlled trial of combined azathioprine and corticosteroids showed no better results than treatment with prednisone alone.[14]

Differential Diagnosis

The differential diagnosis of the recurrent variant is seldom a problem once several episodes have occurred. The characteristic combination of clinical remissions, elevated CSF protein, and profound slowing of motor nerve conduction usually suffice. On occasion, acute intermittent porphyria may mimic this disorder and the appropriate laboratory determinations should be done. Repeated exposure to exogenous toxins should also be ruled out by a careful occupational and enenvironmental history. The differential diagnosis between CRIP and recurrent AIDP has been discussed elsewhere.

The differential diagnosis of the progressive form is frequently very difficult. In our centers this condition is the most common erroneous peripheral neuropathy diagnosis. Meticulous electrophysiology is the key to diagnosis; atypical cases are frequent and are often misidentified as multiple entrapments, mononeuritis multiplex or atypical distal axonopathy by inexperienced or unwary electrophysiologists. Unless the nerve biopsy displays unequivocal evidence of inflammatory cells and demyelinated fibers, this disorder may be clinically indistinguishable from some hereditary neuropathies and the chronic demyelinating neuropathies associated with dysproteinemia and benign gammopathy. The genetic disorders also are characterized by gradual progression, profound but generally *uniform* slowing of motor nerve conduction, and onion-bulb formation (see Chapter 15). A careful family history should always be obtained, especially if the illness begins in childhood or adolescence. The chronic demyelinating neuropathies associated with dysproteinemia and paraproteinemia are often clinically and electrophysiologically identical to CIDP and the diagnosis may be extremely difficult; it depends on detection of the underlying disorder by appropriate laboratory or radiological tests.

Case History: Relapsing Type of CIDP

A 65-year-old woman with a history of chronic bronchitis and pyelonephritis noticed pain and weakness in both hands. During the following month she developed progressive weakness of her legs that led to several falls. She also became aware of pain in both arms, tingling paresthesias in her fingers and toes, and some numbness of her fingers. She was admitted to the hospital 2 months following the appearance of the initial symptoms. Examination on admission revealed obesity, hypertension (250/150), and basal crepitations in both lungs. She showed mild right facial weakness, but no other cranial nerve abnormality. She displayed a generalized weakness of her arms and legs affecting proximal and distal muscles, the arms being more affected than the legs. Her tendon reflexes were all absent and the plantar responses flexor. Apart from loss of vibration sense distally in the legs, no sensory abnormality was detected. A blood count was normal and ESR was 27 mm at 1 hour. The total serum protein level was normal but electrophoresis showed a slight decrease in albumin and beta globulin. The fasting blood sugar level was normal. No evidence of active urinary infection was discovered and radiography of the chest was normal. The CSF contained 120 mg per dl of protein and one lymphocyte per mm^3 3 weeks following admission. Motor nerve conduction velocity in the right ulnar nerve, recording from the abductor digiti minimi, was 30 m/s. Ulnar sensory nerve action potentials were absent in the right hand.

The weakness in her limbs progressively increased following admission. Her hypertension was treated with bethanidine. Prednisone at a dosage of 30 mg daily was commenced after 2 weeks. However, this was discontinued after 6 days because of increasing hypertension and the development of edema. The weakness continued to advance to almost complete paralysis of all four limbs, and because of respiratory distress, she was transferred to the respiratory unit 20 days following admission, although assisted respiration did not become necessary. Over the following month there was some improvement with some return of power in the limbs. A repeat CSF examination, 2 months following admission, revealed a protein content of 3.8 g/L and 2 lymphocytes per mm³. By the end of the third month, her condition was noted to have deteriorated again and she was started on 30 mg daily of prednisolone. A sural nerve biopsy revealed a substantial loss of myelin sheaths in transverse sections. Examination of isolated osmicated nerve fibers showed extensive segmental demyelination and remyelination. Her condition remained unaltered and, toward the end of the fourth month, she suddenly developed chest pain and died. The autopsy revealed a recent red infarct in the right lung and complete blockage of the right pulmonary artery with antemortem thrombus. Otherwise the examination showed bilateral chronic pyelonephritis and left ventricular hypertrophy. No carcinoma was detected.

Portions of cervical, thoracic, and lumbar spinal cord, together with dorsal and ventral spinal nerve roots and dorsal root ganglia from these three levels, were available from the autopsy. The dorsal and ventral spinal roots all showed a moderate loss of myelin sheaths in transverse section, especially in the cervical and lumbar regions. The ventral roots were affected to a greater extent than the dorsal. Little axonal loss was detectable. Single osmicated fibers from a lower lumbar ventral root were examined and showed profuse demyelination and remyelination, with very few normal surviving internodes. Perivascular collections of inflammatory cells, composed predominantly of lymphocytes, were seen at all three levels, most numerous in the ventral roots. The dorsal root ganglia appeared normal apart from some loss of ganglion cells with proliferation of satellite cells. No abnormalities were detected in the spinal cord.

REFERENCES

1. Alter, M: The epidemiology of the Guillain-Barré syndrome. Ann Neurol (Suppl) 27:S7, 1990.
2. Arnason, BG and Chelmica-Szorc, E: Passive transfer of experimental allergic neuritis in Lewis rats by direct injection of sensitized lymphocytes into sciatic nerve. Acta Neuropath 22:1, 1972.
3. Arnason, BG: Inflammatory polyradiculopathies. In Dyck, PJ, Thomas, PK, Lambert, EH, and Bunge, R (eds): Peripheral Neuropathy, ed 2, Vol 2. WB Saunders, Philadelphia, 1984, p 2050.
4. Asbury, AK, Arnason, GB, Karp, HR, et al: Criteria for diagnosis of Guillain-Barré syndrome. Ann Neurol 3:565, 1978.
5. Asbury, AK, Arnason, BG, and Adams, RD: The inflammatory lesion in idiopathic polyneuritis: Its role in pathogenesis. Medicine 48:173, 1969.
6. Brosnan, JV, King, RHM, Thomas, PK, et al: Disease patterns in experimental allergic neuritis (EAN) in the Lewis rat. Is EAN a good model for the Guillain-Barré syndrome? J Neurol Sci 88:261, 1988.
7. Chimowicz, MI, Audet, AJ, Hallet, A, et al: HIV-Associated CIDP. Muscle and Nerve 8:695, 1989.
8. Cornblath, DR: Electrophysiology in Guillain-Barré syndrome. Ann Neurol (Suppl) 27:S17, 1990.
9. Cornblath, DR, McArthur, JC, Kennedy, PGE, et al: Inflammatory demyelinating peripheral neuropathies associated with human T-cell lymphotropic virus III infection. Ann Neurol 21:32, 1987.
10. Cummings, JF and DeLahunta, A: Chronic relapsing polyradiculoneuritis in a dog. A clinical, light and electron-microscopic study. Acta Neuropathol 28:191, 1974.
10a. Cusimano, MD, Bilbao, JM, and Cohen, SM: Hypertrophic brachial plexus neuritis: A pathological study of two cases. Am Neurol 24:615, 1988.
11. Dawson, DM, Samuels, M, and Morris, J: Sensory form of acute polyneuritis. Neurol 38:1728, 1988.
12. Drachman, DA, Paterson, PY, Berlin, BS, et al: Immunosuppression and the Guillain-Barré syndrome. Arch Neurol 23:385, 1970.

13. Dyck, PJ and Arnason, B: Chronic inflammatory polyradiculoneuropathy. In Dyck, PJ, Thomas, PK, Lambert, EH, and Bunge, R (eds): Peripheral Neuropathy, ed 2, Vol 2. WB Saunders, Philadelphia, 1984, p 2101.

14. Dyck, PJ, O'Brien, P, Swanson, C, et al: Combined azathioprine and prednisone in chronic inflammatory demyelinating polyneuropathy. Neurol 35:1173, 1985.

15. Dyck, PJ, Lais, AC, Ohta, M, et al: Chronic inflammatory polyradiculoneuropathy. Mayo Clinic Proc 50:621, 1975.

16. Dyck, PJ, O'Brien, PC, Oviatt, KF, et al: Prednisone improves chronic inflammatory demyelinating polyradiculoneuropathy more than no treatment. Ann Neurol 11:136, 1982.

17. Dyck, PJ, Daube, J, O'Brien, P, et al: Plasma exchange in chronic inflammatory demyelinating polyradiculoneuropathy. N Engl J Med 314:461, 1986.

18. Faed, JM, Day, AB, Pollack, M, et al: High-dose intravenous human gamma-globulin in chronic inflammatory demyelinating polyneuropathy. Neurol 39:422, 1989.

19. Feasby, TE, Gilbert, JJ, Brown, WF, et al: An acute axonal form of Guillain-Barré polyneuropathy. Brain 109:1115, 1986.

20. Feeney, DJ, Pollard, JD, McLeon, JG, Stewart, GJ, and Doran, TJ: HLA antigens in chronic inflammatory demyelinating polyneuropathy. J Neurol Neurosurg Psychiat 53:170, 1990.

21. Genis, D, Busquets, C, Manubens, E, et al: Epidural morphine analgesia in Guillain-Barré syndrome. J Neurol Neurosurg Psychiat 52:999, 1989.

22. Holmes, DF, Schultz, RD, Cummings, JF, et al: Experimental coonhound paralysis: Animal model of Guillain-Barré syndrome. Neurol 29:1186, 1979.

23. Hurwitz, ES, Schoenberger, LS, Nelson DB, et al: Guillain-Barré syndrome and the 1978–1979 influenza vaccine. N Engl J Med 304:1557, 1981.

24. Inoue, A, Tsukada, N, Koh, C-S, et al: Chronic relapsing demyelinating polyneuropathy associated with hepatitis B infection. Neurol 37:1663, 1987.

25. Koski, CL: Characterization of complement fixing antibodies to peripheral nerve myelin in Guillain-Barré syndrome. Ann Neurol (Suppl) 27:544, 1990.

26. Koski, CL, Jungalawala, FB, and Chou, D: Anti-peripheral nerve myelin antibodies in Guillain-Barré syndrome bind a neutral glycolipid of peripheral nerve myelin and cross react with Forssman antigen. Ann Neurol 24:122, 1988.

27. Lewis, RA, Sumner, AJ, Brown, MJ, et al: Multifocal demyelinating neuropathy with persistent conduction block. Neurol 32:958, 1982.

28. Lewis, RA and Sumner, AJ: The electrodiagnostic distinction between chronic familial and acquired demyelinative neuropathies. Neurol 32:592, 1982.

29. McCombe, PA, McManis, PG, Frith, JA, et al: Chronic inflammatory demyelinating polyradiculoneuropathy associated with pregnancy. Ann Neurol 21:102; 1985.

30. McCombe, PA, Pollard, JD, and McLeod, JG: Chronic inflammatory demyelinating polyradiculoneuropathy. Brain 110:1617, 1987.

31. McFarlin, DE: Immunological parameters in Guillain-Barré syndrome. Ann Neurol 27 (Suppl):S25, 1990.

32. McKann, GM, Griffin, JW, Cornblath, DR, et al: Plasmapheresis and Guillain-Barré syndrome: Analysis of prognostic factors and the effect of plasmapheresis. Ann Neurol 23:347, 1988.

33. Mendell, JR, Kollein, S, Kissel, JT, et al: Evidence for central nervous system demyelination in chronic inflammatory demyelinating polyradiculoneuropathy. Neurol 37:1241, 1987.

34. Parry, G and Clarke, S: Pure motor neuropathy with multifocal conduction block masquerading as motor neuron disease. Muscle Nerve 8:617, 1985.

35. Pollard, JD, King, RHM, and Thomas, PK: Recurrent experimental allergic

neuritis: An electron microscope study. J Neurol Sci 24:365, 1975.

36. Prineas, JB: Pathology of the Guillain-Barré syndrome. Ann Neurol (Suppl) 9:6, 1981.

37. Prineas, JB: Acute idiopathic polyneuritis. An electron microscope study. Lab Invest 26:133, 1972.

38. Ropper, A: Campylobacter diarrhea and Guillain-Barré syndrome. Arch Neurol 45:655, 1988.

39. Ropper, A: Unusual clinical variants and signs in Guillain-Barré syndrome. Arch Neurol 43:1150, 1986.

40. Ropper, A: Three patients with Fisher's syndrome and normal MRI. Neurol 38:1630, 1988.

41. Ropper, A and Shahani, B: Pain in Guillain-Barré syndrome. Arch Neurol 41:511, 1983.

42. Rosenberg, NL, Lacy, JR, Kannaugh, RC, et al: Treatment of refractory chronic demyelinating polyneuropathy with lymphoid irradiation. Muscle Nerve 8:223, 1985.

43. Rostami, AM, Burns, JB, Eccleston, PA, et al: Search for antibodies to galactocerebroside in the serum and cerebrospinal fluid in human demyelinating disorders. Ann Neurol 22:381, 1987.

44. Saida, T, Saida K, Silverberg, DH, et al: Experimental allergic neuritis induced by galactocerebroside. Ann Neurol (Suppl)9:87, 1981.

45. Sladky, JT, Brown, M, and Berman, PH: Chronic inflammatory demyelinating polyneuropathy of infancy: A corticosteroid-responsive disorder. Ann Neurol 20:76, 1986.

46. Stevens, JG, Pepose, JS, and Cook, ML: Marek's disease: A natural model for the Landry-Guillain-Barré syndrome. Ann Neurol (Suppl) 9:102, 1981.

47. Stewart, GJ, Pollard, JD, McLeod, JG, et al: HLA antigens in the Landry-Guillain-Barré syndrome and chronic relapsing polyneuritis. Ann Neurol 4:285, 1978.

48. Svennerholm, L and Fredman, P: Antibody detection in Guillain-Barré syndrome. Ann Neurol (Suppl) 27:536–540, 1990.

49. Thomas, PK, Lascelles, RG, Hallpike,

JF, et al: Recurrent and chronic relapsing Guillain-Barré polyneuritis. Brain 92:589, 1969.

50. Thomas, PK, Walker, RHW, Rudge, P, et al: Chronic demyelinating peripheral neuropathy associated with multifocal central nervous system demyelination. Brain 110:53, 1987.

51. vanDoorn, PA, Brand, A, Vermeulen, M, et al: Clinical significance of antibodies against peripheral nerve tissues in inflammatory polyneuropathy. Neurol 37:1798, 1987.

52. vanDoorn, PA, Brand, A, Strengers, PFW, et al: High-dose intravenous immunoglobulin treatment in chronic inflammatory demyelinating polyneuropathy: A double-blind, placebo-controlled, crossover study. Neurol 40:209, 1990.

53. Van Den Bergh, P, Logigan, EL, and Kelley J: Motor neuropathy with multifocal conduction blocks. Muscle Nerve 11:26, 1989.

54. Van der Meche, FGA, Vermeulen, M, and Busch, HFM: Chronic inflammatory demyelinating polyneuropathy. Conduction failure before and during immunoglobulin or plasma exchange therapy. Brain 6:1563, 1989.

55. Wijdicks, EFM and Ropper, AH: Acute relapsing Guillain-Barré syndrome after long asymptomatic intervals. Arch Neurol 47:82, 1990.

56. Winer, JB, Hughes, RAC, and Osmond, C: A prospective study of acute idiopathic neuropathy. I. Clinical features and their prognostic value. J Neurol Neurosurg Psychiat 51:605, 1988.

57. Winer, JB, Hughes, RAC, Anderson, M, et al: A prospective study of acute idiopathic neuropathy. II. Antecedent events. J Neurol Neurosurg Psychiat 51:613, 1988.

58. Winer, JB, Gray, IA, Gregson, NA, et al: A prospective study of acute idiopathic neuropathy. III. Immunologic studies. J Neurol Neurosurg Psychiat 51:619, 1988.

59. Wisniewski, H, Terry, RD, Whitaker, J, et al: The Landry-Guillain-Barré syndrome. A primary demyelinating disease. Arch Neurol 21:269, 1969.

Chapter 6

INFECTIOUS AND GRANULOMATOUS NEUROPATHY*

HERPES ZOSTER
LEPROSY
SARCOID NEUROPATHY
HIV-RELATED PERIPHERAL
  NEUROPATHIES
LYME BORRELIOSIS

HERPES ZOSTER

Definition and Etiology

This condition results from infection of the nervous system with varicella-zoster virus. Peripheral nerve dysfunction stems from involvement of the sensory ganglia, spinal and cranial nerves, and the spinal cord.

Pathology

Inflammation and hemorrhagic necrosis of sensory ganglion neurons is the hallmark of varicella-zoster. Distribution of lesions in any case is usually confined to one or two adjacent dorsal root ganglia, but the trigeminal ganglia and geniculate ganglia of the facial nerve may also be affected. Varicella-zoster virus particles can be seen in the appropriate ganglion neurons, and virus can be recovered from these loci during the active inflammatory phase.[10] Inflammatory changes, usually confined to the dorsal root ganglia and adjacent dorsal roots, occasionally involve the corresponding segments of the spinal cord. Necrotizing myelitis is present in the adjacent dorsal root entry zone and extends into the ventral horns.[55] The segmental nerves and dorsal columns display varying amounts of secondary axonal degeneration.

Pathogenesis

The pathogenesis of herpes zoster is poorly understood. It is generally held that following a generalized varicella-zoster infection, the virus lies dormant in sensory ganglia and becomes activated during various provocative situations (altered immune states, and so forth).[8] There is no direct proof of this notion, and in contrast to herpes simplex, the virus has never been recovered from the ganglia of asymptomatic individuals.

Clinical Features

Herpes zoster can occur at any age, may recur, is more frequent in older individuals, and has an annual incidence of 0.1 to 4.8 per thousand. Malignancy, especially lymphoma, is the most common predisposing factor.[81] Zoster radiculitis occurs in about 5% to 10% of HIV infected patients.

Pain and paresthesias in one sensory dermatome often precede the appearance of vesicles by 1 to 3 weeks and persist throughout the illness.[60] The vesicular rash is usually confined to one spinal segment, most commonly a thoracic dermatome (Fig. 6–1). Zoster infection of the trigeminal and facial nerves may occur. Vesicles become encrusted by 5 to 10 days and disappear, leaving small scars. Within 2 weeks of the rash, rapidly developing segmental paralysis may appear, affecting the identical segments as the preceding

---

*Prepared with the assistance of Dr. Steven Herskovitz.

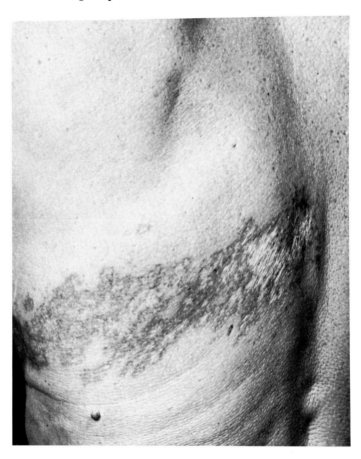

Figure 6–1. Healed lesions of herpes zoster covering adjacent thoracic dermatomes.

sensory involvement. Motor complications most commonly follow lumbosacral and cervical zoster.[104] Usually, only mild radicular deficits occur, but occasionally, marked paralysis and wasting may result, including diaphragmatic paralysis and a neurogenic bladder. Zoster infection of the facial nerve results in facial paralysis identical to idiopathic Bell's palsy. Vesicles are usually noted over the tympanic membrane and external auditory canal but may be absent. Loss of taste over the anterior two thirds of the tongue is common. Partial or complete recovery of facial nerve function is usual.

The CSF displays a variable pleocytosis and moderate increase in protein content. Electromyographic evidence of denervation is present in cases with segmental weakness.

The prognosis for recovery is good in most cases. The most disabling feature of herpes zoster ganglionitis is persistent pain (postherpetic neuralgia). This syndrome of intense segmental pain occurs in about 10% to 20% of cases and may last for months or years.[56] Acyclovir and steroids have proved helpful in shortening the duration of the rash and lessening the acute pain but, to date, have not prevented development of postherpetic neuralgia. Suggested treatments include sympathetic block in the acute phase, local anesthesia and counter irritation, physical therapy, and amitriptyline.[8] None is completely effective.

LEPROSY

Leprosy is the most common treatable neuropathy in the world, although it occurs rarely in Western medicine. At present there are estimated to be 10 to 20 million persons with leprosy.[105]

Definition and Etiology

Leprosy represents a chronic granulomatous infection by *mycobacterium leprae* (Hansen's bacillus) which primarily affects cutaneous nerve, skin, and nasal mucosa. The two major clinical types are lepromatous and tuberculoid; the nature and extent of infection is predominantly determined by the host's immune response. A third type has features of both major varieties. Borderline cases exist between the dimorphous form and the polar lepromatous and tuberculoid forms. Definitive diagnosis is by skin or cutaneous nerve biopsy.

Pathology

The lepromatous and tuberculoid forms represent the polar extremes of the immunologic response to *M. leprae*. The widespread neuropathy of lepromatous leprosy is thought to result from a poor immune response.[5] It is an extensive, diffuse, symmetric neuropathy, featuring many organisms in Schwann cells and macrophages, and in the early stages, by well-preserved nerve architecture and little focal granulomatous inflammatory reaction. PNS involvement extends well beyond the patchy skin lesions. Early stages are characterized by predominant infestation of Schwann cells of unmyelinated fibers by clusters of organisms, occasional foamy macrophages, and little inflammatory response (Fig. 6–2). Eventually there is diffuse, near-total fiber loss; many cutaneous nerves are converted to swollen bundles of connective tissue.[59]

Tuberculoid leprosy, in contrast, is a focal or multifocal condition, and the superficial nerve involvement is often localized near the immediate zone of the skin lesions.[59] The nerve architecture is totally destroyed, even in early

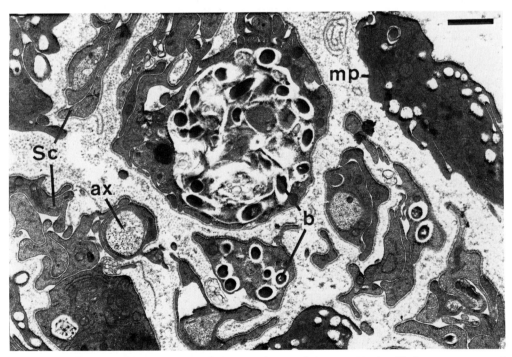

Figure 6–2. Electron micrograph of radial nerve biopsy from a patient with leprous neuropathy. Sc = Schwann-cell processes; b = bacilli; ax = unmyelinated axon; mp = macrophage. Bar = 1 μm.

stages, by an intense inflammatory-granulomatous reaction. Few organisms are present, but *M. leprae* antigen is detectable. The focal nature and severe inflammatory cell reaction of tuberculoid leprosy are thought to result from a strong immunologic response.

Pathogenesis

M. leprae probably enters the body through the skin or nasal mucosa. In lepromatous leprosy, bacillemia follows local infection[37] and organisms settle in stereotyped cool locations (distal extremities, exposed areas of the face, ears, scrotum) that are all several degrees lower than 37°C. This tendency probably explains the initial involvement of superficial cutaneous nerves.[90] Some focal nerve lesions may be related to entrapment secondary to enlargement of nerve trunks. Bloodstream dissemination probably has a much less important role in tuberculoid leprosy, and the initial lesions may be situated close to the portal of entry. It is suggested that there is local spread both from Schwann cell to Schwann cell[33,59] and along axons.[89]

The eventual distribution of destructive nervous system lesions in leprosy thus appears to be determined by the immune response of the host; for example, persons who develop diffuse PNS involvement (lepromatous) probably have poor specific cellular immunity, while individuals with tuberculoid leprosy, whose lesions may be restricted to patchy skin areas with adjacent nerve destruction, are immunologically competent.

Clinical Features

Sensory loss is the cardinal feature of leprosy, regardless of type. Initial sensory loss is due to intracutaneous nerve damage and not in the pattern of individual peripheral nerves or nerve roots. Generally, pain and temperature are the first modalities affected, probably

reflecting early involvement of Schwann cells of small myelinated and unmyelinated fibers in superficial nerves.[91] Loss of sweating in anesthetic areas is also an early feature. Position and vibration sense and tendon reflexes are preserved because they are mediated by larger myelinated fibers in deeply placed (i.e., warm) nerve branches. The overall spatio-temporal pattern of peripheral nervous system involvement is determined by the type of leprosy.

TUBERCULOID LEPROSY

Skin lesions (hypopigmentation) are anesthetic, reflecting concurrence of skin and nerve involvement. Anesthesia rarely extends far beyond the edge of the affected skin.[89] Generally, only one or two lesions are present over the entire body. The most commonly involved sensory nerves are the digital, sural, radial, and posterior auricular. Motor involvement may occur in the distribution of ulnar, median, or peroneal nerves. Affected nerves often are swollen, can be palpated, and usually display slowed nerve conduction from the earliest stages. Autonomic dysfunction of affected nerves is the rule, probably reflecting unmyelinated fiber involvement, so that the cutaneous lesions are anhidrotic.

LEPROMATOUS LEPROSY

Skin lesions are multiple or diffuse and often not anesthetic, in contrast to tuberculoid leprosy. The stereotyped pattern and widespread nature of PNS involvement reflect, respectively, the tendency of *M. leprae* to localize in cool areas and the diffuse hematogenous spread in low-resistance lepromatous leprosy.[51] Cutaneous thermal and pain sensory loss first appear over the ears, the dorsal surface of hands, forearms, feet, and lateral legs. In time—usually years—sensory loss extends to yield a stocking-and-glove distribution with palmar sparing, as well as involving the nasal, malar and eyebrow regions.[89] Sensory loss may result in disabling foot ulcers. At this stage, paralysis of the hands appears secondary to ulnar nerve involvement above the elbow. Other motor nerves are spared at this stage, save for the deep peroneal branch at the ankle. Untreated cases gradually develop sensory loss over the face and palms, paralysis of selected facial muscles, and involvement of median and common peroneal nerves. Widespread anhidrosis is a pronounced feature of the illness. The tendon reflexes are relatively spared. In contrast to tuberculoid leprosy, the enlarged nerves of lepromatous leprosy may function well in the initial stages of the disease.[32]

Prognosis and Treatment

The prognosis for recovery from individual peripheral nerve lesions in tuberculoid leprosy is poor because of the extensive destruction of nerve architecture that characterizes even the early lesions. Early peripheral nerve lesions in lepromatous leprosy may be stabilized by antibacterial treatment and considerable function preserved. Later stages of lepromatous leprosy neuropathy carry a poor prognosis.[89]

The treatment of leprosy is best left to specialists. Specific chemotherapy with sulfones forms the basis for therapy and diamino-diphenylsulfone (dapsone, DDS) is the drug of choice.[57] The daily dose is 50 to 100 mg, and in tuberculoid leprosy is continued for two years after the disease has become inactive. Dapsone itself may cause a reversible motor neuropathy after prolonged use (Chapter 17). Therapy for lepromatous leprosy is continued for 6 to 10 years after bacilli can no longer be detected in skin smears—or perhaps for life. Multiple drug therapy is often used because of drug resistance, and the most common additional drugs are clufazimine (B663)[57] and rifampicin (rifampin).[89]

Reversal reactions are important in the causation of neural damage. An "upgrading" reaction may develop in borderline cases during the first year of treatment, characterized by a heightened cell-mediated response. It may also develop in untreated borderline tuberculoid cases. Pain and swelling of nerve trunks occur with the appearance of sensory and motor deficits in the territory of affected nerves. Tissue necrosis may occur, with the formation of nerve abscesses. The reaction subsides after a few months when many but not all patients are found to have moved toward the tuberculoid form. A downgrading reaction occurs rarely in ineffectually or untreated borderline cases, involving a change in the lepromatous direction.

Erythema nodosum may develop a few months after the initiation of treatment in patients at the lepromatous end of the spectrum. Painful erythematous cutaneous papules appear. These may be accompanied by painful focal nerve lesions. This reaction is associated with the death of large numbers of bacilli. The nerve lesions involve vasculitis and inflammatory infiltration.[52] The process is considered to resemble an Arthus reaction, with the deposition of complement and immune complexes around blood vessels.

A most important aspect in the management of lepromatous neuropathy is the prevention of damage to anesthetic areas by accidental burns, ill-fitting footwear, and so forth. Careful instruction must be given to patients at risk. Reconstructive plastic and orthopedic surgery, proper orthotics, and attention to foot hygiene are helpful in individuals with severe neuropathy. Facial surgery and tendon transplants are among the most common procedures. In many parts of the world, the psychologic trauma that formerly resulted from prolonged hospitalization is now prevented by home treatment.

SARCOID NEUROPATHY

Definition and Etiology

Sarcoidosis is a multisystem granulomatous disorder of unknown etiology, most commonly affecting young adults and presenting frequently with bilateral hilar lymphadenopathy, pulmonary infiltration, and skin or eye lesions.[95] Peripheral neuropathy of uncertain pathogenesis develops in about 5% of cases; both multiple mononeuropathy, involving cranial and limb nerves, and symmetric polyneuropathy occur.[36]

Pathology and Pathogenesis

It is likely that a mixture of localized granulomatous infiltration and vascular compromise play predominant roles in all forms of sarcoid polyneuropathy.[82,83] There are few careful postmortem or nerve-biopsy studies of sarcoid polyneuropathy. Noncaseating granulomas develop within the endoneurium or perineurium of cranial nerve roots, spinal nerve roots, and peripheral nerves. Presumably they displace and compress adjacent fibers with resultant segmental demyelination and axonal degeneration. Granulomas may also infiltrate epineurial arterioles. A nerve-biopsy report from a case of symmetric polyneuropathy describes near-occlusion of arterioles by granulomata, in addition to perivascular inflammation.[22,83] Necrotizing vasculitis and focal areas of ischemic nerve have not been described. Frequently nerve biopsies show nothing other than nerve fiber loss. There is no explanation for the common involvement of the seventh cranial nerve. There is no definitive support for the previous notions that facial neuropathy stems from parotid involvement or basal meningitis.

Clinical Features

MULTIPLE MONONEUROPATHY

There are two distinct types of multifocal neuropathy, one affecting cranial nerves (cranial polyneuritis) and the other, spinal nerves. Each may occur alone, or both patterns may coexist in the same individual.

CRANIAL MONONEUROPATHY

Cranial mononeuropathy in sarcoidosis is almost synonymous with facial palsy.[102] The clinical profile of an episode of facial nerve palsy in sarcoidosis is indistinguishable from severe idiopathic Bell's palsy. Pain behind the ear, sudden onset, and loss of taste are all characteristic. Hyperacusis is rare. Paralysis is usually complete, denervation is common, and incomplete recovery the rule. Although bilateral involvement is frequent, *simultaneous* bilateral involvement is rare; generally one side of the face becomes paralyzed and recovers before the other side is affected. Facial palsy may be the initial sign of sarcoidosis, or may appear after systemic disease is obvious. Other lower cranial nerves (8–12) are less commonly involved.[74,103] CSF examination is usually normal.

SPINAL MONONEUROPATHIES

The spinal nerve mononeuropathies of sarcoidosis have an acute or subacute onset and are generally indistinguishable from other mononeuritides (diabetes, necrotizing vasculitis). Patchy sensory loss over the trunk or abdomen accompanied by pain is particularly characteristic.[22] Rarely, there is predominant involvement of ventral roots, producing a syndrome of diffuse weakness resembling AIDP,[101] but with a more insidious onset.

SYMMETRIC POLYNEUROPATHY

This is a rare complication of sarcoidosis and few cases have been described.[82,83] The condition is a progressive, distal sensorimotor polyneuropathy, clinically indistinguishable from symmetric polyneuropathies caused by toxic or metabolic disease. Progression is either subacute or chronic, and the lower limbs are principally involved. Bilateral facial weakness may coexist. There may be little in the clinical profile to suggest multifocal destruction of nerves, and nerve biopsies have often failed to demonstrate the presence of sarcoid granulomata. Nevertheless, it is still possible that this disorder stems from widespread granulomatous infiltration. Paradoxically, this form of neuropathy is not usually accompanied by obtrusive evidence of systemic disease.

Diagnosis and Prognosis

The diagnosis of sarcoid mononeuropathy is not difficult when accompanied by signs of widespread systemic involvement (pulmonary, skin, ocular). Sural nerve biopsy is usually less helpful than muscle biopsy in mononeuropathy, unless there is clear evidence of sural nerve involvement. The differential diagnosis of the symmetric polyneuropathy is difficult, since most patients do not display overt signs of systemic disease. The presence of facial nerve involvement would heighten suspicion of this disorder, and muscle or nerve biopsy is merited in such cases.

Little is known about the prognosis or natural history of symmetric sarcoid polyneuropathy. Since some cases respond to corticosteroids,[101] a course of treatment is justified, the duration depending on the response. The mononeuropathies generally have a benign prognosis and gradual recovery appears to be the rule. Again, corticosteroid therapy may hasten and improve the degree of recovery. The usual regimen is prednisone 60 mg per day, tapered to 15 mg per day, and continued for the duration of the mononeuropathy. High dose pulse therapy with methyl prednisolone may be of benefit in patients refractory to conventional steroid doses.[3]

HIV-RELATED PERIPHERAL NEUROPATHIES

Definition and Etiology

Peripheral nervous system disease is a frequent complication of infection with the human immunodeficiency virus (HIV). No element of the PNS is immune; virus, neoplasm, or opportunistic infection may affect any structure from anterior horn cell to distal axon. Peripheral neuropathy may complicate any stage of HIV infection and is a source of major morbidity. Figure 6–3 depicts the relative frequency and the time of onset of peripheral nerve disease in HIV infection. Incidence estimates of peripheral neuropathy in the full blown acquired immunodeficiency syndrome (AIDS) range from 9% to 20%.[29,58,67,96] These frequencies probably seriously underestimate the true incidence, as many neuropathies are either subclinical or overshadowed by more pressing medical complications.[109]

Recognition of HIV related peripheral nerve disease is important because neuropathy may: (1) herald HIV seroconversion,[41,87,108] (2) indicate the presence of otherwise asymptomatic HIV infection or mark the transition from an earlier disease stage to more advanced,[26,27] (3) be potentially treatable,[26,31] or (4) predispose to earlier and more severe toxic neuropathies as a side effect of nucleoside therapy.[76] The ability of a pretreatment peripheral neuropathy to potentiate the development of neuropathic side effects from nucleoside anti-viral agents has recently been reported (see Chapter 17). HIV related neuropathies include acute and chronic inflammatory demyelinating neuropathies (AIDP and CIDP),[87,108] cranial neuropathies,[69,78,110] mononeuropathy multiplex,[25,92] progressive polyradiculopathy,[12,40,73] dorsal root ganglioneuritis,[41] distal symmetric sensorimotor neuropathy (DSPN),[7,24,25,66,86,96] and an autonomic neuropathy.[42]

Certain types of HIV-associated neuropathies tend to occur at specific stages of HIV infection (Fig. 6–3).[86] Most strongly associated are the inflammatory demyelinating neuropathies with early stages of relative immunocompetence (up to and including AIDS-related complex-arc) and the later onset DSPN with AIDS.[66,69,78] These pre-

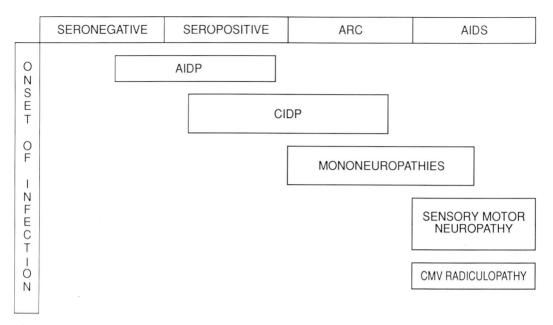

Figure 6–3. A schematic diagram depicting the temporal sequence of the neuropathies associated with HIV infection. Vertical width of each segment is a rough approximation of relative frequency. (Adapted from Johnson, RT, McArthur, JC, and Narayan, O: The neurobiology of human immunodeficiency virus infection. FASEB J 2:2971, 1988, with permission)

dilections are not absolute, and all types of neuropathy can occur at any or all stages of infection. It is alleged that patients who develop inflammatory neuropathies have a better prognosis regarding their ultimate progression to AIDS.[26]

Pathology and Pathogenesis

The precise etiologies and mechanisms for most HIV-related neuropathies remain unclear. It has yet to be shown that the HIV virus itself is directly responsible for any of these varied conditions despite the occasional presence of virus in nerve biopsies, the finding of HIV in infiltrating inflammatory cells,[68] and rare improvement after administration of 3'-azido-2',3'-dideoxythymidine (AZT).[31] Immunopathologic mechanisms are thought to underlie acute and chronic demyelinating neuropathies. Other possible etiologies include coinfection with cytomegalovirus

(CMV),[12,40,73] syphilis,[54] and herpes zoster.[67] CMV has been identified in the Schwann cells of patients with AIDP as well as in the cauda equina of patients with progressive polyradiculopathy and in new biopsies of patients with mononeuropathy multiplex (92A). A recent report claims that the DSPN may result from a CMV dorsal root ganglioneuritis.[15,43] In some instances, neuropathy may be related to or worsened by co-existing metabolic and nutritional factors.

Clinical Syndromes

ACUTE INFLAMMATORY DEMYELINATING NEUROPATHY

A syndrome of subacute generalized weakness, clinically indistinguishable from non-HIV–related AIDP, may rarely occur in patients recently seroconverted or those with asymptomatic infection or ARC.[87] Clinical manifestations and electrophysiology are identical to AIDP as described in Chapter 5. The CSF characteristically includes a mild-moderate lymphocytic pleocytosis[25,26,29] and elevated protein level. This pleocytosis, unusual in non-HIV AIDP, may be the only indication of underlying HIV infection. Positive hepatitis B serology and polyclonal hypergammaglobulinemia may be present.[26] The prognosis is usually good with frequent spontaneous resolution. Plasma exchange is beneficial although the rate of recovery may be slower than that with non-HIV AIDP.[23,26,79]

CHRONIC INFLAMMATORY DEMYELINATING NEUROPATHY

A syndrome clinically identical to CIDP described in Chapter 5 occurs in HIV infected individuals.[18,26,69,79] CIDP occurs early in the course of HIV infections. It is more common than AIDP. Both usually appear during the asymptomatic or ARC stage, although occasionally they may complicate AIDS. Sensorimotor deficits develop over weeks to months in CIDP; they are associated with an elevated CSF protein and electrophysiologic evidence of demyelination and, occasionally, axonal degeneration. CSF pleocytosis and HIV viral antigens in the spinal fluid distinguish HIV related CIDP from the naturally occurring syndrome.[25]

Sural nerve biopsies in patients with AIDP and CIDP occasionally are normal, reflecting the patchy nature of the disorder. In most cases, multiple areas of segmental demyelination associated with varying degrees of axonal degeneration are present, findings similar to the corresponding non-HIV demyelinating neuropathies. In contrast to non-HIV CIDP, however, biopsies from HIV associated CIDP often demonstrate abundant inflammatory cell infiltrates, located in the perineurium, endoneurium, and perivascular areas. HIV is generally not present in nerve biopsy specimens. CMV inclusion bodies have been observed in Schwann cells from spinal roots.[30] Spontaneous recovery

occasionally occurs. Patients with progressive or severe deficits generally respond to human immune gamma-globulin or plasma exchange.[26,43,79] Corticosteroids are best avoided because of the immunocompromised state.

DISTAL SYMMETRIC SENSORIMOTOR POLYNEUROPATHY

The most common neuropathy in HIV infected patients is a distal symmetric sensorimotor polyneuropathy.[7,24,25,66,86,96] It almost always commences in the later stages of AIDS or, less commonly, ARC. DSPN often causes considerable discomfort; clinically, it begins with painful dysesthesias and numbness in the feet with eventual spread to a stocking distribution. Dysesthesia of the soles makes walking difficult and patients commonly wear loose fitting slippers to avoid the unpleasant tactile stimulation of shoes. The hands are rarely involved. Trophic changes, such as loss of hair, thinning of skin, and dependent rubor may be present in the distal legs. There is a variable degree of distal leg weakness, often limited to intrinsic foot muscles. Sensory impairment is usually much more prominent than weakness. The hands usually remain strong. Ankle tendon reflexes are usually absent, the patellar reflexes are diminished but occasionally are brisk, despite prominent sensory loss. Reflexes in the arms are usually normal. A mild CSF pleocytosis and elevated protein level is usual, but less than typically seen with the demyelinating neuropathies.

Electrophysiologic studies indicate an axonal neuropathy characterized by reduced sensory potential amplitudes in the legs and occasionally the arms. Motor potential amplitudes may also be diminished and spontaneous activity is found in intrinsic foot and distal leg muscles. Postmortem examinations and sural nerve biopsies in DSPN demonstrate prominent axonal degeneration with loss of myelinated fibers and varying degrees of segmental demyelination. Inflammatory cell infiltrates are occasionally present but are less conspicuous than with the idiopathic inflammatory demyelinating neuropathies. A necrotizing vasculitis may rarely be present.[44] Prominent degeneration of the gracile tract and loss of dorsal root ganglion cells are described. The neuropathy tends to progress slowly or occasionally plateau with less pain. Spontaneous remissions, although rare, may occur, and in one case treatment with AZT resulted in mild improvement.[31] Therapy for DSPN is usually symptomatic.

PROGRESSIVE POLYRADICULOPATHY

Progressive cauda equina dysfunction characterized by pain, asymmetric flaccid paraparesis, and early sphincter involvement can complicate HIV infection.[12,73] The brunt of the motor and sensory deficits falls on lumbar and sacral radicular segments, but may involve thoracic and cervical roots and even cranial nerves. Sacral sensory loss is prominent with pan-sensory loss in the legs. Pain may be severe. Reflexes are diminished or absent at the ankles and knees. Progressive polyradiculopathy tends to occur in patients with AIDS, although it may also develop during earlier stages of HIV infection. There is a moderate to brisk CSF pleocytosis and markedly elevated CSF protein. Myelography is normal. Electrophysiology demonstrates diffuse axonal degeneration with prominent active denervation in leg and paraspinal muscles. Autopsy examination has demonstrated extensive axonal degeneration within the lumbosacral ventral and dorsal roots with a marked inflammatory cell infiltration. Similar, but less severe changes may occur in thoracic and cervical roots. Inflammatory infiltrates are not present in dorsal root ganglia, lumbosacral plexuses, and peripheral nerves. Numerous intranuclear and in-

tracytoplasmic CMV inclusion bodies are present and lumbosacral anterior horn cells show chromatolysis. The syndrome has a poor prognosis. Treatment, to date, has essentially been unsuccessful. The suggestion that CMV is the responsible agent is somewhat supported by a report of successful treatment with ganciclovir.[80] It is important to exclude syphilitic polyradiculopathy by appropriate serum and CSF serology.[65] Syphilis may complicate HIV infection, and can cause a progressive polyradiculopathy somewhat similar to that ascribed to CMV, but responds to penicillin therapy.

DORSAL ROOT GANGLIONEURITIS

One case of an inflammatory dorsal root ganglioneuritis resulting in an ataxic neuropathy is reported.[41] CSF protein remains normal and antibodies to HIV are present in the spinal fluid. Sural nerve biopsy demonstrates loss of large myelinated fibers without inflammatory cell infiltrates. Autopsy reveals selective loss of sensory neurons in the dorsal root ganglia and inflammatory infiltrates within dorsal root ganglia and dorsal roots. HIV has not been cultured from ganglia or nerve.

MULTIPLE MONONEUROPATHY MULTIPLEX AND CRANIAL NEUROPATHIES

Multifocal involvement of spinal and cranial nerves may occur in HIV-infected patients.[26,69,78,92] The clinical picture may be one of multiple discrete sensorimotor deficits, or a generalized relatively symmetric neuropathy, representing the cumulative deficits of many separate nerve lesions. This multiple mononeuropathy is distinct from multifocal demyelinating lesions occurring in CIDP, radiculitis from herpes zoster,[67] and neoplastic infiltration of peripheral nerves, plexuses, or subarachnoid space. The facial nerve is the most common cranial nerve affected, followed by the trigeminal.[78] Electrophysiology usually shows multifocal axonal degeneration, although mixed axonal and demyelinating lesions may occur. Sural nerve biopsy demonstrates axonal degeneration with inflammatory infiltrates in the endoneurium, epineurium, and perivascular spaces. CMV has recently been described in nerve biopsies from patients with this condition (92A). A frank vasculitis with muscle necrosis is also described.[92] The course is variable; some patients show spontaneous resolution while others progress to multiple deficits and severe disability.

Isolated cranial neuropathies are relatively rare. The majority occur in the earlier stages of HIV infection and result from an aseptic meningitis. Facial palsies and trigeminal neuropathies may appear as a facet of a multifocal neuropathy or with meningeal involvement by neoplasm or infection. Isolated facial palsy occasionally heralds seroconversion. Facial palsies are usually transient and tend to recover spontaneously and completely.[16,110]

AUTONOMIC NEUROPATHY

Occasional patients have orthostasis, impotence, and diarrhea, reflecting a presumed generalized autonomic neuropathy.[42] Autonomic dysfunction may also occur as part of AIDP.

IATROGENIC PERIPHERAL NEUROPATHIES

Peripheral neuropathies may result from administration of various medications including INH, nitrofurantoin, dapsone, and new antiviral agents, such as 2′,3′-dideoxycytidine (ddC) and dideoxyinosine (ddI).[38] The neuropathies from ddC and ddI are noteworthy as they are purely sensory and at times clinically confused with the DSPN occurring during AIDS (see Chapter 17). Patients with pretreatment peripheral nerve disease appear especially vulnerable to the neuropathic side effects of nucleosides such as ddC and ddI.

LYME BORRELIOSIS

Definition and Etiology

Lyme disease, or Lyme borreliosis (LB), is a multisystem infectious disease caused by the spirochete *Borrelia burgdorferi*.[98] It is a vector-borne illness representing a complex interaction between organism, vector (usually the species of tick, *Ixodes dammini*, and similar ticks), reservoir (white-footed mouse; deer), and host (many species of bird and mammal, including man and household pets). Presently, it is the most commonly reported vector-borne disease in the United States, and occurs worldwide. Endemic areas in the United States include the Northeast, upper Midwest, and Pacific Coast. The onset is usually in the spring and summer when ticks are most active, but cases are reported year-round.[17,19,98]

Pathology

There are few histopathologic studies of the PNS. Postmortem samples demonstrate lymphocytes and plasma cells infiltrating the autonomic ganglia and afferent and efferent rootlets.[39] Sural nerve biopsy specimens in patients with radiculoneuropathy or polyneuropathy contain perivascular infiltrates of lymphocytes and plasma cells around small endoneurial, perineurial, or epineurial vessels.[39,48,75,107] There is occasional occlusion of small vessels, but usually no frank necrotizing vasculitis.[39,75] In most studies, axonal degeneration is the predominant pathologic change. Spirochetes are not observed in peripheral nerve, but presumably are there in small numbers; they are identifiable in most other tissues, including the CNS.[39] There is a murine experimental animal model of Lyme arthritis, but not of neuropathy.[9] Muscle biopsies in the few reported patients with clinical myopathy show an inflammatory or noninflammatory necrotizing myositis; the spirochete has been demonstrated in one instance.[6,77,93]

Pathogenesis

Once the spirochete penetrates the skin, there is presumed to be hematogenous spread to both CNS and PNS. The organism can be isolated from blood during the early systemic phase of the illness.[13,39,93] The observation in European series that facial palsy or radiculoneuropathy (Bannwarth's syndrome) generally occurs ipsilateral to the rash suggests that the organism may migrate within the perineurial sheath.[100] Transplacental transmission is documented, resulting in both congenital defects and fetal demise.[71]

Direct tissue invasion is likely responsible for the histopathologic changes.[39] The presence of spirochetes in most tissues and the often clear response of the meningitis and CNS/PNS lesions to antibiotics supports this notion.[98] Vasculopathy is suggested by the presence of multifocal lesions and small vessel occlusion in the PNS, and in the CNS by the presence of occasional transient and permanent cerebrovascular events and a few cases of vasculitis on cerebral angiography.[39,77,106] It is unclear if the immune response is a critical feature. Cross reacting serum IgM antibodies betweeen *B. burgdorferi* and axonal antigens are present; their significance is unknown.[1,94] Studies have failed to show deposition of immunoglobulins or complement in peripheral nerve.[46,48,75]

Clinical Features

SYSTEMIC DISEASE

The clinical features of LB, like those of syphilis, are often described in three stages.[98] **Stage 1:** characteristic rash, erythema migrans (EM), and a flu-like syndrome. **Stage 2:** carditis, arthritis, and meningoradiculitis. **Stage 3:** chronic skin, arthritic, and neurologic manifestations.

The disorder often does not conform to these stages in an orderly fashion. It may present initially with almost any feature weeks to even many years after infection, and elements of different stages may overlap.[47,98] This organism, like *Treponema pallidum*, is capable of prolonged survival in the human host, and may become clinically apparent at variable and unpredictable intervals. An NIH Consensus Conference recommends abandonment of this staging system in favor of one that describes whether the disease is local or disseminated, acute or chronic: stage 1, local; stage 2, acute disseminated; or stage 3, chronic disseminated disease. Some authors use the term early LB for features of stages 1 and 2, and late LB for stage 3.

After inoculation of the organism into the skin by an infected tick, approximately 60% of patients develop the pathognomonic skin rash, EM, within 3 to 30 days.[14,98] This lesion is a centrifugally expanding, erythematous, annular macule or papule, often with a target appearance. Biopsy of the leading edge may reveal the spirochete. Up to 50% of patients may develop multiple satellite skin lesions, representing metastatic foci.[4] Chronic skin manifestations include *Borrelia lymphocytoma*, a tumor-like nodule of lymphocytes occurring on the earlobes or nipples, and *acrodermatitis chronica atrophicans* (ACA), a mixed inflammatory/atrophic lesion of the distal extremities that can persist for decades, with the organism demonstrable on skin biopsy.[4]

Within days to weeks of infection, the spirochete disseminates widely, and can be isolated from blood and multiple tissues.[39,98] This frequently causes systemic symptoms including severe fatigue, headaches, fever, meningismus, arthralgias, myalgias, and lymphadenopathy. These are often intermittent and fluctuating, as are the neurologic symptoms.[98] Approximately 8% of untreated patients develop cardiac abnormalities, including varying degrees of atrioventricular block, and less commonly, myocarditis, pericarditis, or a dilated cardiomyopathy.[61,97,98] Occasionally, a temporary pacemaker is required. Arthritis is common, occurs in

50% to 60% of untreated patients, begins days to years after infection, and is often relapsing-remitting.[98] There may be arthralgia alone, episodic arthritis, or chronic arthritis, occasionally erosive. Usually it is monoarticular or asymmetric oligoarticular, involving large joints, especially the knee.

NEUROLOGIC DISEASE

In stage 1, early LB, there is often headache, mild or severe; usually it is unlike previous headaches. There may be mild meningeal signs, but the CSF is normal. About 15% to 20% of patients later experience neurologic disease.[47,98]

Central Nervous System. Meningitis is common, occurring in about 80% of patients with neurologic involvement.[88] The clinical and CSF picture is of an aseptic meningitis.[85,88,98] Encephalomyelitis may occur, ranging from a mild encephalopathy with subtle cognitive dysfunction, to a severe progressive encephalomyelitis with focal or multifocal features.[2,47,49,62,84] It may mimic many other disorders. Infrequently, transient ischemic attacks, strokes, or seizures are part of the clinical picture. Fatigue is common, and often prominent.[98] It can occur in early or late LB, usually responds to therapy, but occasionally persists for prolonged periods in apparently adequately treated cases and may be addressed symptomatically. A more difficult group are seropositive patients regarded as having a chronic fatigue syndrome but without a clear history or signs implicating LB. The relationship of LB to an isolated chronic fatigue syndrome remains to be established.

Peripheral Nervous System

Cranial Neuropathies. Cranial neuropathies usually occur with associated meningitis, although facial palsy may occur alone. Every cranial nerve (CN), aside from the olfactory nerve, is reportedly involved in LB, but only CN 3, 5, 6, and 7 occur with frequency.[47,98] With trigeminal involvement, patients describe facial paresthesias, numbness or atypical facial pain, invariably have meningitis, and usually respond to antibiotics.[47] Facial palsy is common; it occurs in as many as 10% of patients with untreated LB.[20,47,98] It may be the presenting feature, can be clinically indistinguishable from idiopathic Bell's palsy, and is usually distal to the corda tympani based on clinical symptoms. Up to one third of cases are bilateral.[85] Bilateral facial palsy in endemic areas strongly suggests LB.

Peripheral (Non-Cranial) Neuropathies. The PNS is involved in 30% to 50% of patients with neurologic abnormalities[70,75,88,93,94] (Table 6–1). The process appears as a multifocal neuropathy affecting roots, plexuses, and individual peripheral nerves. The clinical pattern depends on the site of most severe damage. In the acute disseminated stage (stage 2) of LB, PNS involvement may take the form of a monoradiculopathy or polyradiculopathy, plexopathy, mononeuropathy multiplex, or rarely a fulminant, generalized neuropathy that resembles AIDP. Radiculopathy is typically asymmetric, painful, and associated with focal weakness, atrophy, and sensory and reflex loss in varying combinations in the arms, legs, or trunk. It may be clinically indistinguishable from a mechanical radiculopathy or the thoracoabdominal polyradiculopathy of diabetes.[28] The plexopathy pattern is usually brachial, occasionally lumbosacral, and is similar to idiopathic brachial

Table 6–1 PNS SYNDROMES IN LYME BORRELIOSIS

NEUROPATHIES
  Monoradiculopathy/polyradiculopathy
  Brachial/lumbosacral plexopathy
  Multiple mononeuropathy
  Acute motor > sensory neuropathy (AIDP-like)
  Chronic mild sensorimotor neuropathy
  Acrodermatitis chronica atrophicans associated neuropathy
  Carpal tunnel syndrome
  AIDP

CRANIAL NEUROPATHIES
  Facial palsy–common
  All others

neuritis. The mononeuropathy multiplex pattern is uncommon and may be associated with ACA.[63] Rarely, an AIDP-like illness occurs with a generalized, symmetric, predominantly motor polyneuropathy. Nerve conduction values suggest demyelination, but usually there is a vigorous CSF pleocytosis.[99] A few cases of true AIDP are reported in association with LB, but the relationship is unclear.[21,47,53]

It is reported that in the late, chronic, disseminated stage (stage 3) patients may develop a subtle, generalized distal sensorimotor neuropathy.[48] Symptoms include mild, fluctuating, ill-defined paresthesias; signs are minimal. Nerve conduction studies demonstrate a diffuse, patchy axonal polyneuropathy. Carpal tunnel syndrome is described in 25% of patients with LB;[50] LB-induced wrist arthritis may play a role. Myositis is an infrequent feature of LB; when present, it resembles polymyositis.[6,77,93]

Laboratory Studies

CLINICAL LABORATORY FINDINGS

Nonspecific abnormalities include mild elevation of erythrocyte sedimentation rate, serum IgM, cryoglobulins, circulating immune complexes, and liver function tests.[98]

Specific diagnosis by culture or histology is difficult. This probably reflects the small number of organisms present, their fastidious nature, and small size. Serology by IFA or ELISA demonstrates antibodies to *B. burgdorferi* in most patients with well-established LB.[72,98] These tests, however, display considerable intra- and inter-laboratory variability. False positives occur from exposure to other spirochetes, particularly *Treponema pallidum*. False negatives can occur in the first 4 to 8 weeks after infection, or in cases receiving early antibiotic treatment. The CSF may occasionally be positive when the serum is not,[100] and a CSF/serum antibody index may demonstrate intrathecal antibody synthesis.[49,100] Some inves-

tigators use Western blot analysis to confirm or clarify ELISA results. Early antibiotic treatment many abrogate the antibody response; in such patients, an early T-cell response may be demonstrated with a lymphocyte proliferation assay.[35] The availability of monoclonal antibodies against *B. burgdorferi* have made antigen assays in various body fluids possible; their reliability remains to be established at this time. Finally, a diagnostic test based on the polymerase chain reaction promises a powerful method of detecting small amounts of target DNA; its usefulness in LB awaits further analysis.

CEREBROSPINAL FLUID

In cases with meningitis, there is a moderate lymphocytic pleocytosis and elevated protein.[85,88,95] The glucose is usually normal. Patients with PNS syndromes or facial palsy may or may not have abnormal CSF.[47]

ELECTRODIAGNOSTIC STUDIES

Nerve conduction studies and electromyography show focal or multifocal abnormalities of predominantly axonal degeneration in patterns suggesting monoradiculopathy or polyradiculopathy, plexopathy, multiple mononeuropathy, or polyneuropathy.[47,48,85,107] Patients with clinically restricted lesions or minor symptoms, even with facial palsy alone, may have more widespread nerve conduction abnormalities.[48] Uncommonly, demyelinating changes are noted.[45,99] Myopathic motor unit changes and spontaneous activity may be present with myositis.[6,93] Median nerve entrapment is present up to 25% of late LB.[50]

Course, Prognosis, and Treatment

Antibiotics clearly improve symptoms, prevent progression to later stages, and can eradicate the infection in most patients, including those with PNS syndromes.[70,98] Improvement can be documented electrophysiologically.[48] Treatment choice and dosage must be adequate to penetrate the blood-brain and blood-nerve barriers and of sufficient duration to kill this slowly replicating organism. Currently recommended regimens for early stage-1 disease include oral doxycycline, 100 mg, bid, or amoxicillin, 500 mg to 1000 mg, tid for 21 days.[70,109] Whether this regimen for 21 to 30 days is adequate for isolated facial palsy is unclear, since some have abnormal CSF. All other neurologic disease is treated with intravenous ceftriaxone, 2 g per day, or penicillin G, 20 million to 24 million units per day.[70,98] Prolonged treatment with oral antibiotics may prove to be adequate for mild polyneuropathies, but this remains unestablished. A Jarisch-Herxheimer reaction occurs in 10% to 20% of cases.[70] Treatment failures occur with every regimen. Residual or recurrent symptoms and signs may result from true treatment failure, irreversible tissue damage, reinfection, or possibly have an immunologic basis. There is no clear role for corticosteroids in LB, their use may be associated with a poor response to antibiotics.[34]

The prognosis in LB, including the neuropathies and facial palsy is generally favorable with, and surprisingly occasionally without, antibiotic treatment.[11,64,85,98] Most patients make a good-to-excellent recovery over weeks to months, though occasionally it is incomplete. Mortality is unusual, occurring essentially only as a result of fatal arrhythmias or carditis.[98]

Case History: Lyme Borreliosis (LB)

A 25-year-old woman enjoyed gardening in rural Westchester County, New York. She noted a tiny, circular, erythematous rash under the left breast; she recalled no tick bites. The rash was painless but gradually expanded over a few days. As it spontaneously cleared two weeks later, she developed mild flu-like symptoms including headache, fatigue, and myalgias. These fluctuated over

the next few weeks and resolved. One month later there was severe left neck, shoulder, and scapular pain radiating to the lateral hand, along with headache, stiff neck, and fatigue. Within days, she awoke one morning with complete left facial palsy. Examination revealed bifacial palsy, severe on the left and mild on the right.

There was moderate weakness of left C5–C6 segmentally innervated muscles, as well as serratus anterior and trapezius muscles. Sensory examination was normal. The left biceps reflex was depressed. CSF showed 50 lymphocytes, a protein of 85 mg%, and normal glucose. MRI of the cervical spine was normal, as were nerve conduction studies. EMG showed denervation in all left C5, C6, and C7 innervated muscles, as well as multiple cervical paraspinal muscles bilaterally from C4 to C8. Both serum and CSF Lyme titers were strongly positive by ELISA. Treatment was initiated with ceftriaxone, 2 g intravenously, daily. After the first dose, there was mild fever and encephalopathy that lasted 24 to 48 hours. The headache and radicular pain resolved within 1 week; at 2-month follow-up, there was only minimal residual arm and facial weakness. Fatigue lingered for several months.

Comment. This case illustrates several characteristic features of this disease, including the typical rash, aseptic meningitis, bifacial palsy, and cervical polyradiculopathy. Less than 50% of patients recall the usually painless tick bite. A Jarisch-Herxheimer reaction occurred in this case. Despite evidence of axonal degeneration, there was a gratifying recovery.

REFERENCES

1. Aberer, E, Brunner, C, and Suchanek, G: Molecular mimicry and Lyme borreliosis: A shared antigenic determinant between Borrelia burgdorferi and human tissue. Ann Neurol 26:732, 1989.
2. Ackermann, R, Rehse-Kupper, B, and Gollmer, E: Chronic neurologic manifestations of erythema migrans borreliosis. In Benach, JL

and Bosler, EM (eds): Lyme Disease and Related Disorders, Vol 539. The New York Academy of Sciences, New York, 1988, p 16.

3. Allen, RK and Merory, J: Intravenous pulse methyl prednisolone in the successful treatment of severe sarcoid polyneuropathy with pulmonary involvement. Aust N Z Med 15:45, 1985.

4. Asbrink, E and Hovmark, A: Early and late cutaneous manifestations in Ixodes-borne borreliosis (erythema migrans borreliosis, Lyme borreliosis). In Benach, JL and Bosler, EM (eds): Lyme Disease and Related Disorders, Vol 539. The New York Academy of Sciences, New York, 1988, p 4.

5. Asbury, AK and Johnson, PC: Pathology of Peripheral Nerve. WB Saunders, Philadelphia, 1978, p 184.

6. Atlas, E, Novak, SN, and Duray, PH: Lyme myositis: Muscle invasion by Borrelia burgdorferi. Ann Intern Med 109:245, 1988.

7. Bailey, RO, Baltch, AL, Venkatesh, R, Singh JK, and Bishop, MB: Sensory motor neuropathy associated with AIDS. Neurology 38:886, 1988.

8. Baringer, JR and Townsend, JJ: Herpes virus infection of the peripheral nervous system. In Dyck, PJ, Thomas, PK, and Lambert, EH (eds): Peripheral Neuropathy, Vol II, ed 2. WB Saunders, Philadelphia, 1984, p 1941.

9. Barthold, SW, Moody, KD, Terwilliger, GA, et al: An animal model for Lyme arthritis. In Benach, JL and Bosler, EM (eds). Lyme Disease and Related Disorders, Vol 539, The New York Academy of Sciences, New York, 1988, p 264.

10. Bastian, FO, Rabson, AS, and Yee, CL: Herpes virus varicellae. Arch Pathol 97:331, 1974.

11. Bateman, DE, Lawton, NF, and White, JE: The neurological complications of Borrelia burgdorferi in the New Forest area of Hampshire. J Neurol Neurosurg Psychiatry 51:699, 1988.

12. Behar, R, Wiley, C, and McCutchan, JA: Cytomegalovirus polyradiculoneuropathy in acquired immune deficiency syndrome. Neurology 37:557, 1987.

13. Benach, JL, Bosler, EM, and Hannahan, JP: Spirochetes isolated from the blood of two patients with Lyme disease. N Engl J Med 308:740, 1983.

14. Berger, BW: Dermatologic manifestations of Lyme disease. Rev Infect Dis 11, (Suppl 6):1475, 1989.

15. Bishopric, G, Bruner, J, and Butler, J: Guillain-Barré syndrome with cytomegalovirus infection of peripheral nerves. Arch Pathol Lab Med 109: 1106, 1985.

16. Brown, MM, Thompson, A, Goh, BT, Forster, GE, and Swash, M: Bell's palsy and HIV infection. J Neurol Neurosurg Psychiatry 51:425, 1988.

17. Burgdorfer, W: Vector/host relationships of the Lyme disease spirochete, Borrelia burgdorferi. In Johnson, RC (ed): Rheumatic Disease Clinics of North America, Lyme Disease, Vol 15, No 4. WB Saunders, Philadelphia, 1989, p 775.

18. Chimowitz, MI, Audet, AJ, Hallet, A, and Kelly, JJ: HIV-associated CIDP. Muscle Nerve 12:695, 1989.

19. Ciesielski, CA, Markowitz, LE, and Horsley, R: Lyme disease surveillance in the United States, 1983–1986. Rev Infect Dis 11, (Suppl 6):1435, 1989.

20. Clark, JR, Carlson, RD, and Sasaki, CT: Facial paralysis in Lyme disease. Laryngoscope 95:1341, 1985.

21. Clavelou, P, Beytout, J, and Vernay, D: Neurologic manifestations of Lyme disease in the northern part of the Auvergne. Neurology 39, (Suppl 1): 350, 1989.

22. Colover, J: Sarcoidosis with involvement of the nervous system. Brain 71:451, 1948.

23. Cornblath, DR: The treatment of the neuromuscular complications of human immunodeficiency virus infection. Ann Neurol 23:S88, 1988.

24. Cornblath, DR and McArthur, JC: Predominantly sensory neuropathy in

patients with AIDS and AIDS-related complex. Neurology 38:794, 1988.

25. Cornblath DR, McArthur JC, and Griffin JW: The spectrum of peripheral neuropathies in HTLV-II infection. Muscle Nerve 9:S76, 1986.

26. Cornblath, DR, McArthur, JC, Kennedy, GE, Witte, AS, and Griffin, JW: Inflammatory demyelinating peripheral neuropathies associated with human T-lymphocytic virus type III infection. Ann Neurol 21:32, 1987.

27. Crawfurd, EJP, Baird, PRE, and Clark, AL: Cauda equina and lumbar nerve root compression in patients with AIDS. J Bone Joint Surg 69:36, 1987.

28. Daffner, KR, Saver, JL, and Biber, MP: Lyme polyradiculoneuropathy presenting as increasing abdominal girth. Neurology 40:373, 1990.

29. Dalakas, MC: Neuromuscular complications of AIDS. Muscle Nerve 9:92, 1986.

30. Dalakas, MC and Pezeshkpour, GH: Neuromuscular diseases associated with human immunodeficiency virus infection. Ann Neurol 23:S38, 1988.

31. Dalakas, MC, Yarchoan, R, Spitzer, R, Elder, G, and Sever, JL: Treatment of human immunodeficiency virus-related polyneuropathy with 3'-azido-2',3'-dideoxythymidine. Ann Neurol 23:S92, 1988.

32. Dastur, DK: Cutaneous nerve in leprosy: The relationship between histopathology and cutaneous sensibility. Brain 78:615, 1955.

33. Dastur, DK, Ramamohan, Y, and Shah, JS: Ultrastructure of lepromatous nerves. Neural pathogenesis in leprosy. Int J Lepr 41:47, 1973.

34. Dattwyler, RJ, Halperin, SS, and Volkman, DJ: Treatment of late Lyme borreliosis—randomized comparison of ceftriaxone and penicillin. Lancet 1:1191, 1988.

35. Dattwyler, RJ, Volkman, DJ, and Luft, BJ: Seronegative Lyme disease. N Engl J Med 319:1441, 1988.

36. Delaney, P: Neurological manifestations

37. of sarcoidosis: Review of the literature, with a report of 23 cases. Ann Intern Med 87:336, 1977.

37. Drutz, DJ, Chen, TS, and Lu, WH: The continuous bacteremia of lepromatous leprosy. N Engl J Med 287:159, 1972.

38. Dubinsky, RM, Yarchoan, R, Dalakas, M, and Broder, S: Reversible axonal neuropathy from the treatment of AIDS and related disorders with 2',3'-dideoxycytidine (ddC). Muscle and Nerve 12:856, 1989.

39. Duray, PH and Steere, AC: Clinical pathologic correlations of Lyme disease by stage. In Benach, JL and Bosler, EM (eds): Lyme Disease and Related Disorders, Vol 539. The New York Academy of Sciences, New York, 1988, p 65.

40. Eidelberg, D, Sotrel, A, Vogel, H, Walker, P, Kleefield, J, and Crumpacker, CS: Progressive polyradiculopathy in acquired immune deficiency syndrome. Neurology 36:912, 1986.

41. Elder, G, Dalakas, MC, Pezeshkpour, GH, and Sever, JL: Ataxic neuropathy due to ganglioneuronitis after probable acute human immunodeficiency virus infection. Lancet 2:1275, 1986.

42. Evenhouse, M, Haas, E, Snell, E, Visser, J, Pawl L, and Gonzalez, R: Hypotension in infection with the human immunodeficiency virus. Ann Intern Med 107:598, 1987.

43. Fuller, GN, Jacobs, JM, and Guiloff, RJ: Association of painful peripheral neuropathy in AIDS with cytomegalovirus infection. Lancet 937, 1989.

44. Gherardi, R, Lebargy, F, Gaulard, P, Mhiri, C, Bernaudin, JF, and Gray, F: Necrotizing vasculitis and HIV replication in peripheral nerves. N Engl J Med 321:685, 1989.

45. Graf, M, Kristoferitsch, W, and Baumhackl, U: Electrophysiologic findings in meningopolyneuritis of Garin-Bujadoux-Bannwarth. Zbl Bakt Hyg 263-324, 1986.

46. Halperin, JJ, Pass, HL, Anand, AK, et al: Nervous system abnormalities in

Lyme disease. In Benach, JL and Bosler, EM (eds): Lyme Disease and Related Disorders, Vol 539. The New York Academy of Sciences, New York, 1988, p 24.

47. Halperin, JJ: Nervous system manifestations of Lyme disease. In Johnson, RC (ed): Rheumatic Disease Clinics of North America, Vol 15, No 4. WB Saunders, Philadelphia, 1989, p 635.

48. Halperin, JJ, Little, BW, and Coyle, PK: Lyme disease: Cause of a treatable peripheral neuropathy. Neurology 37:1700, 1987.

49. Halperin, JJ, Luft, BJ, and Anand, AK: Lyme neuroborreliosis: Central nervous system manifestations. Neurology 39:753, 1989.

50. Halperin, JJ, Volkman, DJ, and Luft, BJ: Carpal tunnel syndrome in Lyme Borreliosis. Muscle Nerve 12:397, 1989.

51. Hastings, RC, Brand, PN, and Mansfield, RE: Bacterial density in the skin in lepromatous leprosy as related to temperature. Lepr Rev 39:71, 1968.

52. Hastings, RC and Trautman, JR: B663 in lepromatous leprosy: Effect in erythema nodosum leprosum. Lepr Rev 39:3, 1968.

53. Herskovitz, S, Berger, A, and Swerdlow, M: Guillain-Barré syndrome associated with Lyme borreliosis. Neurology 40 (Suppl 1):342, 1990.

54. Ho, DD, Rota, TR, Schooley, RT, Kaplan, JC, Alan, JD, Groopman, JE, Resnick, L, Felsenstein, D, Andrew, CA, and Hirsch, MS: Isolation of HTLV-III from cerebrospinal fluid and neural tissues of patients with neurologic syndromes related to the acquired immunodeficiency syndrome. N Engl J Med 313:1493, 1985.

55. Hogan, EL and Krigman, MR: Herpes zoster myelitis: Evidence for viral invasion of spinal cord. Arch Neurol 29:309, 1973.

56. Hope-Simpson, RE: The nature of herpes zoster: A long term study and a new hypothesis. Proc R Soc Med 58:9, 1965.

57. Jacobson, RR and Trautman, JR: The treatment of leprosy with the sulfones. I. Faget's original 22 patients: A thirty-year follow-up on sulfone therapy for leprosy. Int J Lepr 39:726, 1971.

58. Janssen, RS, Saykin, AJ, Kaplan, JE, Spira, TJ, Pinsky, PF, and Sprehn, GW: Neurologic complications of lymphadenopathy syndrome associated with human immunodeficiency virus infection. Neurology 37:S344, 1987.

59. Job, CK: Pathology of peripheral nerve lesions in lepromatous leprosy: A light and electron microscopic study. Int J Lepr 39:251, 1971.

60. Juel-Jensen, BE and McCallum, BO: Herpes Simplex, Varicella and Zoster. JB Lippincott, Philadelphia, 1972.

61. Kimball, SA, Janson, PA, and LaRaia, PJ: Complete heart block as the sole presentation of Lyme disease. Arch Intern Med 149:1987, 1989.

62. Kohler, J, Kern, U, and Kasper, J: Chronic central nervous system involvement in Lyme borreliosis. Neurology 38:863, 1988.

63. Kristoferitsch, W, Sluga, E, Graf, M, et al: Neurology associated with acrodermatitis chronica atrophicans, In Benach, JL and Bosler, EM (eds): Lyme Disease and Related Disorders, Vol 539. The New York Academy of Sciences, New York, 1988, p 35.

64. Kruger, H, Reuss, K, and Pulz, M: Meningoradiculitis and encephalomyelitis due to Borrelia burgdorferi: A follow-up study of 72 patients over 27 years. J Neurol 236(6):322, 1989.

65. Lanska, MJ, Lanska, DJ, and Schmidley, JW: Syphilitic polyradiculopathy in an HIV-positive man. Neurology 38:1297, 1988.

66. Leger, JM, Bouche, P, Bolgert, F, Chaunu, MP, Rosenheim, M, et al: The spectrum of polyneuropathies in patients infected with HIV. J Neurol Neurosurg Psychiatry 52:1369, 1989.

67. Levy, RM. Bredesen, DE, and Rosen-

blum, ML: Neurological manifestations of the acquired immunodeficiency syndrome (AIDS): Experience of UCSF and review of the literature. J Neurosurg 62:475, 1985.

68. Lin-Greenberg, A, Taneja-Uppal, N: Dysautonomia and infection with the human immunodeficiency virus. Ann Intern Med 106:167, 1987.

69. Lipkin, WI, Parry, G, Kiprov, DD, and Abrams, D: Inflammatory neuropathy in homosexual men with lymphadenopathy. Neurology 35:1479, 1985.

70. Luft, BJ, Gorevic, PC, and Halperin, JJ: A perspective on the treatment of Lyme borreliosis. Rev Infect Dis 11, (Suppl 6):1518, 1989.

71. MacDonald, AB: Gestational Lyme Borreliosis. In Johnson, RC (ed): Rheumatic Disease Clinics of North America, Vol 15, No 4. WB Saunders, Philadelphia, 1989, p 657.

72. Magnarelli, LA: Laboratory diagnosis of Lyme disease. In Johnson, RC (ed): Rheumatic Disease Clinics of North America, Vol 15, No 4. WB Saunders, Philadelphia, 1989, p 735.

73. Mahieux, R, Gray, F, Fenelon, G, et al: Acute myeloradiculitis due to cytomegalovirus as the initial manifestations of AIDS. JNNP 52:270, 1989.

74. Matthews, WB: Sarcoidosis of the nervous system. J Neurol Neurosurg Psychiatry 28:23, 1965.

75. Meier, C, Grahmann, F, and Engelhardt, A: Peripheral nerve disorders in Lyme borreliosis. Acta Neuropathol 79:271, 1989.

76. Merigan, TC, Skowron, G, Bozzette, SA, and Richman, D, et al: Circulating p24 antigen levels and responses to dideoxycytidine in human immunodeficiency virus (HIV) infections. A phase I and II study. Ann Intern Med 110:189, 1989.

77. Midgard, R and Hofstad, H: Unusual manifestations of nervous system Borrelia burgdorferi infection. Arch Neurol 44:781, 1987.

78. Miller, R, Kiprov, D, Parry, G, and Bredesen, D: Peripheral nervous system dysfunction in acquired immunodeficiency syndrome (AIDS). In Rosenblum, ML, Levy, RM, and Bredesen, DE (eds): AIDS and the Nervous System. Raven Press, New York, 1988.

79. Miller, RG, Parry, G, Larry, W, et al: AIDS-related inflammatory polyradiculoneuropathy: Successful treatment with plasma exchange. Neurology 36:S206, 1986.

80. Miller, RG, Storey, JR, and Greco, CM: Ganciclovir in the treatment of progressive AIDS-related polyradiculopathy. Neurology 40:569, 1990.

81. Muller, SA: Association of zoster and malignant disorders in children. Arch Dermatol 96:657, 1967.

82. Nemni, R, Galassi, G, and Cohen, M: Symmetrical sarcoid polyneuropathy: Analysis of a sural nerve biopsy. Neurology (Minneap) 31:1217, 1981.

83. Oh, SJ: Sarcoid polyneuropathy: A histologically proved case. Ann Neurol 7:178, 1980.

84. Pachner, AR, Duray, P, and Steere, AC: Central nervous system manifestations of Lyme disease. Arch Neurol 46:790, 1989.

85. Pachner, AR and Steere, AC: The triad of neurologic manifestations of Lyme disease: Meningitis, cranial neuritis, and radiculoneuritis. Neurology 35:47, 1985.

86. Parry, GH: Peripheral neuropathies associated with human immunodeficiency virus infections. Ann Neurol 23:S49, 1988.

87. Piette, AM, Tusseau, F, Vignon, D, et al: Acute neuropathy coincident with seroconversion for anti-LAV/HTLV-III. Lancet 1:852, 1986.

88. Reik, L, Steere, AC, and Bartenhagen, NH: Neurologic abnormalities of Lyme disease. Medicine 58:281, 1979.

89. Sabin, TD and Swift, TR: Leprosy. In Dyck, PJ, Thomas, PK, and Lambert, EH (eds): Peripheral Neuropathy, Vol II, ed 2. WB Saunders, Philadelphia, 1984, p 1955.

90. Sabin, TD, et al: Temperature along the course of certain nerves affected in

leprosy. Int J Lepr 42:33, 1974.

91. Sabin, TD: Temperature-linked sensory loss: A unique pattern in leprosy. Arch Neurol 20:257, 1969.

92. Said, G, Lacroix, C, Andriev, JN, et al: Necrotizing arteritis in patients with inflammatory neuropathy and human immunodeficiency virus (HIV-III) infection. Neurology 37:176, 1987.

92a. Said, G, Lacroix, C, Chemorilli, P, et al: Cytomegalovirus neuropathy in acquired immunodeficiency syndrome: A clinical and pathologic study. Ann Neurol 29:139, 1991.

93. Schoenen, J, Sianard-Gainko, J, and Carpentier, M: Myositis during Borrelia burgdorferi infection (Lyme disease). J Neurol Neurosurg Psychiatry 52:1002, 1989.

94. Sigal, LH and Tatum, AH: Lyme Disease patients' serum contains IgM antibodies to Borrelia burgdorferi that cross-react with neuronal antigens. Neurology 38:1439, 1988.

95. Siltzbach, LE, James, DG, and Neiville, E: Course and prognosis of sarcoidosis around the world. Am J Med 57:847, 1974.

96. Snider, WD, Simpson, DM, Nielsen, S, Gold, JWM, et al: Neurological complications of acquired immune deficiency syndrome. Ann Neurol 14:403, 1983.

97. Stanek, G, Klein, J, and Bittner, R: Isolation of Borrelia burgdorferi from the myocardium of a patient with longstanding cardiomyopathy. N Engl J Med 322:249, 1990.

98. Steere, AC: Lyme disease. N Engl J Med 321:586, 1989.

99. Sterman, AB, Nelson, S, and Barclay, P: Demyelinating neuropathy accompanying Lyme disease. Neurology 32:1302, 1982.

100. Stiernstedt, G, Gustafsson, R, and Karlsson, M: Clinical manifestations and diagnosis of neuroborreliosis. In Benach, JL and Bosler, EM (eds): Lyme Disease and Related Disorders, Vol 539. New York Academy of Sciences, New York, 1988, p 46.

101. Strickland, GT and Moser, KM: Sarcoidosis with a Landry-Guillain-Barré syndrome and clinical response to corticosteroids. Am J Med 43:131: 1967.

102. Suchenwirth, R: Die Sarkoidose des Nervensystems. Münch Med Wochenschr 110:580, 1968.

103. Tharp, BR and Pfeiffer, JB: Sarcoidosis and the acoustic nerve. Arch Otolaryngol 90:360, 1969.

104. Thomas, JE and Howard, FM: Segmental zoster paresis—a disease profile. Neurology (Minneap) 22:459, 1972.

105. Trautman, JR and Enna, CD: Leprosy. In Tice's Practice of Medicine, Vol III. Harper and Row, New York, 1970, p 1.

106. Uldry, PA, Regli, F, and Bogousslavsky, J: Cerebral angiopathy and recurrent strokes following Borrelia burgdorferi infection. J Neurol Neurosurg Psychiatry 50:1703, 1987.

107. Vallat, JM, Hugon, J, and Lubeau, M: Tick-bite meningoradiculoneuritis: Clinical, electrophysiologic, and histologic findings in 10 cases. Neurology 37:749, 1987.

108. Vendrell, J, Heredia, C, Pujol, M, et al: Guillain-Barré syndrome associated with seroconversion for anti-LAV HTLV-III. Neurology 37:544, 1987.

109. Vishnubhakat, SM and Beresford HR: Prevalence of peripheral neuropathy in HIV disease: Prospective study of 40 patients. Neurology 38:350, 1988.

110. Wechsler, A and Ho, DD: Bilateral Bell's palsy at the time of HIV seroconversion. Neurology 39:747, 1989.

Part III

NEUROPATHY ASSOCIATED WITH SYSTEMIC DISEASE

Chapter 7

DIABETIC NEUROPATHY

SYMMETRIC POLYNEUROPATHY
DIABETIC MONONEUROPATHY
  SYNDROMES
CASE HISTORIES AND COMMENT

Table 7–1 PRINCIPAL TYPES OF DIABETIC NEUROPATHY

SYMMETRIC POLYNEUROPATHY
  Distal sensory/autonomic neuropathy
  Autonomic neuropathy
  Lower limb motor neuropathy
  Acute painful neuropathy
  Rapidly reversible neuropathy

MONONEUROPATHY
  Cranial nerve lesions
  Focal limb nerve and truncal neuropathies
  Asymmetric proximal lower limb motorneu-
    ropathy

A wide range of peripheral nerve disorders may occur in diabetes mellitus; in general, these may be classified into two types: the symmetric polyneuropathies and the mononeuropathies. This classification has little correlation with type of diabetes: most patients with long-standing diabetes, insulin-dependent or non-insulin–dependent, will develop some degree of symmetric polyneuropathy. It should be emphasized that mixed syndromes are common, with isolated mononeuropathies, for example, sometimes occurring on a background of a symmetric sensory polyneuropathy. Table 7–1 is a classification of the principal types of diabetic neuropathy.

SYMMETRIC POLYNEUROPATHY

Pathology, Pathogenesis, and Animal Models

Neuropathologic studies on patients with diffuse diabetic sensory polyneuropathy show a combination of axonal degeneration and segmental demyelination. The axonal degeneration is greater distally and there is often prominent regenerative activity. In early asymptomatic cases, demyelination predominates, the degree of axonal degeneration increasing with advancing neuropathy.[17,57] Multifocal proximal lesions are evident in older diabetic subjects.[64] Minor degrees of loss of dorsal

root ganglion and anterior horn cells may be seen. Microangiopathy is frequently present, especially in older diabetic subjects. This takes the form of reduplication of the basal lamina around endoneurial capillaries (Fig. 7–1) with hyperplasia of their endothelial cells and of the intima of epineural arterioles.[63]

The earliest suggestions for the causation of diabetic polyneuropathy were that it is of vascular origin. Based largely on clinical analogy with the toxic distal axonopathies (Chapter 2) and supported by previous postmortem and nerve biopsy studies,[7,10,43,46,52,57] this view was later supplanted by metabolic hypotheses.

The viewpoint that has attracted most attention links sorbitol accumulation, myo-inositol depletion, and reduced sodium-potassium adenosine triphosphatase (Na^+-K^+-ATPase) activity in nerve.[25] In the presence of hyperglycemia, glucose is converted to sorbitol by aldose reductase. The sorbitol accumulates in nerve and, together with the hyperglycemia, inhibits the uptake of myo-inositol. Na^+-K^+-ATPase is reduced in nerve and there was some evidence

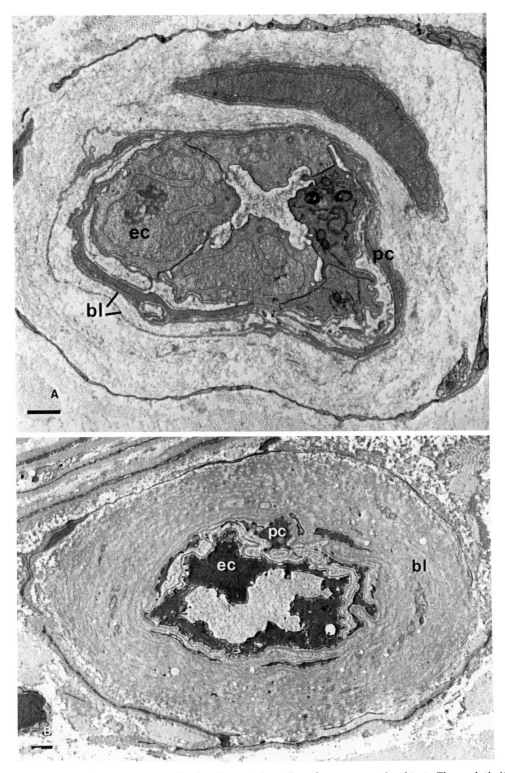

Figure 7–1. (*A*) Electron micrograph of endoneurial capillary from a normal subject. The endothelial cells (*ec*) are surrounded by pericytes (*pc*) and both are enclosed by basal lamina (*bl*). Bar = 1 μm. (*B*) Electron micrograph of endoneurial capillary from a patient with diabetic polyneuropathy. The endothelial cells (*ec*) and pericytes (*pc*) are surrounded by a wide zone of reduplicated basal lamina (*bl*). Bar = 1 μm.

that this was secondary to the myo-inositol depletion. It was postulated that the reduced Na^+-K^+-ATPase activity had various secondary effects that led to the development of neuropathy. Most of these studies were undertaken either in the streptozotocin-diabetic rat or the spontaneously diabetic Biobreeding (BB) rat. These animals develop abnormalities of nerve conduction that are corrected by insulin and dietary myoinositol supplementation as well as[26] diminished axonal nerve fiber caliber, but no convincing evidence of axonal degeneration.[39,49,59] Prevention of sorbitol accumulation by aldose reductase inhibitors improves nerve conduction velocity both in diabetic animals[58] and man,[32] but fails to correct reduced nerve fiber size in animals.[5] Moreover, observations on galactose-fed animals, where the sugar alcohol dulcitol accumulates in nerve, show an *increase* in Na^+-K^+-ATPase activity despite reduced myo-inositol concentrations.[36,61] There is thus no simple relationship between Na^+-K^+-ATPase activity and myo-inositol concentrations.

Sorbitol concentrations are increased in nerve in human diabetic neuropathy,[13] but if this is instrumental in producing the neuropathy, it is probably doing it in some as yet undiscovered way, as myo-inositol concentrations are not reduced.[13] Moreover, dietary myoinositol supplementation fails to improve human diabetic neuropathy.[27] Whether sorbitol accumulation is implicated at all is uncertain as, despite the ability of aldose reductase inhibitors to reduce nerve sorbitol levels, convincing improvement in neuropathy has not been demonstrated.[24,47]

The vascular hypothesis has been revived in recent years. It has long been favored as the explanation for focal diabetic neuropathies but evidence has now been adduced that it may be responsible for diffuse polyneuropathy by the summation of multiple small ischemic lesions.[14,20,31] The evidence, derived both from autopsy and nerve biopsy studies, includes the finding of multifocal fiber loss, capillary closure, and abnormalities in vessel walls.[16,35,63] Some, but not all, of these changes have been correlated with the severity of nerve fiber loss. Nonenzymatic glycation of proteins in connective tissue components could contribute to the vascular abnormalities.[8] Intraneural recordings have also suggested that diabetic nerve is hypoxic.[41] Although an impressive body of evidence has been produced to support the vascular hypothesis for diabetic neuropathy,[15] it has to be pointed out that most was obtained in older individuals in whom concomitant vascular disease would not be unexpected. It is difficult to reproduce these findings in younger individuals with diabetic polyneuropathy,[6] and it has to be concluded that the vascular hypothesis is still sub judice.

There are few studies on the pathology of the autonomic nervous system in patients with diabetic autonomic neuropathy. Enlargement and degeneration of sympathetic chain ganglion cells is described, together with a reduced density of myelinated fibers and segmental demyelination in the greater splanchnic nerve, and a reduced innervation of lower-limb arterioles.[1,28] Studies of the sympathetic nervous system in both the streptozotocin and BB rat models demonstrate axonal dystrophic changes in mesenteric nerves and ganglia.[48,62] Neither neuronal nor axonal loss is described.

Metabolic dysfunction clearly is a prominent feature of the rapidly reversible neuropathy of newly diagnosed diabetics. It may be analogous to the insulin-reversible reduction of nerve conduction in experimental animal studies.[26,49]

Epidemiology

INCIDENCE

Diabetic neuropathy occurs throughout the world equally among men and women. Although it is a common condition, estimates of prevalence vary widely, and most series are confounded by imprecise definition as to what constitutes neuropathy. The recent establishment of clear clinical and neuropathologic criteria for the diagnosis of diabetic neuropathy should ameliorate this problem.[18] Available evidence indicates that there is a prevalence of about 50% for individuals with diabetes of 25 years' duration.[44] Neuropathy is present in less than 10% of individuals at the time of the initial diagnosis of diabetes and is uncommon in diabetic children.

PREDISPOSING FACTORS

The premise that neuropathy develops more readily and recovery is less likely in individuals with poorly controlled blood sugar levels is widely held but as yet, unproved.[19] There is considerable variation in individual susceptibility and there are many instances of severe neuropathy in apparently well-controlled diabetic subjects. Genetic factors may be important.

Distal Sensory Polyneuropathy

Table 7–2 outlines the clinical features of the symmetric polyneuropathy syndromes.

A distal, predominantly sensory polyneuropathy is the commonest type of diabetic peripheral nerve disorder. It may be asymptomatic, with abnormal signs first detectable on routine neurologic examination, or it may present with a variety of symptoms. In most instances, symptoms and quantitative sensory testing indicate involvement of sensory fibers of all diameters; autonomic dysfunction gradually emerges and, in general, correlates with the severity of somatic neuropathy.[56] Small fiber abnormalities, somatic and autonomic, are invariable features of most long-duration cases. The striking small or large fiber patterns of some cases (see further on) formerly held as evidence for selective vulnerability of different classes of sensory neurons, may merely reflect extremes of a diffuse spectrum of nerve fiber loss.[20]

Table 7–2 CLINICAL FEATURES OF SYMMETRIC POLYNEUROPATHY SYNDROMES

DISTAL SENSORY POLYNEUROPATHY
  Large-fiber type
    paresthesias in feet, loss of vibration and
      position senses
  Small-fiber type
    pain, loss of thermal and pain sense; ulcers
      on feet

AUTONOMIC NEUROPATHY
  Cardiovascular
    hypotension, tachycardia
  Genitourinary
    bladder, impotence
  Gastrointestinal
    gastroparesis, nocturnal diarrhea

SYMMETRIC PROXIMAL LOWER LIMB MOTOR NEUROPATHY
  Painful wasting of thighs in elderly, sensation
    spared

ACUTE PAINFUL NEUROPATHY
  Weight loss, intense foot pain, good response
    to diabetic control

RAPIDLY REVERSIBLE NEUROPATHIES
  Hyperglycemia type
    Untreated diabetics with reduced nerve
      conduction velocity
  Treatment-induced type
    Paresthesias following initial treatment with
      insulin

The onset of symptoms may be insidious in individuals with long-standing diabetes mellitus, or may lead to discovery of the condition in maturity-onset cases, in which undetected diabetes has been present for a considerable time. Symptoms may also arise in a more precipitate manner, sometimes following an episode of severe ketosis, or after initiation of insulin treatment. Sensory dysfunction displays a typical length-related pattern, with earliest manifestations in the toes, eventually spreading up to the feet, legs, and finally the hands and anterior abdominal wall.

SYMPTOMS AND SIGNS

While most patients display dysfunction in all fiber types, two striking patterns occasionally appear.

In the large-fiber pattern, symptoms

present as paresthesias in the feet and lower legs.[56] Prominent signs are absent ankle jerks, impaired appreciation of light touch in a stocking distribution, and quantitative sensory testing revealing reduction or loss of vibration sense in the feet with relative preservation of thermal sense. In more severe cases, joint position sense is impaired, the hands are involved, and slight distal weakness appears. Morphologic confirmation of this large-fiber hypothesis has yet to be obtained.

The small-fiber pattern has also been recognized[7,46] and usually presents with pain. Dull aching may be experienced in the feet or the patient may experience deep pain, sometimes described as "in the bones" in the legs. Burning sensations in the soles are common, and are especially troublesome at night. Prominent signs include impairment of cutaneous and deep pain and temperature sensation in the legs. Autonomic nervous system involvement is especially associated with this variety. Strength, tendon reflexes, and the sensations of light touch, vibration and position are relatively spared. When neuropathic foot ulceration and joint degeneration are associated, this is sometimes referred to as the pseudo-syringomyelic form. Electrophysiologic and pathologic studies on nerves from these patients have supported the notion of predominant *small-fiber* involvement.[7,46]

The *pseudotabetic form* is now fortunately rare. The Romberg sign is present and tendon reflexes are absent in the legs. Ulceration of the feet and joint deformation (neuropathic arthropathy) are hallmarks of this condition, developing in individuals with long-standing diabetes mellitus associated with profound sensory loss. Ulcers are most often situated on the ball of the foot over the heads of the medial metatarsal bones, but may occur at other pressure sites, such as on the toes or heel (Fig. 7–2). Diabetic neuropathic arthropathy generally involves the distal joints, in contrast to the Charcot joint of tabes dorsalis, which commonly affects the

knees, hips, and ankles. Autonomic dysfunction is usually associated with this variety. It is suggested that loss of nociceptive C-fiber function contributes to foot complications in this pattern.[46]

LABORATORY STUDIES

The CSF protein may be moderately elevated in most varieties of diabetic neuropathy. Levels as high as 5 g/L may rarely occur. For undetermined reasons, many diabetic subjects with no evidence of neuropathy have elevated CSF protein levels.

QST Studies. Quantitative studies of vibratory and thermal sensation, even in asymptomatic diabetics, commonly reveal diminished sensory perception.[9] In symptomatic sensory neuropathy, QST studies are almost always abnormal.

Electrodiagnostic Studies. In patients with the above types of neuropathy, changes in sensory electrophysiology, especially sural sensory potential amplitude, are a consistent abnormaltiy.[18] An exception is the small-fiber type, in which sensory conduction studies may be normal. Reduction in motor potential amplitudes and abnormalities of peroneal and tibial late responses may also be present in cases with marked sensorimotor deficit. Electromyography often reveals signs of denervation in distal leg muscles even when these muscles are clinically spared. Abnormalities of cardiac vagal innervation (R-R interval) and sudomotor function (sympathetic skin response) are often present even when there are no autonomic symptoms.

COURSE AND PROGNOSIS

There are no large-scale prospective studies that address the prognosis of this common condition. In the authors' experience, some individuals with the large-fiber variety of sensory-motor polyneuropathy have a steady, insidiously progressive neurologic decline. A few will fluctuate, improve transiently, and

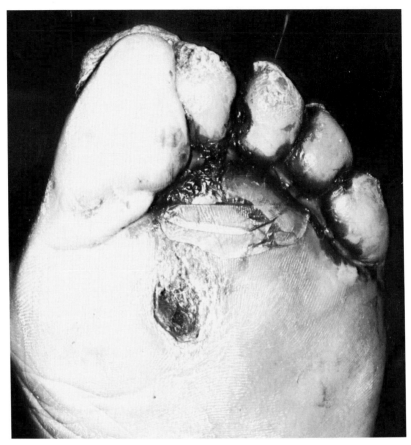

Figure 7–2. Ventral view of swollen, infected foot with ulcer on plantar surface in a patient with diabetic neuropathy.

then slip back. Most will plateau and experience moderately annoying sensory symptoms and mild distal lower-extremity weakness. Severe disability and limitation of activities may occasionally occur but are the exception, unless an associated autonomic or mononeuropathy syndrome also develops. The painful small-fiber type is often self-limited, but the pseudotabetic variety generally progresses relentlessly, and is much more disabling. This is partly due to the associated joint deformities and foot ulceration.[42]

GENERAL TREATMENT

Diabetic neuropathies of all types appear more likely to develop if the meta-bolic state is poorly controlled. Strict control of blood sugar concentration is therefore mandatory whether or not neuropathy is present. When neuropathy has developed, rigid control of glycemia with an insulin pump has been shown to retard deterioration in comparison with conventional insulin treatment. A recent study has shown that pancreatic transplantation may slow the progression of diabetic neuropathy.[34]

Therapeutic trials with myo-inositol,[27] gangliosides,[29] and aldose reductase inhibitors[24,27] have for the most part been inconclusive or unrewarding. The common practice of administering vitamins does not alter the neuropathy.

PAIN TREATMENT

Sensory neuropathy is frequently accompanied by persistent, severe pain which can, on occasion, be incapacitating. Simple analgesics rarely help and the chronicity of the problem contraindicates the use of opiates. Phenytoin or carbamazepine are often given, but the response is rarely satisfactory. The authors initially employ a clinical trial of 3 weeks of phenytoin at 300 mg per day or a gradually increasing dose of carbamazepine to 200 mg 3 times per day for 2 weeks to determine efficacy. Intravenous lidocaine may produce relief for several weeks following infusion,[33] and there is one report of relief following oral mexilitine.[11]

Amitriptyline, sometimes in combination with phenothiazines, has been reported to relieve the chronic pain syndrome of diabetic neuropathy in patients with normal or depressed mood.[38] It requires a 6-week course at between 25 mg to 125 mg per day to determine efficacy, and is probably the most effective form of therapy currently available.

Lancinating pain sometimes responds to 0.5 mg tid clonazepam. Transcutaneous nerve stimulation may also produce some relief of chronic distal extremity pain. We have no experience with the use of acupuncture in diabetic neuropathy. Pain may be relieved in some patients by cold, so much so that cold injury has been described as a result of excessive self-treatment.

Autonomic Neuropathy

Diabetic autonomic neuropathy generally is associated with the symmetric sensory neuropathy, although there are instances where impairment of autonomic function constitutes the major neurologic disturbance. It is probably widely underdiagnosed, since the signs are often subtle and best detected by special tests not routinely employed in clinical practice. The course of autonomic neuropathy varies from benign imperceptible progression to an incapacitating disability.

Three types of autonomic dysfunction are prominent in diabetics: cardiovascular, genitourinary, and gastrointestinal.

CARDIOVASCULAR

Asymptomatic abnormalities of cardiovascular reflexes can be detected at the time of diagnosis in many diabetics. Impairment of vasomotor reflexes are present in most patients with evidence of sensory neuropathy, even in mildly affected cases. Postural hypotension is rare, but when present, can be extremely disabling. It is aggravated by the administration of insulin, tricyclic antidepressants, diuretics, and phenothiazines. In mild cases of postural hypotension, simple measures may be helpful, such as sleeping with the feet slightly elevated, rising gradually from a seated or supine position, and elastic support stockings. In more severe cases, an elasticized body stocking[51] or treatment with plasma expanders such as fludrocortisone are advocated.[59]

Cardiac denervation, either vagal alone or in combination with sympathetic denervation, is well described. Vagal denervation is characterized by a high resting heart rate and loss of the sinus arrythmia that occurs with respiration. Some instances of sudden death in young diabetics may be attributable to impaired respiratory reflexes.

GENITOURINARY

Voiding dysfunction is uncommon but when present has an insidious onset and is characterized by a progressive failure of bladder emptying. In addition, sensory deafferentation causes a poor sensation of bladder filling. Symptoms include increased intervals between voiding, difficulty in the initiation of micturition, an intermittent stream, and post voiding dribbling. The principal goal of treatment is to improve bladder emptying and prevent infection. In patients with bladder atony, regular voiding according to a time schedule is advisable, with micturition assisted by suprapubic pressure. Urinary infections require prompt treatment. Resection of the bladder neck is sometimes helpful if there is severe difficulty in voiding, but dribbling incontinence can follow if the distal sphincter is impaired.[65] For similar reasons, urethral dilatation in females may be hazardous.

Impotence is sometimes the initial manifestation of autonomic neuropathy; it usually steadily worsens and rarely, if ever, is improved by control of hyperglycemia. It is characterized initially by loss of erections with preserved ejaculation and orgasm.[22] Retrograde ejaculation may occur, when the seminal fluid passes backwards into the bladder instead of down the urethra because of failure of closure of the internal sphincter of the bladder. Diabetic impotence also may be caused by sensory and vascular dysfunction. This complaint requires skilled urologic evaluation since efficacious therapy is now available, but varies with the cause.[40] Once present, autonomic impotence remains constant and there is no nocturnal tumescence. Local papaverine injection has proved successful as a diagnostic maneuver in provoking an erection,[60] and it has emerged as a popular treatment alternative to the penile prosthesis.[66] Vasculogenic impotence is common in diabetics, secondary to small vessel disease, and does not respond to reconstructive vascular surgery.

GASTROINTESTINAL

In general, gastrointestinal symptoms from diabetic autonomic neuropathy are uncommon and surprisingly episodic. Asymptomatic disturbances of esophageal motility are sometimes demonstrable by cineradiography. Gastroparesis and delayed emptying of the stomach, probably secondary to vagal denervation, are treated by agents that increase gastric motility and emptying, either metoclopramide or domperidone.[53] A recent report suggests that erythromycin may have therapeutic value.[30] Persistent vomiting is rare, may require surgery, and is best managed by an expert gastroenterologist.

Nocturnal diarrhea, frequently apoplectic, usually episodic and severe, is the best known diabetic alimentary disturbance. The feces are watery and the patients experience urgency and incontinence. One or two 250 mg doses of tetracycline usually abort an attack if given at the onset. Its mechanism of action is uncertain; the rapidity of relief makes an antibacterial effect unlikely. Both codeine phosphate and loperamide also may be helpful. Constipation, associated with colonic dilatation, is the most common gastrointestinal symptom in diabetic autonomic neuropathy. It usually responds to simple symptomatic therapy.

Other prominent autonomic features are abnormalities of pupillary function, thermoregulation, and sweating. Gustatory sweating may be an especially troublesome complication, with profuse facial sweating usually beginning shortly after eating and extending to the scalp and shoulders symmetrically. It can be prevented by self-administration of anticholinergic drugs, such as propantheline hydrochloride, before meals. Nocturnal total body sweating, not related to hypoglycemia, also occurs. Diminished sweating of the distal extremities may lead to hot, dry skin that fissures easily.

Acute Painful Neuropathy (Diabetic Neuropathic Cachexia)

Clinical features of this striking condition are precipitous weight loss followed by severe burning or stabbing pain on the soles of the feet.[21] Often likened to walking on burning sand, pain is disabling, associated with contact discomfort and intense hypersensitivity of the skin. Sensory loss is usually slight compared to the magnitude of the pain and is accompanied by cutaneous hyperesthesia (allodynia). The upper limbs are usually spared and weakness is not a feature of this condition. Only modest electrophysiologic changes indicating large myelinated axonal degeneration are described in the few nerve conduction studies available. Improved diabetic control with insulin is associated with weight gain and diminution in severity of symptoms, usually within 6 months. This condition occurs early or late in diabetes and there appears little correlation with other diabetic complications. This suggests a metabolic basis and not microvascular disease. A careful nerve biopsy study of three cases found normal vasa nervorum, supporting this notion.[2]

Symmetric Proximal Lower Extremity Motor Neuropathy

For nearly a century the syndrome of proximal lower limb weakness has been recognized as a type of diabetic neuropathy. Formerly termed diabetic amyotrophy,[23] it is now recognized that two distinct, similar-appearing neuropathic conditions are associated with proximal weakness in diabetics—a slowly progressive *symmetric* condition, the subject of the present section, and a rapidly progressive *asymmetric* type, discussed later as a mononeuropathy syndrome.[4]

Symmetric proximal motor neuropathy (SPMN) is most common in older diabetics but may occur at any stage; sometimes it is the initial manifestations of diabetes. Initial symptoms of low back or thigh pain may erroneously suggest a rheumatologic or mechanical disorder. Pain is soon followed by slowly progressive weakness of quadriceps and iliopsoas muscles, with eventual atrophy and loss of patellar reflexes. Involvement of glutei and thigh adductors may cause a waddling gait. Sensation is surprisingly spared given the degree of weakness and wasting. Electrodiagnos-

tic findings vary with the evolution of the condition. The initial abnormality is reduced motor-unit recruitment, several weeks later, evidence of denervation appears. Increased conduction time in the femoral nerve may be demonstrable electrophysiologically. Coexisting distal symmetric neuropathy may produce changes in distal sensory nerves. There are no satisfactory histopathological studies of this entity and its pathogenesis is unclear. Aside from control of hyperglycemia, there is no specific therapy. In cases of insidious onset, recovery is often disappointing. In those patients with a more rapid evolution, satisfactory recovery may occur following good diabetic control.

Rapidly Reversible Neuropathies

"HYPERGLYCEMIC" NEUROPATHY

Newly diagnosed, untreated diabetics may demonstrate reduced nerve conduction velocities on electrodiagnostic testing without clinical signs or symptoms of peripheral nerve involvement. The reduced conduction velocity is rapidly reversed by lowering blood sugar concentration to near normal levels. Occasional untreated diabetics likewise manifest distal, painful paresthesias that respond promptly to control of the diabetic state. It seems unlikely that these rapidly reversible phenomena are associated with structural breakdown in peripheral nerve fibers and it is not clear whether their occurrence correlates with later development of persistent neuropathy.

TREATMENT-INDUCED NEUROPATHY ("INSULIN NEURITIS")

Distal lower limb paresthesias and pain occasionally follow commencement of insulin therapy or treatment with hypoglycemic drugs. Nerve fiber loss is present on sural nerve biopsy; slow recovery and disappearance of sensory symptoms follows continued glycemic control.[37]

Differential Diagnosis

Diabetes mellitus is a common condition and other causes for neuropathy may be coincidental.

PARANEOPLASTIC SENSORY NEUROPATHY

This purely sensory condition usually has a subacute onset, predominantly involves the lower extremities, affects all modalities, and may be associated with signs of CNS dysfunction. In these aspects, it differs from most cases of diabetic sensory or sensorimotor neuropathy and, when present, such findings should initiate a search for an occult carcinoma or lymphoma.

COMBINED SYSTEM DISEASE (B$_{12}$ DEFICIENCY)

This condition shares many features with diabetic sensorimotor neuropathy, namely, gradual onset in later life, abnormal sensations in the lower limbs, "large-fiber" pattern of sensory loss, and absent ankle jerks. Hyperreflexic knee jerks, positive Babinski's responses, psychiatric disturbance, and anemia may suggest this condition, and a low serum B$_{12}$ level establishes the diagnosis.

PHARMACEUTICAL AND INDUSTRIAL TOXINS

Certain of the toxic neuropathies (e.g., cisplatin, nitrous oxide, pyridoxine, acrylamide, and dimethylaminoproprionitrile) share features with diabetic sensorimotor neuropathy. These diagnoses may be suggested by a careful history and established by recovery following removal of the toxin.

MISDIRECTED INVESTIGATION AND THERAPY

Patients in whom the presence of mild diabetes is overlooked may be subjected to extensive and discomforting investigation or misdirected therapy. Three frequent examples of this phenomenon are: (1) myelography in cases of symmetric proximal lower extremity or genitourinary autonomic neuropathy, and in cases of diabetic amyotrophy. Diabetic amyotrophy is often a diagnosis of exclusion and radiographic imaging of the area is usually needed. (2) Extensive evaluation for occult carcinoma in individuals with a symmetric sensory neuropathy or for lumbar plexus neoplasm in cases of asymmetric diabetic amyotrophy. (3) Unwarranted psychotherapy or androgen therapy in males with impotence secondary to autonomic neuropathy.

DIABETIC MONONEUROPATHY SYNDROMES

The mononeuropathies are less common, more often occur in the elderly, and may appear earlier in the disease course than the symmetric polyneuropathies. Table 7–3 outlines the clinical features of the mononeuropathy syndromes.

Table 7–3 CLINICAL FEATURES OF MONONEUROPATHY SYNDROMES

CRANIAL NERVE LESIONS
Oculomotor nerve commonly affected, abrupt, painful, pupil spared, good prognosis

FOCAL LIMB AND TRUNCAL LESIONS
Limb type has abrupt, painful onset; good prognosis with distal lesion
Truncal type has acute, often nonradicular thoracoabdominal pain

ASYMMETRIC PROXIMAL LOWER LIMB MOTOR NEUROPATHY
Sudden onset of unilateral thigh pain, weakness, wasting, partial recovery.

Pathology, Pathogenesis and Animal Models

It is widely held that isolated peripheral nerve lesions in diabetics have a vascular basis. Three clinical facts support this notion: they are most common in the elderly in whom vascular disease is likely to be frequent, they may have an abrupt onset, and they tend to improve spontaneously.

Two autopsy studies of diabetics who died shortly following the onset of oculomotor palsy describe focal lesions in the intracavernous portion of the corresponding third nerve.[3,12] Destruction of myelin and, to a lesser extent, of axons, was prominent in the central portion of these nerves. No vascular occlusion was demonstrated, but in one case, there was pronounced narrowing and hyalinization of adjacent arterioles.[3] A postmortem study of the lumbar nerves of a patient with asymmetric lower limb motor neuropathy also disclosed multiple changes in vessels and nerve.[45]

In sum, vascular factors appear to have a significant role in the pathogenesis of focal diabetic peripheral nerve lesions. The clinical pattern—abrupt onset of painful, focal nerve lesions—is strikingly analogous to the pattern encountered in other vascular neuropathies (see Chapter 10). Vascular occlusions have not been adequately documented in humans with mononeuropathy, and further investigation of this condition is required to support this attractive, but still unproved, assumption.

Clinical Features

CRANIAL NERVE LESIONS

Isolated or multiple palsies of the nerves to the extraocular muscles are specific to older diabetics.[3,12] Frequently there is no other evidence of neuropathy and, on occasion, this may be the first indication of diabetes in an otherwise asymptomatic individual. The third nerve is most commonly affected, the sixth less frequently. Fourth nerve lesions are rare. Involvement may be unilateral or bilateral. The onset is usually abrupt and associated with an intense, retro-orbital aching sensation. Occasionally, paralysis evolves in a subacute, painless manner. Sparing of the pupillomotor fibers is frequent with diabetic third nerve palsy and, when present, helps differentiate this entity from lesions that compress the nerve, such as an aneurysm of the posterior communicating artery. The probable explanation is that the pupilloconstrictor fibers, situated around the periphery of the nerve, are not affected by ischemic demyelination that disrupts the central part of the nerve.[12] Rare cases of internal ophthalmoplegia have been reported in combination with third nerve pareses. Such cases are indistinguishable from oculomotor palsy due to posterior communicating artery aneurysm. Satisfactory recovery of nerve function is usual in diabetic oculomotor palsy occurring within several weeks in most cases.

Isolated lesions of other cranial nerves, especially the seventh, probably occur more frequently in diabetics than in the general population. The course of these disorders is characterized by rapid development of severe dysfunction of involved nerves, followed by a satisfactory recovery. Other cranial nerves subsequently may be involved. Repeated palsies of the same cranial nerve may occur but are rare. Cranial neuropathies that affect extraocular muscles may be accompanied by severe pain. Since they are self-limited conditions, opiates may be judiciously used with little risk of addiction. Facial palsy with paralysis of the orbicularis oculi may lead to exposure keratitis and require tarsorrhaphy.

FOCAL LIMB NERVE AND TRUNCAL LESIONS

Focal Limb Nerve Lesions. Almost every major peripheral nerve can be affected by diabetic mononeuropathy. Lesions of the ulnar, median, radial, lateral cutaneous nerve of the thigh, sciatic, and peroneal nerves have been described. Since diabetic nerves are especially vulnerable to compression, the most common focal neuropathies are those due to compression at commonly entrapped sites (e.g., median at the wrist, ulnar at elbow, and peroneal at the fibular head). Noncompressive diabetic mononeuropathies usually have an abrupt onset of symptoms and are frequently painful. Recovery is often satisfactory if the lesion is sited distally. Proximal lesions may be associated with prolonged deficits, presumably because of the greater distance required for axonal regeneration. Rapid recovery of a peripheral mononeuropathy probably indicates a demyelinating lesion. Electrodiagnostic abnormalities tend to be confined to affected nerves and their innervated muscles. Electrodiagnostic evaluation commonly demonstrates localized slowing at common sites of nerve entrapment, such as the ulnar nerve in the cubital tunnel or the median nerve in the carpal tunnel, despite the absence of clinical symptoms.

The prognosis of mononeuropathies involving limb nerves is better in cases with an acute onset. In general, the more distal the lesion, the more rapid and complete the recovery. Exceptions to this rule are the median nerve lesions that occur at the wrist, reinforcing the notion that diabetic nerves respond less favorably to surgical decompression than normal nerves. In general, treating the limb mononeuropathies utilizes repetitive physical therapy. Anterior tibial weakness from a common peroneal palsy usually requires foot support, and persistent wrist drop from a radial nerve palsy occasionally necessitates tendon-transfer surgery. Chapter 21 contains suggestions for management of specific limb nerve dysfunction.

Truncal Neuropathy. The syndrome of acute, unilateral pain in the distribution of one or more thoracic nerves is most common in patients over 50 years of age. It can occur in any diabetic state and may be the initial finding in an unsuspected diabetic. Because of the sudden onset and dermatomal distribution, it is generally believed to be a vascular thoracic mononeuropathy or radiculopathy, but there is no histopathologic confirmation of this concept. This condition is usually heralded by significant weight loss, followed by sudden onset of pain or dysesthesias in the chest or abdomen. Pain appears in many varieties: deep, aching, sharp, stabbing, and burning are described. Occasionally, a radicular pattern is apparent but more often pain is not dermatomally localized.[54] Examination commonly reveals hypersensitive areas, described as *scalding* or *sunburn-like*, or loss of pin and cold. Distribution of sensory abnormality is unilateral but variable: some involve entire thoracic or abdominal dermatomes in a clear radicular pattern; others display patchy areas in the distribution of dorsal or ventral areas and their branches. Involvement of lower thoracic motor segments may result in abdominal muscle paralysis and bulging. Electrodiagnostic studies may reveal denervation in the intercostal, rectus abdominis, or paraspinal muscles. The prognosis is generally favorable, pain gradually dissipates within two years, and hypersensitive zones gradually disappear. Management of pain is a major concern and is often unsatisfactory. High levels of amitriptyline (100 mg to 150 mg daily) have been most successful. Accurate diagnosis and reassurance are as important as pharmacotherapy. It is the authors' experience that this disorder is commonly misdiagnosed. A combination of abdominal-thoracic pain and weight loss in a middle-aged person suggests an occult malignancy; frequently an extensive evaluation, occasionally with laparotomy, is undertaken before diabetologic consultation. The term *diabetic truncal neuropathy* has in the past been erroneously applied to the bilateral symmetric midline thoracic sensory loss (shield distribution) in persons with advanced sensory polyneuropathy.

ASYMMETRIC PROXIMAL LOWER LIMB MOTOR NEUROPATHY

The syndrome of proximal limb weakness, formerly termed diabetic amyotrophy, encompasses the symmetric type (SPMN), described on page 99, and an asymmetric condition (ASPMN).[4] ASPMN is most common in patients over 50 years of age and can occur at any stage of diabetes. It is usually heralded by weight loss. Onset of pain and weakness is typically sudden, in contrast to SPMN, and evolves over 1 or 2 weeks. Pain is on the same side as weakness. Usually a deep aching discomfort in hip, buttock or thigh appears; rarely is it dysesthetic. Significant sensory loss or marked limitation of straight leg raising is atypical. There may be few mechanically induced symptoms. Pain is often most prominent at night, preventing sleep or awakening the patient. Weakness of quadriceps femoris, thigh adductors, and iliopsoas is the rule; occasionally, the glutei, hamstrings, and gastrocnemius are involved. The patellar reflex is usually absent or strikingly diminished at the onset, and wasting of the appropriate muscles follows within weeks. The deficit is usually best localized to the lumbosacral plexus or upper lumbar roots. Electrodiagnostic findings demonstrate axon loss affecting those nerves supplied by the lower thoracic and upper lumbar roots. As for the chronic symmetric form, femoral nerve conduction time is often increased. One study suggests that ASPMN is not as commonly associated with distal sensory neuropathy as SPMN.[55] A postmortem neuropathological study strongly supports the notion that ASPMN is an ischemic process of the lumbosacral plexus,[45] although electrodiagnostic studies demonstrate profuse paraspinal denervation suggesting additional radicular or spinal nerve involvement. There is no specific therapy other than treatment of pain and the diabetic state. Most patients improve slowly; frequently recovery is partial and it may take as long as 2 years for pain to subside. Rarely the disorder may recur on the same or the contralateral side.

Differential Diagnosis

Diabetes mellitus is a common condition and other causes for neuropathy may be coincidental. There are numerous causes for focal and multifocal neuropathies; the following examples are cited.

POLYARTERITIS NODOSA

Mononeuropathy from this condition differs in few clinical respects from that due to diabetes mellitus. Polyarteritis should be suspected if the individual is young and has evidence of systemic illness (fever, elevated ESR, elevated blood urea). Nerve biopsy is often diagnostic.

LEPROSY

Thickening of nerve trunks, the distribution of sensory loss (ears, malar eminences, forearms, and lateral legs) and the coexistence of cutaneous lesions in advanced cases of leprosy should suggest this diagnosis, especially in individuals from endemic areas. Detection of early cases, in which one nerve is prominently involved, may require nerve biopsy.

ANEURYSM OF INTERNAL CAROTID OR POSTERIOR COMMUNICATING ARTERY WITHOUT SUBARACHNOID HEMORRHAGE

Like diabetic oculomotor neuropathy, nerve compression from this aneurysm occurs with abrupt onset and orbital pain. Sparing of the pupillomotor fibers strongly suggests a diabetic basis and such patients should not undergo invasive neuroradiologic procedures. Pupillary dilatation is often sufficient indication for carotid angiography.

HERNIATED LUMBAR DISC OR INFILTRATIVE LESION OF LUMBOSACRAL PLEXUS

These conditions may be extraordinarily hard to differentiate from diabetic lower limb asymmetric wasting. Neuroimaging studies are usually decisive.

CASE HISTORIES AND COMMENTS

Case History: Symmetric, Primarily Sensory Neuropathy with Subsequent Cranial Mononeuropathy

A 32-year-old female was discovered to have diabetes mellitus at a pre-employment physical examination. Treatment with diet and tolbutamide adequately controlled hyperglycemia until she reached age 45, when she additionally required 10 units of NPH insulin twice a day. At the age of 47, unpleasant burning pain appeared over the soles of both feet. This steadily worsened until she reached age 50, when it began to disturb her sleep. Wearing shoes produced great discomfort and she preferred to walk barefoot about the house. Outpatient evaluation at that time revealed no abnormalities other than symmetric, lower-extremity sensory and reflex changes. Strength and sweating were normal. Knee jerks were normal; ankle jerks were absent. Vibratory stimuli (128 cps) were not felt at toes or ankles, but were appreciated normally at the knees. Position sense was moderately diminished at the toes. Pinprick and light touch sensation were slightly diminished over both feet and lower legs to a level just below the knee, and both stimuli elicited an abnormal "burning" sensation. Electromyography of the leg muscles was normal, as was motor conduction in the peroneal nerves. (Sensory conduction in the peroneal nerves was profoundly

slowed to 20 m per sec—normal 40 to 60—and sensory conduction in the median nerves was at the lower limit of the normal range—50 m per sec.) Treatment with diphenylhydantoin failed to alleviate the pain, and she declined further medication. During the following 10 years, she scrupulously maintained dietary and insulin control of hyperglycemia. In that time, there was gradual progression of sensory loss in the legs and occasional numb sensations of the fingers. The spontaneous burning pain did not worsen and did not limit her activities.

At age 61 years, while she was on vacation, she awakened one morning with severe throbbing pain in the right eye and vomited repeatedly. Double vision appeared by noon and the right lid was half-closed. The following day she was admitted to the hospital with continued orbital pain and vomiting. On examination there was complete ptosis of the right lid and paralysis of all extraocular muscles innervated by the oculomotor nerve. The pupils were equal, and reacted normally to light, and to accommodation-convergence. Vision was 20/30 in both eyes. Facial sensation was normal as was other cranial nerve function. In the upper extremities there was no perception of vibration over the fingers or wrists and pinprick was poorly perceived over the finger tips. Other sensory modalities were intact, and no dysesthesias were reported. Tendon reflexes were normal in the arms. In the lower extremities, vibration sensation was absent except at the iliac crests. Pinprick and touch sense were severely diminished up to the mid-shin, and were perceived better on moving up the proximal limb, becoming normal at mid-thigh. The skin was cool and moist in the feet, sweating appeared normal and pedal pulses were present. Joint position sense was poor at the toes but normal at the ankles. Rubbing her soles elicited only mild discomfort. There was slight (4/5) weakness of dorsiflexion of ankles and toes. Gait was slow, deliberate, narrow-based, and required no support. Romberg's sign was not present. Toe walking was performed well, but she could not walk on her heels. Tendon jerks were absent in the lower limbs.

The orbital pain was relieved by large doses of codeine and subsided on the third hospital day. The CSF was clear, contained no cells, and showed a protein level of 1.3 g/L. CT scan was unremarkable. She refused electrodiagnostic studies and was discharged after 1 week. After 2 months, strength returned to the right eyelid, and in the subsequent month extraocular muscles regained a full range of motion.

Comment. This case demonstrates that several types of neuropathy may occur in the same diabetic individual. Despite excellent control of hyperglycemia, a primarily large-fiber type, dysesthetic, symmetric sensory neuropathy occurred. Initially, it was featured by distal loss of vibration sense and tendon reflexes. Typically, progression of neuropathy was slow and the patient was still functional after 10 years without therapy. The abnormal sensations either subsided or she accommodated to them. Definite weakness was present in the legs after 10 years but was not apparent to the patient. Autonomic dysfunction was not symptomatic or detected by bedside maneuvers, indicating that unmyelinated fibers were relatively spared. A painful cranial mononeuropathy, typically involving the third nerve, developed in later life and had a benign course. Sparing of the pupillomotor fibers strongly argued against an aneurysm compressing the third nerve and, together with the negative CT scan, obviated the need for carotid angiography.

Case History: Symmetric Proximal Lower Limb Motor Neuropathy

A 54-year-old female was diagnosed as diabetic and treated by diet alone. She did well for 5 years and then began to lose weight in the 6th year, losing nearly 40 pounds in a 9-month period. The weight loss was accompanied by pain in the legs and steadily increasing leg weakness over a 5-month period. She required assistance in walking on admission to the hospital. There was severe muscle wasting in the lower limbs, more marked in proximal than distal muscles, and more on the right side than the left. Tendon reflexes were absent in the legs, de-

pressed in the arms, and a right Babinski's response was present. Sensation was preserved throughout. The CSF protein was 2.4 g/L. Rigorous control of hyperglycemia was attained by insulin administration. The wasted muscles improved dramatically within 20 months and after 3 years, there was only minimal weakness in the lower limbs.

Case History: Asymmetric Proximal Lower Limb Motor Neuropathy

A 65-year-old male, diabetic for 10 years and well-controlled by diet and tolbutamide, suddenly developed deep pain in the left anterior thigh. Weakness of the proximal left lower limb developed within the day; the pain progressed for 1 week, then subsided, leaving a slight burning sensation above the knee. On examination, there was 2/5 weakness of the left iliopsoas, quadriceps femoris, and adductor muscles, the left knee jerk was absent and there was a poorly defined area of diminished pinprick appreciation over the mid-anterior thigh. Cerebrospinal fluid protein was 1 g/L; 8 weeks later there was moderate atrophy of the left thigh, strength was unchanged, and electromyographic signs of denervation were present in the above-named muscles. Motor conduction in the peroneal and median nerves was within normal limits. Six months following the original episode, 4/5 strength in the proximal muscles was recovered, sensory loss was not detected, and he was able to walk unaided. The left patellar reflex remained absent.

One year following the original episode, persistent dull pain developed in the right buttock. Two weeks later, incapacitating weakness (2/5) appeared in most muscles of the right lower limb. The gluteus maximus, hamstrings, gastrocnemius, anterior tibial, peroneal muscles, quadriceps femoris, and iliopsoas were involved. Tendon reflexes were absent in the right leg. Pain persisted for one month, occasionally requiring opiates for relief. No sensory loss was detected. Four months later there was atrophy and no improvement in strength; the other muscles remained at 2/5. Two years later, he remains unchanged and walks with a walker and a short right leg brace.

Comment. These two patients displayed evidence of lesions that could be localized to multiple levels of lower limb nerves and were accompanied by pain. The two cases differ strikingly in rate of progression and symmetry, and constitute good examples of the clinical extremes and pathogenetic dilemmas surrounding proximal lower limb motor neuropathy. The first case features gradual onset of a relatively symmetric pure motor syndrome with good recovery and suggests a metabolic axonopathy (see Chapter 2), save for the predominantly proximal distribution and absence of sensory findings. The second case initially features abrupt onset of asymmetric combined femoral and obturator nerve deficits. Later, a similar affliction additionally involved the contralateral sciatic nerve. This clinical profile strongly suggests a vascular multiple mononeuropathy syndrome.

REFERENCES

1. Appenzeller, O and Richardson, EP: The sympathetic chain in patients with diabetic and alcoholic polyneuropathy. Neurology (Minneap) 16:1205, 1966.
2. Archer, AG, Watkins, PJ, Thomas, PK, et al: The natural history of acute painful neuropathy in diabetes mellitus. J Neurol Neurosurg Psychiatry 46:491, 1983.
3. Asbury, AK, Aldredge, H, Hershberg, R, et al: Oculomotor palsy in diabetes mellitus: A clinicopathological study. Brain 93:555, 1970.
4. Asbury, AK: Proximal diabetic neuropathy. Ann Neurol 2:179, 1977.
5. Bhoyrul, S, Sharma, AK, Stribling, D, et al: Ultrastructural observations on myelinated fibers in experimental diabetes: Effect of the aldose reductase inhibitor ponalrestat given alone or in conjunction with insulin therapy. J Neurol Sci 85:131, 1988.
6. Bradley, JL, Thomas, PK, King, RHM, et al: Morphometry of endoneurial capillaries in diabetic sensory and autonomic neuropathy. Diabetologia 33:611, 1990.
7. Brown, MJ, Martin, JR, and Asbury, AK: Painful diabetic neuropathy: A morphometric study. Arch Neurol 33:164, 1976.
8. Brownlee, M, Cerami, A, and Vlassara, H: Advanced glycosylation end products in tissue and the biochemical basis of diabetic complications. N Engl J Med 318:1315, 1988.
9. Conomy, JP, Barnes, KL, and Conomy, JM: Cutaneous sensory function in diabetes mellitus. J Neurol Neurosurg Psychiatry 42:656, 1969.
10. De La Monte, S: Case records of the Massachusetts General Hospital. N Engl J Med 316:1325, 1987.
11. Dejgard, A, Petersen, P, and Kastrup, J: Mexiletine for treatment of chronic painful diabetic neuropathy. Lancet 2:9, 1988.
12. Dreyfus, PM, Hakim, S, and Adams, RD: Diabetic ophthalmoplegia: Report of a case, with postmortem study and comments on vascular supply of human oculomotor nerve. AMA Arch Neurol Psychiatry 77:337, 1957.
13. Dyck, PJ, Zimmerman, BR, Vilen, TH, et al: Nerve glucose, fructose, sorbitol, *myo*-inositol, and fiber degeneration and regeneration in diabetic neuropathy. N Engl J Med 319:542, 1988.
14. Dyck, PJ, Karnes, JL, O'Brien, P, et al: The spatial distribution of fiber loss in diabetic polyneuropathy suggests ischemia. Ann Neurol 19:425, 1986.
15. Dyck, PJ: Hypoxic neuropathy: Does hypoxia play a role in diabetic neuropathy? Neurology (Minneap) 39:111, 1989.
16. Dyck, PJ: Capillary number and percentage closed in human diabetic sural nerve. Proc Nat Acad Sci 82:2513. 1985.
17. Dyck, PJ, Sherman, WR, Hallcher, LM, Service, FJ, O'Brien, PC, Grina, LA, Palumbo, PJ, and Swanson, CJ: Human diabetic endoneurial sorbitol, fructose and *myo*-inositol re-

lated to sural nerve morphometry. Ann Neurol 8:590, 1980.

18. Dyck, PJ, Karnes, JL, Daube, J, et al: Clinical and neuropathological criteria for the diagnosis and staging of diabetic polyneuropathy. Brain 108:861, 1985.

19. Dyck, PJ, Brown, M. Greene, D, et al: Does improved control of glycemia prevent or ameliorate diabetic polyneuropathy? Ann Neurol 19:288, 1986.

20. Dyck, PJ, Lais, A, Karnes, JL, et al: Fiber loss is primary and multifocal in sural nerves in diabetic polyneuropathy. Ann Neurol 19:425, 1986.

21. Ellenberg, M: Diabetic neuropathic cachexia. Diabetes 23:418, 1974.

22. Fairburn, CG, McCulloch, DK, and Wu, FC: The effects of diabetes on male sexual function. Clin Endocrinol Metab 11:749, 1982.

23. Garland, HT: Diabetic amyotrophy. Br Med J 2:1287, 1955.

24. Gillon, KRW, King, RHM, and Thomas, PK: The pathology of diabetic neuropathy and the effects of aldose reductase inhibitors. Clinics in Endocrinology and Metabolism 15:837, 1986.

25. Green, DA, Lattimer, SA, and Sima, AF: Sorbitol, phosphoinositides and sodium-potassium-ATPase in the pathogenesis of diabetic complications. N Engl J Med 316:599, 1987.

26. Greene, DA, De Jesus, Jr, PV, and Winegrad, AI: Effects of insulin and dietary *myo*-inositol on impaired peripheral motor nerve conduction velocity in acute streptozotocin diabetes. J Clin Invest 55:1326, 1975.

27. Gregersen, G, Bertelsen, B, Harbo, H, et al: Oral supplementation of myoinositol: Effects on peripheral nerve function in human diabetics and on the concentration in plasma, erythrocytes, urine, and muscle tissue in human diabetics and normals. Acta Neurol Scand 67:164, 1983.

28. Grover-Johnson, NM, Baumann, FG, Imparato, AM, Kim, GE, and Thomas, PK: Abnormal innervation

of epineural arterioles in human diabetes. Diabetologia 20:31, 1981.

29. Hallett, M, Flood, T, Slater, N, et al: Trial of ganglioside therapy for diabetic neuropathy. Muscle and Nerve 10:822, 1987.

30. Janssens, J, Peeters, TL, Vantrappen, G, et al: Improvement of gastric emptying in diabetic gastroparesis by erythromycin. Preliminary studies. N Engl J Med 322:1028, 1990.

31. Johnson, PC: Pathogenesis of diabetic neuropathy. Ann Neurol 19:450, 1986.

32. Judzewitsch, RG, Jaspan, JB, Polonsky, KS, et al: Aldose reductase inhibition improves nerve conduction velocity in diabetic patients. N Engl J Med 308:119, 1983.

33. Kastrup, J, Petersen, P, Dejgard, A, et al: Treatment of chronic painful diabetic neuropathy with intravenous lidocaine infusion. Br Med J 292:173, 1986.

34. Kennedy, WR, Navarro, X, Goetz, FC, et al: Effect of pancreatic transplantation on diabetic neuropathy. N Engl J Med 322:1031, 1990.

35. Korthals, JK, Gieron, MA, and Dyck, PJ: Intima of epineural arterioles is increased in diabetic polyneuropathy. Neurology (Minneap) 38:1582, 1988.

36. Llewelyn, JG, Patel, NJ, Thomas, PK, and Stribling, D: Sodium, potassium adenosine triphosphatase activity in peripheral nerve tissue of galactosaemic rats. Effects of aldose reductase inhibition. Diabetologia 30:971, 1987.

37. Llewelyn, JG, Thomas, PK, Fonseca, V, King, RHM, and Dandona, P: Acute painful diabetic neuropathy precipitated by strict glycaemic control. Acta Neuropathol (Berl) 72:157, 1986.

38. Max, MB, Culnane, M, Schafer, SC, et al: Amitriptyline relieves diabetic neuropathy pain in patients with normal or depressed mood. Neurol (Minneap) 37:589, 1987.

39. Medori, R, Autilio-Gambetti, L, Jenick, H, et al: Changes in axon size and

slow axonal transport are related to experimental diabetic neuropathy. Neurology 38:597, 1988.

40. Melman, A: The evaluation of erectile dysfunction. Urol Rad 10:119, 1988.

41. Newrick, PG, Wilson, AJ, Jakubowski, J, Boulton, AJ, and Ward, JD: Sural nerve oxygen tension in diabetes. Br Med J 293:1053, 1986

42. Parkhouse, N and LeQuesne, PM: Impaired vasogenic vascular response in patients with diabetes and neuropathic foot lesions. N Engl J Med 318:1306, 1988.

43. Pirart, J: Diabetic neuropathy: A metabolic or a vascular disease? Diabetes 14:1, 1965.

44. Pirart, J: Diabetes mellitus and its degenerative complications: A prospective study of 4,400 patients observed between 1947 and 1973. Diabetes Care 1:168, 1978.

45. Raff, MC and Asbury, AK: Ischemic mononeuropathy and mononeuropathy multiplex in diabetes mellitus. N Engl J Med 279:17, 1968.

46. Said, G, Slama, G, and Selva, J: Progressive centripetal degeneration of axons in small fibre diabetic polyneuropathy. Brain 106:791, 1983.

47. Schaumburg (personal observation).

48. Schmidt, RE and Plurad, SB: Ultrastructural and biochemical characterization of autonomic neuropathy in rats with chronic streptozotocin diabetes. J Neuropath Exp Neurol 45:525, 1986.

49. Sharma, AK and Thomas, PK: Animal Models. In Dyck, PJ, Thomas, PK, Asbury, AK, Winegrad, AI, and Porte, D. Diabetic Neuropathy. WB Saunders, Philadelphia, 1987, p 237.

50. Sharma, AK, Bajada, S, and Thomas, PK: Influence of streptozotocin-induced diabetes on myelinated nerve fiber maturation and on body growth in the rat. Acta Neuropathol 53:257, 1981.

51. Sheps, SG: Use of an elastic garment in the treatment of orthostatic hypotension. Cardiol 61 (Suppl 1):271, 1976.

52. Slager, U: Diabetic myelopathy. Arch Pathol Lab Med 102:467, 1978.

53. Snape, WJ, Battle, WM, Schwartz, SS, et al: Metoclopramide to treat gastroparesis due to diabetes mellitus. Ann Intern Med 96:444, 1982.

54. Stewart, JD: Diabetic truncal neuropathy: Topography of the sensory deficit. Ann Neurol 25:233, 1989.

55. Subramony, SH and Wilbourn, AJ: Diabetic proximal neuropathy. Clinical and electromyographic studies. J Neurol Sci 53:293, 1982.

56. Thomas, PK and Brown, MJ: Diabetic polyneuropathy. In Dyck, PJ, Thomas, PK, Asbury, AK, Winegrad, AI, and Porte, D (eds). Diabetic Polyneuropathy. WB Saunders, Philadelphia, 1987, p 57.

57. Thomas, PK and Lascelles, RG: The pathology of diabetic neuropathy. QJ Med 35:489, 1966.

58. Tomlinson, DR, Moriarty, RJ, and Mayer, H: Prevention and reversal of defective axonal transport and motor nerve conduction velocity in rats with experimental diabetes by treatment with the aldose reductase inhibitor Sorbinil. Diabetes 33:470, 1984.

59. Watt, SJ, Tooke, JE, Perkins, CM, et al: The treatment of idiopathic orthostatic hypotension: A combined fludrocortisone and flurbiprofen regime. QJ Med 50:205, 1981.

60. Weiske, WH and Weidner, W: Papaverine-injection and Doppler sonographic examination in the diagnosis of vascular erectile dysfunction. Urol Int 42:61, 1967.

61. Willars, GB, Calcutt, NA, and Tomlinson, DR: Reduced anterograde and retrograde accumulation of axonally transported phosphofructokinase in streptozotocin-diabetic rats: Effects of insulin and the aldose reductase inhibitor "Statil". Diabetologia 30:239, 1987.

62. Yagihashi, S and Sima, A: Neuroaxonal and dendritic dystrophy in diabetic autonomic neuropathy. Classification and topographic distribution in the BB-rat. J Neuropath Exp Neurol 45:545, 1986.

63. Yasuda, H and Dyck, PJ: Abnormalities of endoneurial microvessels and sural nerve pathology in diabetic neuropathy. Neurology 37:20, 1987.

64. Yasuda, H and Dyck, PJ: Abnormalities of endoneurial microvessels and sural nerve pathology in diabetic neuropathy. Neurology 37:20, 1987.

65. Zincke, H, Campbell, JT, Palumbo, PJ, et al: Neurological vesical dysfunction in diabetics: Another look at vesical neck resection. J Urol 111: 488, 1974.

66. Zorgniotti, AU and Lefleur, RS: Auto-injection of the corpus cavernosum with a vasoactive drug combination for vasculogenic impotence. J Urol 133:39, 1985.

Chapter 8

ENDOCRINE NEUROPATHIES OTHER THAN DIABETES

HYPOTHYROIDISM

Two types of peripheral nerve disorders occur in hypothyroidism: symmetric polyneuropathy[1,3,4,5,8,12,14,15,18,20,22,26] and mononeuropathy.[1,13,19] The latter usually consists of the carpal tunnel syndrome. Sensorineural hearing loss is also associated with myxedema, but its etiology is obscure and it may not reflect primary alterations in the PNS.[1,4,8,10,13,19,20,22]

Pathology

MONONEUROPATHY

The carpal tunnel syndrome in hypothyroidism is probably secondary to compression of the median nerve by swollen tendon sheaths synovial membranes, and other engorged connective tissues.[13] That myxedematous deposits occur within the perineurium and compress nerve fibers seems unlikely.[1] There are no contemporary morphologic studies of this entity; presumably the pathologic changes are similar to those described in entrapment neuropathy from other causes (see Chapter 16).

SYMMETRIC POLYNEUROPATHY

There are no credible postmortem descriptions of the PNS changes in hypothyroid neuropathy. Contemporary studies, utilizing sural nerve biopsy specimens, indicate both axonal loss and segmental demyelination.[1,22] This conclusion is based both on teased fiber and ultrastructural examination. It appears likely that axonal change is primary, but this notion remains unproved. Neither deposits of foreign material nor excessive thickening of intraneural connective tissues are described. Two studies document Schwann cells with excessive deposition of glycogen and nonspecific mitochondrial changes.[8,18] A recent study describes endoneurial vessel wall changes including cytoplasmic inclusions, endothelial cell protrusion into and obliteration of the lumen, disruption of junctions, and basement membrane thickening similar to that described in diabetic neuropathy.[3]

Pathogenesis

MONONEUROPATHY

As already noted, the pathogenesis of the carpal tunnel syndrome is best understood as chronic compression of the median nerve by swollen connective tissue (see Chapter 16). Its severity does not correlate with that of the metabolic defect.

SYMMETRIC POLYNEUROPATHY

It appears that metabolic derangement secondary to thyroid hypofunction may be responsible for the symmetric polyneuropathy, giving rise to axonal degeneration with secondary or concomitant Schwann-cell dysfunction.

The nature of the metabolic disturbance is unknown. A decrease in the transport of slow component a, as demonstrated in hypothyroid rats, has been proposed as a possible mechanism.[23] The finding of endoneurial vessel wall changes, as described above, raises the question of whether the nerve fiber involvement may be secondary to vascular abnormalities.[3]

Clinical Features

INCIDENCE

There are no reliable estimates for the incidence of either a symmetric polyneuropathy or mononeuropathy.[1] Paresthesias of the fingertips (acroparesthesias) are common and usually reflect the carpal tunnel syndrome, the commonest peripheral nerve disturbance in hypothyroidism. Symmetric polyneuropathy is rare.[8,22]

SYMPTOMS AND SIGNS

Mononeuropathy. Intermittent paresthesias in the hands, especially at night, are the main feature of the carpal tunnel syndrome.[13] The clinical manifestation of the hypothyroidism-associated type are identical to those usually found in this disorder, and are described in Chapter 16. Other entrapment neuropathies may occur, for example, meralgia paresthetica.

Symmetric Polyneuropathy

Symptoms. Distal-extremity paresthesias and muscle cramps are frequent in hypothyroidism,[5] and it has been suggested that they reflect subclinical polyneuropathy.[1] This notion is debated, as these symptoms are not always accompanied by signs of PNS dysfunction, but there may be electrophysiologic abnormalities in such individuals.[9] Most patients with hypothyroid neuropathy complain initially of numbness in the feet. Within a few months, the hands are also affected. Leg weakness accompanied by unsteady gait develops subsequently.

Symptoms of polyneuropathy are often accompanied or obscured by other manifestations of hypothyroidism such as constipation, scaly skin, fatigue, cold intolerance, hearing loss, and hoarseness.[1]

Signs. Touch, vibration, and position sense are decreased in the distal extremities in all instances; diminished pain or thermal sense is rare.[8] Tendon reflexes are usually absent in the legs and depressed in the arms. Mild distal weakness of the legs is frequent, atrophy and fasciculations are rare. Proximal limb weakness, if present, is usually due to coexisting hypothyroid myopathy. Clumsiness and limb ataxia occasionally occur and are commonly attributed to cerebellar degeneration.[4] Prolonged relaxation of the tendon reflexes is characteristic of hypothyroidism, and does not reflect peripheral neuropathy.

Laboratory Studies

CLINICAL LABORATORY FINDINGS

Values of most routine laboratory tests are normal but the serum cholesterol and creatine kinase concentrations may be elevated. Serum thyroxine levels are decreased in all varieties of hypothyroidism. Serum tri-iodothyronine levels and radioactive iodine uptake values are less predictable.

CEREBROSPINAL FLUID

The CSF protein is frequently mildly elevated in severe hypothyroidism, independent of malfunction of the peripheral nervous system.

Electrodiagnostic Studies

MONONEUROPATHY

Electrodiagnostic studies of individuals with the carpal tunnel syndrome are valuable, both in establishing the site of the lesion and in excluding a more diffuse neuropathy. Prolonged latency of motor and sensory impulses at the wrist and diminished amplitude of sensory action potentials are usually present. Electromyography of the thenar muscles reveals denervation changes in advanced cases (see Chapter 16).

SYMMETRIC POLYNEUROPATHY

Nerve conduction studies of motor and sensory nerves show mild-to-moderate slowing of conduction velocities and prolonged distal latencies, absent or reduced sensory action potential amplitudes, and occasionally reduced motor action potential amplitudes.[8,12,18,26] Electromyography sometimes reveals signs of denervation in distal muscles.[22] Electrodiagnostic abnormalities may appear before the clinical symptoms of polyneuropathy become manifest.[26] Co-existent carpal tunnel syndrome may complicate interpretation of upper-limb electrodiagnostic studies in individuals with symmetric neuropathy.

Course, Treatment, and Prognosis

MONONEUROPATHY

The severity of the carpal tunnel syndrome correlates poorly either with the degree or the duration of hypothyroidism. In general, the untreated course is a gradual increase in paresthesias, followed by sensorimotor loss in the distribution of the distal median nerve. Most cases improve following hormone replacement therapy; surgical section of the transverse carpal ligament usually is unnecessary.[1,19]

SYMMETRIC POLYNEUROPATHY

The untreated course is relentless progression from initially mild sensory impairment to a severe distal symmetric sensorimotor neuropathy. The natural history of untreated neuropathy is un-

known, since published reports all deal with treated cases. Symmetric polyneuropathy responds well to hormone replacement therapy; usually there is dramatic improvement in signs, symptoms, and electrophysiologic function within 6 months.[1,8,22]

Differential Diagnosis

MONONEUROPATHY

The diagnosis of hypothyroidism should be entertained in individuals with the carpal tunnel syndrome, even if they appear euthyroid and the condition is unilateral. In blatantly hypothyroid patients with the carpal tunnel syndrome, it is unlikely that other factors are operant, but it is prudent to rule out chronic arthritic compression of the median nerves.

SYMMETRIC POLYNEUROPATHY

It is not difficult to identify hypothyroid neuropathy when it is accompanied by florid myxedema, but the clinical features are nonspecific and the condition cannot be distinguished from many other symmetric polyneuropathies. Fortunately, symptoms and signs of clinical hypothyroidism are clearly apparent in most cases and recovery after hormonal replacement therapy will confirm the diagnosis.

ACROMEGALY

Acromegaly is usually caused by growth hormone-secreting tumors developing in the pituitary gland. Two types of PNS involvement may develop: an entrapment mononeuropathy, usually the carpal tunnel syndrome,[16,17] and a distal symmetric polyneuropathy.[6,11] Proximal muscle weakness (acromegalic myopathy) occurs independently of PNS changes, and may confuse the clinical profile.[17]

Pathology and Pathogenesis

MONONEUROPATHY

There is no contemporary description of morphologic changes in the acromegalic carpal tunnel syndrome. Presumably compression results from combined acral soft-tissue hyperplasia, synovial edema, bony overgrowth, and osteoarthritis.[2,16,17,21] The pathology and pathogenesis of this condition are discussed in Chapter 16.

SYMMETRIC POLYNEUROPATHY

There are widely differing descriptions of the peripheral nerves in acromegaly. It is also possible that some early reports describe diabetic changes in peripheral nerve, since diabetes is a well-known complication of acromegaly. One study examines sural nerve biopsies from four nondiabetic acromegalic subjects and describes a reduction in the number of myelinated and unmyelinated fibers, segmental demyelination, and occasional onion-bulb formations.[11] An extensive investigation of nine nondiabetic acromegalics suggests that the primary pathologic abnormality is demyelination with hypertrophic onion-bulb formation.[7]

The pathogenesis of acromegalic neuropathy is obscure. Although no direct metabolic effect of excessive growth hormone is proved to be responsible, a temporal relationship appears to exist between the degree of elevation of growth hormone levels and the polyneuropathy.

Clinical Features

INCIDENCE

The carpal tunnel syndrome is a well-known complication of acromegaly and can be detected in more than one third of cases.[17] Estimates of the incidence of symmetric polyneuropathy vary; it is probably uncommon.

SYMPTOMS AND SIGNS

Mononeuropathy. The manifestations of the carpal tunnel syndrome in acromegaly do not differ from those usually encountered in this condition (see Chapter 16). It seems likely that acroparesthesias of the fingers, which occur in one third of all acromegalics, arise from this cause.

Symmetric Polyneuropathy. Symptoms of polyneuropathy may develop at any time in the course of acromegaly, but mostly appear late in the illness.[11,17] Initial symptoms are paresthesias in the feet and hands, followed by the insidious development of weakness.[25]

Decreased touch, vibration, and position sense in the lower extremities is characteristic. Tendon reflexes are usually absent in the legs and depressed in the arms. Mild distal weakness of the legs is frequent and atrophy and fasciculations are rare. Thickening of peripheral nerve trunks has been described but is an inconsistent finding.[24]

Values of most routine laboratory tests are normal, although the plasma glucose may be elevated. Serum growth hormone is usually elevated and its level is not suppressed following glucose ingestion. The CSF is usually normal.

Electrodiagnostic Studies

MONONEUROPATHY

Nerve conduction studies of individuals with the carpal tunnel syndrome are diagnostically helpful, both in establishing the site of damage and in excluding more widespread nerve dysfunction (see Chapter 16).

SYMMETRIC POLYNEUROPATHY

Motor and sensory nerve conduction in limb nerves is abnormal, with moderately reduced motor conduction velocity and depressed or absent sensory action potentials.[11]

Course, Treatment, and Prognosis

MONONEUROPATHY

The untreated course is a gradual increase in paresthesias, followed by sensorimotor loss in the distribution of the distal median nerve. Most cases improve following removal of the pituitary adenoma, and surgery of the transverse carpal ligament usually is unnecessary.[17]

SYMMETRIC POLYNEUROPATHY

The course is gradual worsening of distal motor and sensory manifestations. There are no studies of the effect of removal of pituitary adenomas on the course of the neuropathy.[7,11,17]

Differential Diagnosis

There is usually little difficulty establishing the diagnosis, since mononeuropathy or symmetric polyneuropathy usually occur after the somatic manifestations of acromegaly have become obvious. This is not always true for the carpal tunnel syndrome, as the acromegalic changes develop insidiously and may at first be overlooked.

REFERENCES

1. Bastron, JA: Neuropathy in diseases of the thyroid. In Dyck, PJ, Thomas, PK, and Lambert, EH (eds): Peripheral Neuropathy, Vol II, ed. 2 WB Saunders, Philadelphia, 1974, p 1833.
2. Biglieri, EG, Waltington, CO, and Forsham, PH: Sodium retention with human growth hormone and its subfractions. J Clin Endocrinol Metab 21:361, 1961.
3. Cavaletti, G, Guazzi, M, Marcucci, A, et al: Endoneural vessel involvement in hypothyroidism. Ital J Neurol Sci 8:259, 1987.
4. Cremer, GM, Goldstein, NP, and Paris, J: Myxedema and ataxia. Neurology (Minneap) 19:37, 1969.
5. Crevasse, LE and Logue, RR: Peripheral neuropathy in myxedema. Ann Intern Med 50:1433, 1959.
6. Dinn, JJ: Schwann-cell dysfunction in acromegaly. J Clin Endocrinol Metab 31:140, 1970.
7. Dinn, JJ and Dinn, EI: Natural history of acromegalic peripheral neuropathy. Q J Med 57:833, 1985.
8. Dyck, PJ and Lambert, EH: Polyneuropathy associated with hypothyroidism. J Neuropathol Exp Neurol 29:631, 1970.
9. Fincham, RW and Cape, CA: Neuropathy in myxedema: A study of sensory nerve conduction in the upper extremities. Arch Neurol 19:464, 1968.
10. Greene, R: The thyroid gland: Its relationship to neurology. In Vinken, PJ and Bruyn, GE (eds): Handbook of Clinical Neurology, Vol 27, Metabolic and Deficiency Diseases of the Nervous System, Part I. Elsevier/North Holland, New York, 1976, p 255.
11. Low, PA, McLeod, JG, Turtle, P, et al: Peripheral neuropathy in acromegaly. Brain 97:139, 1974.
12. Martin, J, Tomkin, GH, and Hutchinson, M: Peripheral neuropathy in hypothyroidism—an association with spurious polycythaemia (Gaisbrock's syndrome). J R Soc Med 76:187, 1983.
13. Murray, IPC and Simpson, JA: Acroparaesthesia in myxoedema. A clinical and electromyographic study. Lancet 1:1360, 1958.
14. Nickel, SN and Frame, B: Neurologic manifestations of myxedema. Neurology (Minneap) 8:511, 1958.
15. Nickel, SN, Frame, B, Bebin, J, et al: Myxedema neuropathy and myopathy. A clinical and pathologic study. Neurology (Minneap) 11:125, 1961.
16. O'Duffy, JD, Randall, RV, MacCarty, CS: Median neuropathy (carpal tunnel syndrome) in acromegaly. A sign of endocrine overactivity. Ann Intern Med 78:379, 1973.
17. Pickett, JBE, Layzer, RB, Levin, SR, et

al: Neuromuscular complications of acromegaly. Neurology (Minneap) 25:638, 1975.

18. Pollard, JD, McLeod, JG, Angel Honnibal, TG, et al: Hypothyroid polyneuropathy. Clinical, electrophysiological and nerve biopsy findings in two cases. J Neurol Sci 53:461, 1982.

19. Purnel, DC, Daly, DD, and Lipscomb, PR: Carpal-tunnel syndrome associated with myxedema. Arch Intern Med 108:751, 1961.

20. Sanders, V: Neurologic manifestations of myxedema. N Engl J Med 266:547, 1962.

21. Schiller, F, and Kolb, FO: Carpal tunnel syndrome in acromegaly. Neurology (Minneap) 4:271, 1954.

22. Shirabe, T, Tawara, S, Terao, A, et al: Myxoedematous polyneuropathy: A light and electron microscopic study of the peripheral nerve and muscle. J Neurol Neurosurg Psychiatry 38:241, 1975.

23. Sidenius, P, Nagel, P, Larsen, JR, et al: Axonal transport of slow component in sciatic nerves of hypo- and hyperthyroid rats. J Neurochem 49:1790, 1987.

24. Stewart, BM: The hypertrophic neuropathy of acromegaly. Arch Neurol 14:107, 1966.

25. Woltman, HW: Neuritis associated with acromegaly. Arch Neurol Psychiatry 45:680, 1941.

26. Yamamoto, K, Saito, K, Takai, T, et al: Unusual manifestations in primary hypothyroidism. Prog Clin Biol Res 116:169, 1983.

Chapter 9

NEUROPATHY ASSOCIATED WITH ALCOHOLISM, NUTRITIONAL DEFICIENCIES, AND MALABSORPTION*

ALCOHOL–NUTRITIONAL DEFICIENCY
POLYNEUROPATHY SYNDROME
SPECIFIC VITAMIN DEFICIENCY
POLYNEUROPATHY SYNDROMES
MALABSORPTION NEUROPATHIES:
POSTGASTRECTOMY AND SPRUE-
RELATED DISORDERS

Nutritional neuropathy in North America and Europe is almost synonymous with the multiple vitamin-deficiency polyneuropathy associated with alcoholism. The selective vitamin B_{12} and pyridoxine (B_6) deficiency polyneuropathies have become rare, as has peripheral nerve damage secondary to malabsorption syndromes. While it is acknowledged that polyneuropathy occurs with selective deficiencies of specific B vitamins, disorders such as beriberi and pellagra are now medical curiosities in Western medical practice.

Most deficiency syndromes affecting the nervous system are now held to result from lack of multiple vitamins, frequently combined with other dietary imbalance. Rarely can an instance of polyneuropathy be defined in terms of a single vitamin deficiency.[61]

ALCOHOL–NUTRITIONAL DEFICIENCY POLYNEUROPATHY SYNDROME

Definition and Etiology

This disorder is defined most simply as the polyneuropathy associated with chronic abuse of alcohol. In North America, it occurs in association with selective nutritional deprivation.[60]

The precise dietary imbalance in most cases of alcohol–nutritional deficiency polyneuropathy is usually not identified. Thiamine deficiency may have a dominant role, but other vitamins and nutrients may also be lacking.[62] In addition, alcohol itself may be neurotoxic. Thus, while most cases of alcoholic neuropathy in North America are associated with excessive intake of carbohydrate relative to thiamine and other vitamins, neuropathy also is reported to occur occasionally in apparently well-nourished alcoholics.[7] Blood levels of thiamine may be normal in patients with alcoholic polyneuropathy;[15,47] decreased transketolase activity suggesting a defect in thiamine utilization has been reported in some.[47]

Pathology

Postmortem examination of the nervous system from cases of beriberi and alcohol–nutritional deficiency neurop-

*Prepared with the assistance of Dr. Jerry Kaplan

athy consistently reveals severe degenerative changes in distal limb peripheral nerves and in the cervical portion of the gracile fasciculi. Degeneration of the distal vagus and recurrent laryngeal nerves, and "axonal reaction" of anterior horn and dorsal root ganglion cells are present in advanced cases.[61]

Several studies of nerve biopsies from human alcohol–nutritional deficiency neuropathy strongly indicate that initial axonal changes are followed by alterations in myelin.[7,64] These findings, taken in concert with the distal distribution of changes, are evidence that this condition is a distal axonopathy. Confirmation of this notion awaits definitive experimental animal studies. Unfortunately, there is neither a satisfactory experimental mammalian model for many nutritional deficiency neuropathies, nor firm evidence that ingested ethyl alcohol results in PNS degeneration.

Pathogenesis

The pathogenesis of the presumed alcohol–nutritional deficiency distal axonopathy is widely held to be similar to that of other metabolic axonopathies (see Chapter 2). Supposedly, lack of thiamine and other vitamins—cofactors for several key biochemical reactions—disrupts both axonal and nerve cell body metabolism, triggering the eventual breakdown of distal axonal integrity.[56] It is possible that alcohol itself has some role in this process.

In vitamin-deficient alcoholics, at least four additional factors have been proposed in the pathogenesis of polyneuropathy: one is decreased intake of vitamins as the alcoholic loses interest in nourishment,[60] second is malabsorption associated with alcohol-induced pancreatic damage,[64] third is an increased metabolic need for thiamine as alcohol and carbohydrates (pretzels, pasta, etc.) become the major foods,[61] and fourth is that alcohol, whether in the blood or the intestinal lumen, interferes with absorption of thiamine.[25]

It has been suggested that polyneuropathy can occur as the result of direct toxicity of alcohol on peripheral nerves, independent of nutritional deficiencies.[7] The evidence for this derives from reports of neuropathy in apparently well-nourished alcoholics[7,15] and laboratory animals.[11] Further, a study correlating alcoholic neuropathy and cerebral atrophy claims a direct toxic effect of alcohol on peripheral nerves. Carefully controlled experiments in monkeys have failed to demonstrate peripheral neurotoxicity[24] and peripheral neuropathy has been shown to be independent of cerebellar degeneration in alcoholics.[52] Proper evaluation of nutrition requires repeated observations over time,[24] and the criteria used in many studies may not be sufficiently strict.[54] Clinical evidence of malnutrition, usually substantial weight loss (15 to 20 kg), is present in the majority of alcoholics prior to or accompanying the evolution of the peripheral neuropathy.[54] These facts suggest that the role of direct neurotoxicity in the pathogenesis of alcoholic polyneuropathy is limited; evaluation of this concept awaits further studies.

Clinical Features

INCIDENCE

Alcohol-nutritional deficiency polyneuropathy is unusual in North American general practice, but is common in some municipal institutions. In the Boston City Hospital series, 9% of all alcoholics had evidence of polyneuropathy;[61] the disorder appears more frequently among women.

SYMPTOMS AND SIGNS

Symptoms of mild aching of the calf muscles and discomfort over the soles are present in about one half of the patients with alcoholic polyneuropathy. Commonly, patients are asymptomatic.[40] Examination of such individuals usually reveals depression or loss of Achilles reflexes, with impairment of pain and touch sensation over the feet, and thinned legs.[10]

Symptoms in more severely involved cases usually consist of distal lower-limb paresthesias, pain, and weakness. The legs are always affected before the arms, and often are involved exclusively.[62] Mixed sensorimotor neuropathy is the rule. In the early stages, only sensory symptoms and signs may be present, but eventually weakness appears. Pure motor neuropathy is not described. Distressing symptoms of pain and paresthesias of the feet occur in about one quarter of cases, and some experience the fully developed "burning feet" syndrome. This condition, also associated with some toxic (e.g., thallium) and nonalcohol-related malnutrition states, develops gradually over weeks or months. The soles are initially affected with aching pain that may become pricking, electric, or stabbing, sometimes evolving into an intense burning sensation.[10] The entire plantar surfaces become exquisitely sensitive to touch and the pressure of shoes and stockings cannot be tolerated. These sensations spread to the dorsum of the feet, eventually may cover the entire leg, and usually are accompanied by excessive sweating.[61]

Weakness is initially distal. Foot and wrist drop may occur and when accompanied by atrophy, become disabling. Prominent weakness of the thigh muscles can be evident. Symptoms of lower cranial nerve involvement may appear in advanced cases, and include difficulty in swallowing, hoarseness, and impaired phonation. Postural hypotension occasionally occurs, while other symptoms of autonomic involvement are extremely rare.[61]

Clear signs of PNS impairment are present in cases with only minimal symptoms.[18] The Achilles reflexes are diminished or absent in individuals with any degree of weakness. Tenderness upon squeezing the calf muscles is characteristic. Sensory impairment generally involves all modalities, and a symmetric glove-and-stocking pattern is the rule. Motor and sensory loss in extreme cases may extend to the proximal limbs. Severe dysesthesias of the soles are always accompanied by sensory loss.[61]

Signs of nutritionally-induced CNS impairment may accompany the polyneuropathy. Gait and lower-limb ataxia, formerly attributed exclusively to PNS involvement, probably primarily reflects degeneration in the anterior cerebellar vermis. Some evidence of Wernicke's disease (nystagmus, oculomotor palsy, confusion, Korsakoff's psychosis) is occasionally detectable.[62]

Laboratory Studies

CLINICAL LABORATORY FINDINGS

Abnormal values of routine clinical laboratory tests are common and usually reflect nutritional anemia or alcohol-induced liver disease. Special laboratory tests have been advocated as helpful in establishing the presence of vitamin deficiencies. Blood pyruvate and thiamine levels are less reliable. The best index of thiamine deficiency is low serum or erythrocyte transketolase activity, and the restoration of normal activity during treatment with vitamin B_1.[20,65]

CEREBROSPINAL FLUID

The CSF is acellular and the protein usually normal. A modest elevation in protein may be present in advanced cases.

ELECTRODIAGNOSTIC STUDIES

Nerve conduction and late response abnormalities are present even in the early stages of alcoholic polyneuropathy. Sural nerve action potential amplitude and late responses are abnormal in over 90% of alcoholics with neuropathy, even when patients are asymptomatic.[40] Abnormalities of H reflex and F response studies are present in about three fourths of these patients. Conventional motor nerve conduction studies are frequently normal[14,40] consistent with the notion that alcoholic polyneuropathy is a distal axonopathy with the longest sensory fibers most affected. Late response studies are more sensitive than conventional motor nerve conduction studies because they measure conduction over a greater length of nerve and because they exhibit a narrower range of normal values than motor conduction velocities.[40] Studies that emphasize the sensitivity of ulnar and peroneal studies[53] should be viewed with caution as these nerves are frequent sites of entrapment or compression in alcoholics, even in the absence of generalized peripheral neuropathy.[36] It is claimed that nerve conduction studies may be sensitive indicators of recovery in alcoholics with neuropathy who remain abstinent.[31]

The role of needle electromyography in the evaluation of alcoholic polyneuropathy is less clear. Most electrophysiological studies of this disorder have examined only nerve conduction. It is suggested that electromyography is a sensitive indicator of motor nerve dysfunction in alcoholic polyneuropathy[54] and that motor unit potential size increases with continued alcoholic intake.[28] Single fiber electromyography is reported to be useful in studying early alcoholic neuropathy.[37]

The use of experimental techniques to monitor refractory periods,[5] autonomic function,[34,42] and vestibular function[51] awaits clinical correlation. In the future, quantitative sensory threshold testing may prove especially helpful in the early detection and monitoring of alcoholics with neuropathy.

Course, Prognosis, and Treatment

The prognosis of untreated alcohol–nutritional deficiency neuropathy is poor.[28] Axonal degeneration continues and leg weakness becomes extreme. Weakness in the upper extremities is not as profound, and proximal strength and tendon reflexes remain relatively spared. Signs of Wernicke's encephalopathy or delirium tremens may overshadow the polyneuropathy, and can constitute a life-threatening situation.[62]

With treatment, the prognosis is excellent if the polyneuropathy is still in an early stage.[31] Treatment begun in the later stages of neuropathy may rapidly alleviate paresthesias, but objective improvement in strength may not appear until much later. Proximal muscles usually regain bulk and strength within several months. A year may elapse before foot-drop lessens. Rarely, distal atrophy, contractures, and sensory loss are permanent.

Treatment includes abstinence from alcohol, a high calorie, protein-rich diet, and multiple vitamin supplements.[61] Thiamine initially should be administered by intramuscular injection of 50 mg per day for one week.[64] Multiple-vitamin therapy should also be given. Physical therapy should stress range-of-motion exercises, and splints may be necessary to avoid contractures in severe cases.

Differential Diagnosis

The diagnosis of alcohol–nutritional deficiency neuropathy is straightforward in an individual with evidence of liver disorder, malnutrition, and a history of alcoholism. Paraneoplastic neuropathies (see Chapter 11) may share common clinical and neurologic features (e.g., weight loss, paresthesias) and should be considered in cases refractory to nutritional therapy and abstinence from alcohol.

Alcoholism is a common condition, and its associated polyneuropathy is clinically indistinguishable from many other distal axonopathies. Therefore, the diagnosis of alcoholic polyneuropathy should be entertained with caution in individuals with coexistent metabolic disorders, or in those who also may be exposed to neurotoxic agents (see Chapters 17 and 18).

Disulfiram causes a peripheral neuropathy in abstinent alcoholics.[8,45,67] This may be mistaken for alcoholic neuropathy unless a careful history is taken.

Case History: Alcohol–Nutritional Deficiency Polyneuropathy

A 41-year-old bachelor and construction worker had consumed a pint of whiskey, but otherwise a balanced diet, almost daily for 10 years, without apparent ill effect. He sustained a fractured humerus in early June and was unable to work. His daily whiskey consumption increased to a quart, and he gradually lost interest in preparing meals or shopping. In addition to alcohol, his diet consisted mainly of spaghetti, pretzels, potato chips, and pizza. By Christmas, he appeared tremulous and seemed unsteady. In April, he fell down the stairs, was knocked unconscious and was admitted to the hospital. General physical examination disclosed no evidence of weight loss, but an enlarged liver and rhinophyma were present. He was alert, cooperative, and oriented. Memory was intact. Cranial nerves and upper limb strength and sensation were normal. The gait was broad-based and he lost balance frequently. Coordination of the arms was normal, the heel-knee-shin maneuver was irregular, and toe-tapping was jerky. Tendon reflexes were absent throughout, save for weak biceps and triceps jerks. Strength was normal proximally in the legs. The dorsiflexors of ankles and toes could be overcome with ease, and he was unable to stand on toes or heels. Gentle pressure on the calf muscles elicited severe pain, as did squeezing the anterior thighs. Pain and touch sensation were markedly impaired over the feet, and gradually shaded into normal appreciation at knee level. Both pinprick and touch elicited an unpleasant "stabbing" sensation over the soles. Vibration sense was diminished over the toes and ankles, while position sense was intact. Routine laboratory tests disclosed a mild hypochromic microcytic anemia. Liver function tests were unremarkable.

The day following admission he became tremulous and, in the ensuing 3 days, he developed visual hallucinations and florid delirium tremens. He was hospitalized for three weeks and given a high protein diet with vitamin supplements. He refused electrodiagnostic studies or lumbar puncture, and was discharged to live with a sister. She supervised his diet and general well being, but could not limit his alcohol consumption. Six months later, gait was still broad-based and tendon reflexes were unchanged. Strength was nearly normal, and there was no muscle tenderness or dysesthetic response in the legs. Sensation to pin and touch was still impaired over the feet and ankles.

Comment. This case demonstrates many of the salient clinical features of alcohol–nutritional deficiency neuropathy. It also illustrates the role of increased carbohydrate consumption in its genesis, and the ability of therapy rapidly to reverse some of the changes. The patient shows evidence of cerebellar dysfunction from which he may never recover.

SPECIFIC VITAMIN DEFICIENCY POLYNEUROPATHY SYNDROMES

Thiamine (B₁)

Although thiamine-deficiency polyneuropathy in North America is generally associated with alcoholism, in other areas of the world a thiamine-poor diet alone causes nervous system damage, such as polyneuropathy ("dry" beriberi) resulting from a diet exclusively of polished rice. Recurrent vomiting, independent of alcoholism, is a well-documented gastrointestinal cause of Wernicke's encephalopathy (cerebral beriberi). One of Wernicke's original cases was a young woman with an esophageal stricture.[66] Thiamine deficiency may follow gastric resection surgery for morbid obesity.[2] Cardiac failure with peripheral edema ("wet" beriberi) is a further consequence of deficiency of this vitamin.

STRUCTURE, SOURCE, REQUIREMENT, ABSORPTION

Thiamine is a water-soluble vitamin, formed by a pyrimidine ring and a thiazole moiety linked by a methylene bridge. It is synthesized by many plants and bacteria, and is present in most vegetable and nonfatty animal tissues. Most of the thiamine in cereal grains is in the outer layers, hence the occurrence of beriberi in individuals eating only milled rice. The minimal daily requirement for an adult is 0.7 mg to 0.9 mg. Daily gastrointestinal absorption capacity is limited to 5 mg to 10 mg. Only approximately 25 mg is stored in the body—excessively administered oral thiamine being excreted in the feces. These two factors emphasize the need for daily parenteral treatment of the thiamine-deficiency neurologic syndromes and may explain the rapidity with which symptoms develop.[62] Certain bacteria and raw fish contain thiaminases, and a diet containing raw carp may contribute to the development of beriberi. Thiamine-deficiency neurologic syndromes are generally associated with diminished intake rather than malabsorption, although, as has been noted, recurrent vomiting may precipitate Wernicke's disease.

MECHANISM OF ACTION

Thiamine diphosphate is a coenzyme for several important biochemical reactions, including the oxidative decarboxylation of pyruvate and ketoglutarate and the transketolase reaction of the pentose phosphate pathway.[62] It is also alleged that thiamine has a specific role in nerve conduction, independent of its coenzyme function in general metabolism.[33]

PATHOLOGY AND PATHOGENESIS

The pathology and pathogenesis of thiamine-deficiency polyneuropathy are widely held to be identical to those described in the preceding section.[59]

There is no satisfactory mammalian experimental model of thiamine-deficiency neuropathy.[50] Degeneration of long and large myelinated nerve fibers is a consistent feature in pigeons made thiamine deficient.[58] An unpublished study of this avian model (by B.G. Cragg and P.K. Thomas) found that no electrophysiologic or morphologic changes were identifiable in acutely thiamine-deficient birds or chronic thiamine-deficient ataxic pigeons. Brainstem structures seem much more vulnerable than peripheral nerve.

GENERAL CLINICAL FEATURES OF THIAMINE DEFICIENCY

The most devastating neurologic effects of thiamine deficiency are Wernicke's disease and its associated syndrome, Korsakoff's psychosis. Cerebellar degeneration and polyneuropathy often accompany Wernicke's disease.[62] The signs, symptoms, course, and prognosis of thiamine-deficiency polyneuropathy ("dry" beriberi) are indistinguishable from those described for alcohol-nutritional deficiency. Cardiac failure ("wet" beriberi) may be associated with neuropathy.

DIAGNOSIS AND TREATMENT

The most reliable biochemical test of thiamine deficiency is the measurement of whole-blood or erythrocyte transketolase activity before and following the addition of thiamine diphosphate in vitro. An increase in enzyme activity following treatment further supports this diagnosis.[20,65] Treatment consists of intramuscular injection of 50 mg thiamine daily for 2 weeks, followed by 5 mg per day orally. Since these individuals often suffer multiple vitamin deficiencies, other water-soluble vitamins should also be given.[61] In individuals suspected of having Wernicke's disease, treatment is urgently indicated and may dramatically reverse extraocular dysfunction.[62]

Riboflavin (B₂)

Riboflavin deficiency almost invariably occurs in combination with other vitamin deficiencies. It is usually secondary to a riboflavin-deficient diet, but has also been described following gastrectomy and attributed to malabsorption. Riboflavin deficiency is alleged to be one of the factors responsible for the "burning feet" syndrome seen in certain malabsorptive and undernutrition states, but the administration of riboflavin alone does not relieve the neurologic symptoms.[61]

Niacin (Nicotinic Acid, B₃)

Deficiency of niacin or its amino acid precursor, tryptophan, precipitates pellagra. Pellagra may result from a diet poor in these substances (primary pellagra) or from gastrointestinal malabsorption (secondary pellagra). Degeneration of CNS neurons and coexistent dementia are well documented in pellagrins, and clearly are secondary to niacin-tryptophan depletion. Claims of a niacin-deficiency peripheral neuropathy are viewed with skepticism, and this entity remains sub judice. It is generally held that the peripheral neurologic manifestations of pellagra can be attributed to coexisting deficiencies of other vitamins.

Pyridoxine (B₆)

Naturally occurring, selective pyridoxine deficiency is extremely rare, largely because this vitamin is present in many foods; the condition may be induced both because the vitamin is destroyed during food processing and because certain drugs act as pyridoxine antagonists. Isoniazid antituberculous therapy accounts for almost all instances of human pyridoxine-deficiency polyneuropathy. Isoniazid toxicity and its pathogenesis are described in Chapter 17.

Paradoxically, the daily administration of excessive amounts of pyridoxine to experimental animals results in degeneration of dorsal root ganglion cells and the appearance of a permanently disabling toxic sensory neuronopathy syndrome.[38] Peripheral neuropathy has been reported in humans with a daily intake of greater than 200 mg. Ironically, this is often iatrogenic, as this vitamin is mistakenly dispensed for the premenstrual and carpal tunnel syndromes (see Chapter 17).

STRUCTURE, SOURCE, REQUIREMENT, ABSORPTION

Three closely related compounds—pyridoxine, pyridoxal, and pyridoxamine—are designated as vitamin B₆. The coenzyme form is pyridoxal-5-phosphate and the three compounds owe their enzymatic activity to tissue conversion to this moiety. The vitamin is widely and uniformly distributed in all foods. The minimum daily requirement for an adult is 0.6 mg to 1.3 mg. Ingested ethyl alcohol interferes with the metabolism of pyridoxal phosphate, possibly providing an additional factor in the alcohol–nutritional deficiency neuropathy syndrome. Gastrointestinal absorption is rapid and occurs by passive diffusion; pyridoxine-deficiency neurologic syndromes rarely result from malabsorption even with extensive postsurgical loss of small intestine.[48]

MECHANISM OF ACTION

Pyridoxal phosphate acts as a cofactor for many enzymes in amino acid metabolism (transaminases, synthetases, hydroxylases), and is especially important in the metabolism of tryptophan, glycine, serine, and glutamate. Pyridoxine has a vital, poorly understood role in neuronal excitability, possibly related to gamma-aminobutyric acid metabolism.

PATHOLOGY AND PATHOGENESIS

See Isoniazid Toxicity in Chapter 17.

GENERAL CLINICAL FEATURES OF PYRIDOXINE DEFICIENCY

Pyridoxine-deficiency neurotoxicity produced in human volunteers by administering desoxypyridoxine, a pyridoxine antagonist, includes generalized seizures as well as peripheral neuropathy of the type seen in isoniazid toxicity.[63] Some genetic conditions characterized by abnormalities in B_6 metabolism feature generalized seizures and CNS damage. Those can be prevented only by giving large amounts of pyridoxine.[32]

DIAGNOSIS AND TREATMENT

The most common diagnostic index is the measurement of tryptophan metabolites, particularly xanthurenic acid, following tryptophan loading.[26] Also useful is the in vitro assessment of erythrocyte glutamic pyruvate transaminase in the presence and absence of pyridoxine.[48]

Oral treatment with 30 mg of pyridoxine is advocated for prophylaxis in pregnancy, and for individuals taking isoniazid.

Pantothenic Acid

Selective pantothenic acid deficiency is extremely rare because of the ubiquitous occurrence of this vitamin. Since pantothenic acid is not stored in the body in any great amount, severely malnourished individuals rapidly deplete their reserve of this and other vitamins. Pantothenic acid–deficiency syndromes secondary to malabsorption are not known.

Experimental human pantothenic acid deficiency is alleged to result in the "burning feet" syndrome and one report describes relief of this condition in malnourished individuals by the administration of pantothenic acid.[22] These observations were not confirmed by a subsequent investigation of human pantothenic acid deficiency.[61] The existence of a polyneuropathy caused by selective pantothenic acid deficiency appears moot.

Vitamin B_{12}

Vitamin B_{12} deficiency is generally related to malabsorption. Dietary inadequacy of B_{12} is rare and encountered solely in strict vegetarians. Malabsorption usually results from either inadequate gastric production of intrinsic factor (pernicious anemia, gastrectomy) or from disorders of the terminal ileum (celiac disorders, intestinal resection). The prominent clinical features common to all conditions from which B_{12} deficiency may develop are hematologic, gastrointestinal, and neurologic (both CNS and PNS)[48] though neurologic dysfunction may occur in the absence of anemia.[9,21] There exists an enormous clinical and experimental body of knowledge of the hematologic and CNS effects[49] of B_{12} deficiency that is beyond the scope of this review. In contrast, remarkably little is known about the peripheral neuropathy.

STRUCTURE, SOURCE, REQUIREMENT, ABSORPTION

Vitamin B_{12} is a complex organometallic compound characterized by a cobalt atom sited within a corrin ring. It cannot be synthesized in the human body, but can be formed by some bacteria normally present in the intestine. The principal dietary sources of vitamin B_{12} are meat and dairy products. The minimum adult daily requirement is 3 μg. Dietary vitamin B_{12} combines with a glycoprotein intrinsic factor (IF) that is produced by the parietal cells of the stomach. The B_{12}-IF complex travels to the distal ileum where B_{12} is absorbed into the blood. Thus, disorders of the stomach or lower intestine may result in B_{12} deficiency. Normally, about 4 mg of B_{12} are stored in the body. In view of the extremely low minimum daily requirement, it takes 3 to 4 years of malabsorption to produce a human deficiency state. This may explain the repeated failures of earlier investigators to produce nervous system degeneration in short-term animal experiments.

MECHANISMS OF ACTION

Methylcobalamin is necessary for the demethylation of methyltetrahydrofolate (TFH). This involves the methylation of homocysteine to methionine. TFH is required for the production of the active folate coenzymes that are necessary for DNA synthesis. Nitrous oxide is known to inactivate methylcobalamin, but not adenosylcobalamin, and gives rises to megaloblastic changes in bone marrow. Reports suggest that prolonged exposure to N_2O gives rise to a myelopathy with combined posterior and lateral column degeneration, possibly similar to human combined-system disease.[19,30]

PATHOLOGY AND PATHOGENESIS

Little is known about either the pathology or pathogenesis of the PNS lesions of human B_{12} deficiency.[3,61] This contrasts strikingly with the extensive studies of the human CNS lesions[49] and the findings of an ultrastructural investigation of similar CNS lesions appearing in primates following prolonged, carefully controlled B_{12} deficiency.[3,4] Human and primate CNS lesions are confined to the white matter, are prominent in the dorsal and lateral funiculi of the spinal cord and peripheral visual pathways, and are scattered in the cerebrum. Vacuolation of myelin with relative sparing of axons appears to be the initial change, while at later stages, there is loss of axons and gliosis.

The PNS lesions have not been extensively studied morphologically. Most reports are limited to descriptions of nerve biopsies processed by conventional histologic methods.[23] Degeneration of both myelin[43,44] and axons is described but it is difficult to determine whether the polyneuropathy is axonal or demyelinative. There is a manifest lack of detailed study of the extent of the human PNS lesions utilizing modern histologic techniques and no suitable animal model of the PNS lesion of B_{12} deficiency exists. Curiously, the B_{12}-deficient primate, which displays profound CNS demyelinative lesions closely mimicking those of humans, does not become anemic or develop peripheral neuropathy.[3,4]

Theories concerning the biochemical basis underlying the nervous system lesions that accompany B_{12} deficiency have included occult cyanide poisoning and lipid metabolic abnormalities resulting from impairment of the methylmalonyl CoA mutase reactions. These theories have been thoroughly reviewed elsewhere.[3] More recently, methyl group deficiency has been considered.[19]

GENERAL CLINICAL FEATURES OF B_{12} DEFICIENCY POLYNEUROPATHY

The frequency of peripheral neuropathy among the neurologic complications of vitamin B_{12} deficiency is controversial. Two reviews of this issue yield these diametrically opposing statements: "In our opinion peripheral neuropathy is probably the commonest neurological complication of vitamin B_{12} deficiency,[61] and, One may conclude that the neurological manifestations of pernicious anemia are due primarily to the spinal-cord lesions. In the course of the myelopathy the peripheral nerves may also be involved, but the latter affection is less frequent than the former and of less clinical significance."[48]

There is overwhelming evidence that PNS changes occur in individuals with severe B_{12} deficiency. Clinical evidence of neuropathy initially appears in the legs and consists of early absent or diminished Achilles reflexes, distal leg sensory deficits for multiple modalities, and occasionally, distal leg weakness that exceeds the accompanying pyramidal signs. Since myelopathy accompanies the PNS degeneration, in most cases, it is extraordinarily difficult to determine whether the paresthesias in the feet are secondary to changes in the dorsal columns or in the peripheral nerves.[61] Signs of pyramidal tract involvement (Babinski's sign) usually appear early, and often accompany the diminished or absent Achilles reflexes. Electrophysiologic investigations have generally supported the existence of lower-limb peripheral nerve involvement but have not helped in elucidating its nature.[43]

DIAGNOSIS AND TREATMENT

Neuropsychiatric disorders may develop in the absence of anemia or macrocytosis.[41] Specific assays for serum B_{12} are now generally available in developed countries. The normal range of vitamin B_{12} in serum is 200 pg to 900 pg per ml; values less than 200 pg per ml merit suspicion, and those less than 100 pg per ml indicate clinically significant deficiency.[41] Measurements of serum methylmalonic acid and total homocysteine are useful in the diagnosis of cobalamin deficiency.[41]

Treatment consists of 100 µg of vitamin B_{12} (as hydroxocobalamin) intramuscularly on alternate days for a week, then weekly for 3 months. Since the defect is almost always one of absorption, replacement is given parenterally and must be maintained for life, usually at a level of 1000 µg every 3 months.

Vitamin E (α-Tocopherol)

Vitamin E deficiency, from malabsorption due to chronic cholestasis, fibrocystic disease, abetalipoproteinemia, and blind loop syndromes, causes a peripheral neuropathy often overshadowed by central nervous system and ocular abnormalities.[6,12,27] Electrophysiologic and morphologic[39] studies suggest it is distal axonopathy, but Schwann cell inclusions have been noted.[13,39] This neuropathy is at least partially reversible if treated early.[13] The pathogenesis of this disorder remains unclear: one study has demonstrated depressed intraneural levels of tocopherol in vitamin E deficient patients with peripheral neuropathy.[55] It is probably related to free-radical damage.[55a]

MALABSORPTION NEUROPATHIES: POSTGASTRECTOMY AND SPRUE-RELATED DISORDERS

Postgastrectomy Neuropathy

Total gastrectomy abolishes the absorption of vitamin B_{12} due to lack of intrinsic factor and if untreated, a fully developed vitamin B_{12}-deficiency neurologic syndrome (optic neuropathy, myelopathy, peripheral neuropathy) would seem likely. Myelopathy has been demonstrated in such individuals and paresthesias occur, but there are few reports of polyneuropathy.

Partial gastrectomy or plication for morbid obesity is clearly associated with peripheral neuropathy or myelopathy, or both—usually from 10 to 20 years after surgery. Although it is widely held that malabsorption of vitamin B_{12} underlies these disorders, serum B_{12} levels are sometimes normal and replacement therapy elicits a variable response. It appears likely that some cases of postgastrectomy neuropathy are unrelated to vitamin B_{12} deficiency.[21,55]

Sprue-Related Disorders (Tropical Sprue, Gluten-Induced Enteropathy, Celiac Disease, and Other Malabsorption-Associated Enteropathies)

These disorders, of varied etiology and pathogenesis, result in impairment of the absorptive function of the small intestine.[48] They are characterized by varying degrees of fatty diarrhea (steatorrhea), weight loss, disturbance of calcium and protein metabolism, and anemia. The neurologic syndromes associated with these disorders include, in varying degrees, peripheral neuropathy, myelopathy, encephalopathy, and myopathy. It is often stated that these neurologic disorders, like those following gastrectomy, are somehow related to selective malabsorption of vitamin B_{12}. Analysis of the few well-studied cases of sprue-related neuropathy, myelopathy, or both does not strongly support this assumption.[48] Treatment with vitamin B_{12} has not always ameliorated the symptoms, and individuals have recovered without such therapy. It has been suggested that the malabsorptive neurologic disorders possibly result from multiple nutritional deficiencies. Vitamin E deficiency has been advanced as a possible cause in some instances.[9,27,46]

Gluten-induced enteropathy has been associated with distal axonopathy. Serum levels of pyridoxine, thiamine, B_{12}, folic acid, and vitamin E may be normal in this disorder. The relationship between enteropathy and neuropathy is unclear, but both appear to respond to dietary gluten restriction.[35]

REFERENCES

1. Abarbanel, JM, Berginer, VM, Osimani, A, et al: Neurologic complications after gastric resection for morbid obesity. Neurology 37:196, 1987.
2. Abarbanel, JM, Frisher, S, Osimani, A, et al: Vitamin B_{12} deficiency neuropathy: Sural nerve biopsy study. Isr J Med Sci 22:909, 1986.
3. Agamanolis, DP, Chester, EM, Victor, M, et al: Neuropathology of experimental vitamin B_{12} deficiency in monkeys. Neurology (Minneap) 26: 905, 1976.
4. Agamanolis, DP, Victor, M, Harris, JW, et al: An ultrastructural study of subacute combined degeneration of the spinal cord in vitamin B_{12} deficient rhesus monkeys. J Neuropathol Exp Neurol 37:273, 1978.
5. Alderson, M and Petajan, J: Relative re-

6. Alvarez, F, Landrieu, P, Laget, P, et al: Nervous and ocular disorders in children with cholestasis and vitamin A and E deficiencies. Hepatology 3:410, 1983.
7. Behse, F and Buchthal, F: Alcoholic neuropathy: Clinical, electrophysiological and biopsy findings. Ann Neurol 2:95, 1977.
8. Bergouignan, FX, Vital, C, Henry, P, et al: Disulfiram neuropathy. J Neurol 235:382, 1988.
9. Binder, HJ, Solitare, GB, and Spiro, HM: Neuromuscular disease in patients with steatorrhea. Gut 11:549, 1970.
10. Bischoff, A: Die alkoholische polyneuropathie: Klinische, ultrastructurelle und pathogenetische aspekte. Dtsch Med Wochenschr 96:317, 1971.
11. Bosch, EP, Pelham, RW, Rasool, CG, et al: Animal model of alcoholic neuropathy: Morphologic electrophysiologic and biochemical findings. Muscle Nerve 7:133, 1979.
12. Brin, M, Pedley, TA, Lovelace, RE, et al: Electrophysiologic features of abetalipoproteinemia: Functional consequences of vitamin E deficiency. Neurology 36:669, 1986.
13. Burck, U, Goebel, HH, Kuhlendahl, HD, et al: Neuromyopathy and vitamin E deficiency in man. Neuropediatrics 12:267, 1981.
14. Casey, EB and LeQuesne, PM: Electrophysiological evidence for a distal lesion in alcoholic neuropathy. J Neurol Neurosurg Psychiatry 35: 624, 1972.
15. Claus, D, Eggers, R, Engelhardt, A, et al: Ethanol and polyneuropathy. Acta Neurol Scand 72:312, 1985.
16. Coers, C and Hildebrand, J: Latent neuropathy in diabetes and alcoholism. Neurology (Minneap) 15:19, 1965.
17. Coers, C and Woolf, AL: The Innervation of Muscle. Blackwell Scientific Publications, Oxford, 1959.

5. fractory period: A measure to detect early neuropathy in alcoholics. Muscle Nerve 10:323, 1987.

18. Cooke, WT and Smith, WT: Neurological disorders associated with adult coeliac disease. Brain 88:683, 1966.

19. Dinn, JJ, Wilson, P, and Weir, DG: Animal model for subacute combined degeneration. Lancet 11:1154, 1978.

20. Dreyfus, PM: Clinical application of blood transketolase determinations. N Engl J Med 267:596, 1962.

21. Feit, H, Glasberg, M, Ireton, C, et al: Peripheral neuropathy and starvation after gastric partitioning for morbid obesity. Ann Intern Med 96:453, 1982.

22. Gopalan, C: The "burning-feet" syndrome. Indian Medical Gazette 81:22, 1946.

23. Greenfield, JG and Carmichael, EA: The peripheral nerves in cases of subacute combined degeneration of the cord. Brain 58:483, 1935.

24. Hallett, M, Fox, JG, Rogers, AE, et al: Controlled studies on the effects of alcohol ingestion on peripheral nerves of Macaque monkeys. J Neurol Sci 80:65, 1987.

25. Halsted, CH, Robles, EA, and Mazey, E: Decreased jejunal uptake of labeled folic acid (3H-PGA) in alcoholic patients: Roles of alcohol and nutrition. N Engl J Med 285:701, 1971.

26. Hansson, O: Tryptophan loading and pyridoxine treatment in children with epilepsy. Ann NY Acad Sci 166:306, 1969.

27. Harding, AE, Matthews, S, Jones, S, et al: Spinocerebellar degeneration associated with a selective defect of vitamin E absorption. N Engl J Med 313:32, 1985.

28. Hawley, R, Kurtzke, JF, Armbrustmacher, VW, et al: The course of alcoholic-nutritional peripheral neuropathy. Acta Neurol Scand 66:582, 1982.

29. Hensing, JA: Subacute combined degeneration, neutrophilic hypersegmentation and the absence of anemia. A case report. Ariz Med 38:768, 1981.

30. Heyer, EJ, Simpson, DM, Bodis-Wollner, I, et al: Nitrous oxide: Clinical and electrophysiological investigation of neurologic complications. Neurology 36:1618, 1986.

31. Hillbom, M and Wennberg, A: Prognosis of alcoholic peripheral neuropathy. J Neurol Neurosurg Psychiatry 47:699, 1984.

32. Hunt, AD, Stokes, J, McCrory, WW, et al: Pyridoxine dependency: A report of a case of intractable convulsions in an infant controlled by pyridoxine. Pediat 13:140, 1954.

33. Itokawa, Y and Cooper, JR: Ion movements and thiamine. II. The release of the vitamin from membrane fragments. Biochim Biophys Acta 196:274, 1970.

34. Jensen, K, Andersen, K, Smith T, et al: Sympathetic vasoconstrictor nerve function in alcoholic neuropathy. Clinic Physiol 4:253, 1984.

35. Kaplan, JG, Pack, D, Horoupian, D, et al: Distal axonopathy associated with gluten enteropathy: A treatable disorder. Neurology 38:642, 1988.

36. Kemppainen, R, Juntunen, J, Hillbom, M, et al: Drinking habits and peripheral alcoholic neuropathy. Acta Neurol Scand 65:11, 1982.

37. Kontouris, D: Superiority of single fiber electromyography to conventional EMG in early diagnosis of alcoholic neuropathy. Electromyog Clin Neurophysiol 25:295, 1985.

38. Krinke, G, Schaumburg, HH, Spencer, PS, et al: Pyridoxine megavitaminosis produces degeneration of peripheral sensory neurons (sensory neuronopathy) in the dog. Neurotoxicology 2:13, 1981.

39. Landrieu, P, Selva, J, Alvarez, F, et al: Peripheral nerve involvement in children with chronic cholestasis and vitamin E deficiency: A clinical, electrophysiological and morphological study. Neuropediatrics 16:194, 1985.

40. Lefebvre-D'Amour, M, Shahani, BT, Young, RR, et al: The importance of studying sural nerve conduction and late responses in the evaluation of alcoholic subjects. Neurology 29:1600, 1979.

41. Lindenbaum, J, Healton, EB, Savage, DG, et al: Neuropsychiatrics disorders caused by cobalamin deficiency in the absence of macrocytosis. N Engl J Med 318:1720, 1988.

42. Matikainen, E, Juntunen, J, Salmi, T, et al: Autonomic dysfunction in long-standing alcoholism. Alcohol-Alcohol 21(1):69, 1986.

43. Mayer, RF: Peripheral nerve function in vitamin-B_{12} deficiency. Arch Neurol 13:335, 1965.

44. McCombe, PA and McLeod, JG: The peripheral neuropathy of vitamin B_{12} deficiency. J Neurol Sci 66:117, 1984.

45. Mokri, B, Ohnishi, A, Dyck, PJ, et al: Disulfiram neuropathy. Neurology 31:730, 1981.

46. Nelson, JS, Fitch, CD, Fischer, VW, et al: Progressive neuropathologic lesions in vitamin E-deficient rhesus monkeys. J Neuropathol Exp Neurol 40:166, 1981.

47. Paladin, F and Russo-Perez, G: The haematic thiamine level in the course of alcoholic neuropathy. Eur Neurol 26:129, 1987.

48. Pallis, CA and Lewis, PD: The Neurology of Gastrointestinal Disease. WB Saunders, Philadelphia, 1974.

49. Pant, SS, Asbury, AK, and Richardson EP: The myelopathy of pernicious anemia. A neuropathological reappraisal. Acta Neurol Scand 44 (Suppl 35):1, 1968.

50. Prineas, J: Peripheral nerve changes in thiamine-deficient rats. Arch Neurol 23:541, 1970.

51. Sasa, M and Takaori, S: Peripheral and central vestibular disorders in alcoholics. A three-year followup study. Arch Otolaryngol 230:93, 1981.

52. Scholz, E, Diener, HC, Dichgans, J, et al: Incidence of peripheral neuropathy and cerebellar ataxia in chronic alcoholics. J Neurol 233:212, 1986.

53. Shankar, K, Maloney, FP, and Thompson, C: An electrodiagnostic study in chronic alcoholic subjects. Arch Phys Med Rehab 68:803, 1987.

54. Shields, RN: Alcoholic polyneuropathy. Muscle Nerve 8:183, 1985.

55. Somer, H, Bergstrom, L, Mustajoki, P, et al: Morbid obesity, gastric plication and a severe neurological deficit. Acta Med Scand 217:575, 1985.

55a. Southam, E, Thomas, PK, King, RHM, et al: Experimental vitamin E deficiency in rats. Morphological and functional evidence of abnormal axonal transport secondary to free radical damage. Brain 114:1991, (in press).

56. Spencer, PS, Sabri, MI, Schaumburg, HH, et al: Does a defect in energy metabolism in the nerve fiber underlie axon degeneration in polyneuropathies? Ann Neurol 5:501, 1979.

57. Steiner, I, Kidron, D, Soffer, D, et al: Sensory peripheral neuropathy of vitamin B_{12} deficiency: A primary demyelinative disease? J Neurol 235:163, 1988.

58. Swank, RL: Avian thiamine deficiency. J Exp Med 71:683, 1940.

59. Takahashi, K: Thiamine deficiency neuropathy: A reappraisal. Int J Neurol 15:245, 1981.

60. Victor, M, Adams, RD, and Collins, GH: The Wernicke-Korsakoff Syndrome, ed 2. FA Davis, Philadelphia, 1990.

61. Victor, M: Polyneuropathy due to nutritional deficiency and alcoholism. In Dyck, PJ, Thomas, PK, and Lambert, EH. Peripheral Neuropathies, Vol 2, ed 2 WB Saunders, Philadelphia, 1984, p 1899.

62. Victor, M and Adams, RD: On the etiology of the alcoholic neurologic diseases. With special reference to the role of nutrition. Am J Clin Nutr 9:379, 1961.

63. Vilter, RW, Mueller, JF, Glazer, HS, et al: The effect of vitamin B_6 deficiency induced by desoxypyridoxine in human beings. J Lab Clin Med 42:335, 1953.

64. Walsh, JC and McLeod, JG: Alcoholic neuropathy: An electrophysiological and histological study. J Neurol Sci 10:457, 1970.

65. Warnock, LG: Transketolase activity of blood hemolysate, a useful index for

diagnosing thiamine deficiency. Clin Chem 21:432, 1975.

66. Wernicke, C: Lehbuch der Gehirnkrankenheiten fur Aerzte und Studirende, Vol 2. T Fischer. Kassel, 1881, p 229.

67. Wormer, TM: Peripheral neuropathy after disulfiram administration: Reversibility despite continued therapy. Drug Alcohol Depend 10:199, 1982.

Chapter 10

ISCHEMIC NEUROPATHY (PERIPHERAL VASCULAR-OCCLUSIVE DISEASE AND VASCULITIS)

NEUROPATHY OF PERIPHERAL
  VASCULAR DISEASE
NEUROPATHY DUE TO NECROTIZING
  ANGIITIS

Clinicians most commonly equate ischemic disease of the PNS with diabetic mononeuropathy, with nerve compression, or with trauma. Peripheral neuropathy associated with chronic peripheral vascular-occlusive disease is probably frequent, although the manifestations are only mild; it has been little studied. In contrast, neuropathology associated with necrotizing angiopathy is rare but well described.

NEUROPATHY OF PERIPHERAL VASCULAR DISEASE

It is generally held that peripheral nerve is relatively resistant to large vessel occlusive disease, and only widespread, severe involvement of small blood vessels produces ischemic nerve damage. Experimental studies that fail to produce ischemic lesions in leg nerves by ligating single large limb vessels support this notion,[1] and emphasize the abundant collateral sources of blood flow in peripheral nerve. It is usually necessary to occlude[26-28] or microembolize[24] multiple small vessels or the aorta[20] to produce limb nerve ische-

mia in an experimental animal. One recent experimental study suggests that chronic endoneurial ischemia is associated with axonal atrophy.[32] Nevertheless, several clinical reports have documented neurologic abnormalities in the extremities of individuals with chronic severe peripheral vascular disease or as a complication of aortic manipulation.[12]

Studies of individuals with chronic obliterative arterial disease of the lower limbs describe evidence of sensory peripheral neuropathy in 50%[14] to 88%.[11] The extent of the deficit appears proportional to the degree of ischemia. Many had no tendon reflexes. Nerve biopsy reveals loss of myelinated fibers with evidence of segmental demyelination, wallerian degeneration, and abundant occlusion of small epineurial vessels. The latter finding suggests the possibility that the arteriosclerotic process extends into many of the arterioles of nerve, resulting in a "diffuse-small-vessel" neuropathy. It appears likely that mild neuropathy is a common feature of chronic vascular insufficiency, but goes unrecognized, since the more obvious effects of ischemia on skin and muscle dominate the clinical picture.

NEUROPATHY DUE TO NECROTIZING ANGIITIS

No fewer than eleven systemic disorders are associated with vasculitic neu-

Table 10–1 DISORDERS ASSOCIATED WITH VASCULITIC NEUROPATHY

SYSTEMIC DISORDERS
  Polyarteritis nodosa
  Rheumatoid arthritis
  Systemic lupus erythematosus (SLE)
  Hypersensitivity angiitis
  Allergic granulomatosis (Churg-Strauss)
  HIV infection
  Systemic sclerosis
  Sjögren's syndrome
  Wegener granulomatosis
  Giant-cell arteritis (temporal arteritis)
  Spanish oil syndrome

NONSYSTEMIC VASCULITIC NEUROPATHY

ropathy and a primary, nonsystemic neuropathy has been recently described[9,18] (Table 10–1). Although there is considerable variety among these conditions, they present similar clinical and pathologic profiles of vasculitic peripheral nerve involvement. Only polyarteritis nodosa, rheumatoid arthritis, lupus erythematosus, and Sjögren's syndrome are encountered with any frequency in clinical practice and these conditions are discussed individually. The Spanish oil syndrome is covered in Chapter 18 and HIV neuropathies in Chapter 6. Angiitis characteristically gives rise to a focal or multifocal neuropathy. Other types of polyneuropathy or neuronopathy may be encountered in relation to SLE, rheumatoid arthritis, and HIV infection,[31] and a subacute motor polyneuropathy in SLE.[3] They likely have a differing pathogenesis.

Polyarteritis Nodosa

PATHOLOGY AND
PATHOGENESIS

Spinal or cranial nerves display vascular lesions in 75% of cases.[21] Nerves may be focally swollen and hemorrhagic. In polyarteritis the necrotizing inflammation involves medium-sized and small arteries and, occasionally, veins. Capillaries and venules are spared. Thus, the epineurial vessels are

heavily involved while endoneurial capillaries are preserved. Segments of vessels are affected and there appears to be a predilection for bifurcation points where small, fragile aneurysms occur. Acute lesions feature fibrinoid necrosis of the vessel wall accompanied by focal accumulations of polymorphonuclear leukocytes, chronic (or healed) lesions, by modest numbers of mononuclear cells, granulation tissue, and collagen.[11,31] An experimental study utilizing local injections of arachidonic acid suggests that unmyelinated fibers are most vulnerable following acute small-vessel occlusion of peripheral nerve,[28] but this is not reflected in the pattern of sensory loss observed in humans.

The pathogenesis of nerve fiber destruction presumably is related to focal ischemia resulting from arteriolar occlusion. Nerve biopsy usually reveals abundant vascular change and teased nerve fiber preparations a combination of axonal degeneration and segmental demyelination.[3] An experimental study utilizing microembolization of capillaries demonstrates central fascicular lesions, axonal attenuation and degeneration, and secondary demyelination.[24] Postmortem sampling of nerves clearly demonstrates distally accentuated loss of nerve fibers. Focal infarcts of nerve have not been convincingly demonstrated in polyarteritis. It is proposed that circulating antigen-antibody complexes are deposited in vessel walls with subsequent recruitment of leukocytes that release enzymes causing vessel damage.[8] A recent study implicates a T-cell dependent, cell-mediated process as a key component in this process.[19]

CLINICAL FEATURES

Polyarteritis nodosa is rare, but clinically detectable neuropathy occurs in almost two thirds of cases at some stage in the illness and may be the presenting feature.[17] Neuropathy usually occurs subsequent to other manifestations of generalized disease, such as fever, pericarditis, hypertension, renal failure, or abdominal pain. Occasionally, peripheral neuropathy may be the first clear indication of specific organ involvement in an individual with puzzling, nonspecific, debilitating illness. Polyarteritic neuropathy is usually heralded by dysfunction of one or more scattered peripheral nerves (multiple mononeuropathy).

Branches of the radial, median, and sciatic nerves are especially frequent early sites for mononeuropathy in this condition. The onset of dysfunction is abrupt, accompanied by pain and numbness in the distribution of the affected nerve, and may be followed in days by near-total motor and sensory loss. Pain is sometimes experienced locally at the site of ischemia in the nerve (nerve trunk pain). In time, other nerves often become involved, eventuating in a severe, symmetric, predominantly distal polyneuropathy affecting arms as well as legs. This profile is so well recognized that the occurrence of a multiple mononeuropathy in a nondiabetic, chronically ill person immediately suggests a polyarteritic basis. Over one third of polyarteritic neuropathy cases display a distal symmetric pattern from the outset, clinically indistinguishable from the toxic-metabolic neuropathies.[4,18,31] This probably is secondary to an especially widespread and fulminant vasculitic involvement of many peripheral nerves. Nerve biopsy is diagnostic in about one half of cases of neuropathy from systemic vasculitis of all types. One recent study suggests that muscle biopsy increases the yield to 80%.[31]

The prognosis is poor for polyarteritis nodosa with multiple organ involvement. Treatment with corticosteroids (60 mg prednisolone on alternate days) is customary and may produce rapid, albeit temporary, improvement. Higher doses may be necessary. Refractory cases may require treatment with cyclophosphamide. Occasional cases may have vasculitis that predominantly affects the nervous system and spares the heart, abdominal viscera, and kidney. In our experience, such individuals tend to a more prolonged survival, and occasionally display a remarkable recovery from severe generalized polyneuropathy following prolonged corticosteroid therapy. A variety of neuropathic syndromes, probably secondary to angiitis, occur in 14% of patients with giant cell (temporal) arteritis. They are usually ameliorated by corticosteroid therapy.[5]

Rheumatoid Arthritis

COMPRESSION SYNDROMES

Pressure palsy and entrapment mononeuropathies are common in rheumatoid arthritis, reflecting prolonged immobilized postures and compression of nerve by articular deformity.[25] The carpal tunnel syndrome is especially frequent in advanced cases. The pathology, pathogenesis, and clinical features of these conditions are discussed in Chapter 16.

CHRONIC SYMMETRIC SENSORY NEUROPATHY

A few individuals with long-standing, moderately severe rheumatoid arthritis develop a mild, distal symmetric sensory neuropathy. All modalities of sensation are equally affected. It is likely that some weakness accompanies this predominantly sensory neuropathy, and electrodiagnostic studies support this view;[6] however, strength is difficult to evaluate in patients with prolonged immobilization and severe deformity. This symmetric polyneuropathy is generally a benign condition, spontaneous improvement is the rule, and corticosteroid treatment is not indicated. Occasionally, severe pain may accompany this condition, even in its mild form. The symmetric sensory neuropathy, in itself, is rarely debilitating. A careful histologic study of nerve biopsies from three individuals with mild sensory neuropathy demonstrated epineurial vessels occluded by nonspecific thickening, large-fiber loss, and axonal degeneration.[34]

SUBACUTE MULTIPLE MONONEUROPATHY DUE TO NECROTIZING ANGIITIS

This rare condition is the most severe peripheral nervous system complication of rheumatoid arthritis. The clinicopathologic profile closely resembles that of polyarteritis nodosa.[7,25] Most cases occur after many years of chronic arthritis with destructive joint change and rheumatoid nodules, but may be encountered in comparatively mild cases. Mononeuritis is usally part of an overall changing clinical picture reflecting systemic vasculitis (Raynaud's phenomenon, nail bed infarctions, skin ulcers, coronary and mesenteric artery occlusions). The clinical profile is an acute onset of painful paresthesias within the distribution of one peripheral nerve, followed within days by appropriate motor and sensory dysfunction. Usually, multiple nerves in several limbs become involved within a few months and the prognosis is poor. There is a high incidence of life-threatening complications from coronary artery disease, mesenteric artery occlusion, and septicemia. The pathophysiology of this conditon clearly is related to ischemia secondary to necrotizing vasculitis. High-dose corticosteroid treatment in combination with cyclophosphamide is merited, but the response is often disappointing. A meticulous postmortem study of this condition has highlighted several features of the PNS pathology including focal areas of nerve fiber degeneration.[10] Changes appear mostly in the center of fascicles and focal lesions in major limb nerves. The focal lesions begin at mid-humeral and midfemoral levels, suggesting these areas are watersheds of poor perfusion.[7] Endoneurial infarction is not described in this study.

Systemic Lupus Erythematosus (SLE)

Approximately 10% of SLE cases develop peripheral neuropathy in one of several varieties: a subacute illness of the AIDP type, chronic demyelinating type, a diffuse distal sensory or sensorimotor neuropathy, mononeuropathy and multiple mononeuropathy.[3,16,22,30] None of these conditions has been thoroughly characterized morphologically. Presumably the AIDP and chronic demyelinating pictures result from immune-mediated demyelination and resemble the conditions discussed in Chapter 5. It is generally held that both diffuse sensorimotor neuropathy and multiple mononeuropathy somehow result from ischemia secondary to vasculitis, since arteritis is clearly associated with other manifestations of SLE. Severe multifocal neuropathy carries a bad prognosis and is best treated with high-dose corticosteroids. The cases of symmetric sensory neuropathy also appear to have a vascular basis.[22] The symmetric sensorimotor neuropathy may be mild and is not inevitably an ominous prognostic sign. Nerve biopsies from such cases demonstrate increased expression of Class II (Ia) antigen within nerve fascicles and endothelial cells.[22]

Sjögren's Syndrome

Sjögren's syndrome (SS) is a chronic inflammatory disorder of exocrine glands primarily affecting middle-aged women; it is defined as a triad of keratoconjunctivitis sicca (dry eyes), xerostomia (dry mouth), and arthritis. The sicca complex of ocular and oral dryness results from an autoimmune exocrinopathy characterized by lymphocyte and plasma cell infiltration of glandular structures. Diagnosis is established by combination of Schirmer test (for tearing), sialogram, minor salivary gland (lip) biopsy, and appropriate serologic tests. SS has long been associated with trigeminal neuropathy, sensorimotor neuropathy, and multiple mononeuropathy, all presumed due to vasculitis.[2,16,23,29] SS also can cause an acute or chronic sensory and autonomic neuronopathy characterized by T-cell infiltration of dorsal root and autonomic ganglia and neuronal degeneration, without evidence of necrotizing arteritis.[13]

Nonsystemic Vasculitic Neuropathy

Several studies have suggested that peripheral neuropathy may be the sole manifestation of vasculitis.[18,33] This notion is strongly supported by a recent review of 20 well-studied cases of long duration.[9] These cases have few systemic symptoms or serologic abnormalities. The condition is indolent and not life-threatening, and there is a wide range in severity and nature of the neurologic deficits, including symmetric sensory neuropathy. This vascular disease process appears to affect smaller epineurial anterioles to a greater extent than do the systemic vasculitides. Muscle biopsy in such cases, however, indicates that the vasculitic process may be more widespread[31] and systemic spread can eventually occur. Focal conduction block may occur, erroneously suggesting a diagnosis of inflammatory demyelinating neuropathy. Transient focal conduction block is also described in one experimental study of nerve ischemia.[26]

REFERENCES

1. Adams, WE: The blood supply of nerves. II. The effects of exclusion of its regional sources of supply on the sciatic nerve of the rabbit. J Anat 77:243, 1943.
2. Alexander EL, Provost, TT, Stevens, MB, et al: Neurologic complications of primary Sjögren syndrome. Medicine 61:247, 1982.
3. Asbury, AK and Johnson, PC: Pathology of Peripheral Nerve. WB Saunders, Philadelphia, 1978, p 110.
4. Bouche, P, Leger, JM, Travers, HP, et al: Peripheral neuropathy in systemic vasculitis: Clinical and electrophysiologic study of 22 patients. Neurology 36:1598, 1986.
5. Caselli, RJ, Daube, JR, Hunder, GG, et al: Peripheral neuropathic syndromes in giant cell (temporal) arteritis. Neurology 38:685, 1988.
6. Chamberlain, MA and Bruckner, FE: Clinical and electrophysiological features of rheumatoid neuropathy. Ann Rheum Dis 29:609, 1970.
7. Conn, DL and Dyck, PJ: Angiopathic neuropathy in connective tissue diseases. In Dyck, PJ, Thomas, PK, Lambert, EH, and Bunge, R (eds) Peripheral Neuropathy, Vol II, ed. 2. WB Saunders, Philadelphia, 1984, p 2027.
8. Cupps, TR and Fauci, AS: Pathophysiology of vasculitis. In The Vasculitides. WB Saunders, Philadelphia, 1981.
9. Dyck, PJ, Benstead, TJ, Conn, DL, et al: Nonsystemic vasculitic neuropathy. Brain 110:843, 1987.
10. Dyck, PJ, Conn, DL, and Okazaki, H: Necrotizing angiopathic neuropathy. Mayo Clin Proc 47:461, 1972.
11. Eames, RA and Lange, LS: Clinical and pathological study of ischemic neuropathy. J Neurol Neurosurg Psychiatry 30:215, 1967.
12. Ettinger, A, Wolfson, L, Kaplan, R, et al: Causalgia complicates ischemic neuropathy following use of an intra-aortic balloon pump (IABP). Neurology 37 (Suppl 1):261, 1987.
13. Griffin, JW, Cornblath, DR, Alexander, E, et al: Sensory ganglionitis in Sjögren syndrome. Ann Neurol 27:304, 1990.
14. Hutchinson, EC and Liversedge, LA: Neuropathy in peripheral vascular disease. Q J Med 25:267, 1956.
15. Johnson, RT and Richardson, EP: The neurologic manifestations of lupus erythmatosus: A clinical-pathological study of 24 cases and review of the literature. Medicine 47:337, 1968.
16. Kennett, R and Harding, AE: Peripheral neuropathy associated with sicca syndrome. J Neurol Neurosurg Psychiatry 49:90, 1986.
17. Kernohan, JW and Woltman, HW: Periarteritis nodosa: A clinico-pathologic study with special reference to

the nervous system. Arch Neurol 39:655, 1938.

18. Kissell, JT, Slivka, AP, Warmoltz, JR, et al: The clinical spectrum of necrotizing angiopathy of the peripheral nervous system. Ann Neurol 18: 251, 1985.

19. Kissell, JT, Ritchman, JL, Omerza, J, et al: Peripheral nerve vasculitis: Immune characterization of the vascular lesions. Ann Neurol 25: 291, 1989.

20. Korthals, JK and Wisniewski, HM: Peripheral nerve ischemia. Part I. Experimental model. J Neurol Sci 24: 65, 1975.

21. Lovshin, LL and Kernohan, JW: Peripheral neuritis in periarteritis nodosa. A clinicopathologic study. Arch Intern Med 82:321, 1948.

22. McCombe, PA, McLeod, JG, Pollard, JD, et al: Peripheral sensorimotor and autonomic neuropathy associated with systemic lupus erythmatosus. Brain 110:533, 1987.

23. Molina, R, Provost, TT, and Alexander, EL: Peripheral inflammatory vascular disease in Sjögren syndrome. Arthritis Rheum 28:1341, 1985.

24. Nukada, H and Dyck, PJ: Acute ischemia causes axonal stasis, swelling, attenuation and secondary demyelination. Ann Neurol 22:311, 1987.

25. Pallis, CA and Scott, JT: Peripheral neuropathy in rheumatoid arthritis. Br Med J 1:1141, 1965.

26. Parry, G and Linn, D: Transient focal conduction block following experimental occlusion of the vasa nervorum. Muscle Nerve 9:345, 1986.

27. Parry, GJ and Brown MJ: Arachidonate-induced experimental nerve infarction. J Neurol Sci 50:123, 1981.

28. Parry, GJ and Brown, MJ: Selective fiber vulnerability in acute ischemic neuropathy. Ann Neurol 11:147, 1982.

29. Peyronnard, JM, Charron, L, Beaudet, F, et al: Vasculitic neuropathy in rheumatoid disease and Sjögren syndrome. Neurology 32:839, 1982.

30. Rechthand, E, Cornblath, DR, Stern, BJ, et al: Chronic demyelinating polyneuropathy in systemic lupus erythmatosus. Neurology 34:1375, 1984.

31. Said, G, Lacroix-Ciaudo, C, Fujimura, H, et al: The peripheral neuropathy of necrotizing arteritis: A clinicopathological study. Ann Neurol 23: 461, 1988.

32. Sladky, JT, Tschoepe, BA, Greenberg, JH, and Brown, MJ: Peripheral neuropathy after endoneurial ischemia. Ann Neurol 29:272, 1991.

33. Torvik, A and Berntzen, AE: Necrotizing vasculitis without visceral involvement. Postmortem examination of three cases with affection of skeletal muscles and peripheral nerves. Acta Med Scand 184:69, 1968.

34. Weller, RO, Bruckner, FE, and Chamberlain, MA: Rheumatoid neuropathy: A histological and electrophysiology study. J Neurol Neurosurg Psychiatry 33:592, 1970.

Chapter 11

NEUROPATHIES ASSOCIATED WITH PARAPROTEINEMIA, DYSPROTEINEMIA, AND MALIGNANCY

PARAPROTEINEMIC NEUROPATHY
DYSPROTEINEMIC NEUROPATHIES
MALIGNANCY-ASSOCIATED
NEUROPATHIES

PARAPROTEINEMIC NEUROPATHY (BENIGN MONOCLONAL GAMMOPATHY)

Neuropathy related to benign monoclonal gammopathies has become recognized as an important category of late-onset nervous system disease.[8,28,33] Most patients have IgM paraproteinemias, usually with a kappa light chain. Because the distinction between benign and malignant paraproteinemias may be difficult to make and the progression to malignancy slow, the designation monoclonal gammopathies of undetermined significance (MGUS) has been recommended.[32] Benign monoclonal gammopathies develop in a small number of individuals in later life, but it is clear that the association with neuropathy is not a chance effect.[28]

Pathology and Pathogenesis

Nerve biopsies characteristically show evidence both of axonal loss and extensive demyelination and remyelination with hypertrophic "onion bulb" changes.[53,54] Electron microscopy in patients with IgM paraproteins frequently shows myelin that is abnormally widely spaced[53,54] (Figs. 11–1 and 11–2). Focal myelin thickenings, similar to those that occur in patients with hereditary liability to pressure palsies, may also be encountered.[44]

Immunohistochemical studies demonstrate the presence of the paraprotein on surviving myelin sheaths;[33,53] it is deposited in the region of the widely spaced myelin.[37] The IgM paraprotein reacts with myelin-associated glycoprotein (MAG) in about half of these cases.[5,29] Recent observations suggest that this leads to complement-mediated demyelination.[38] Intraneural injection of serum from patients into animals has produced demyelination.[21]

Not all neuropathies related to IgM paraproteinemia are demyelinating in type; those that are not may have a different mechanism or be chance associations. Paraproteins may react with other peripheral nerve antigens[29] and produce either axonal neuropathies[50] or other lower motor neuron syndromes.[18]

Neuropathies associated with IgG and IgA paraproteinemia are also usually demyelinating, but deposition of the paraprotein on myelin is not usually demonstrable, nor is widely spaced myelin encountered. It has been suggested that some of these patients may represent examples of chronic inflammatory demyelinating polyneuropathy.[29]

Clinical Features

Neuropathy associated with IgM paraproteinemia mainly occurs in later life and is more common among males.[53]

136

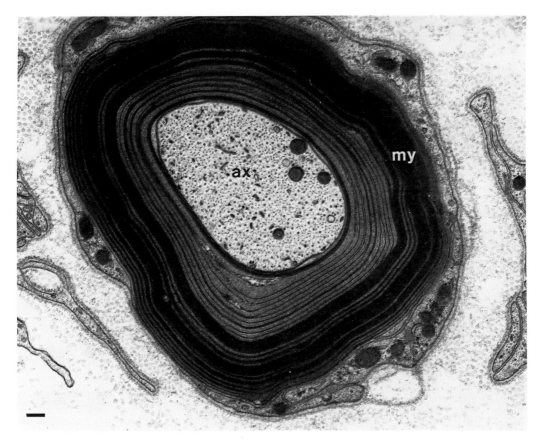

Figure 11—1. Electron micrograph of a transverse section through a myelinated nerve fiber from a patient with an IgM kappa paraproteinemic polyneuropathy. The axon (*ax*) is surrounded by a myelin sheath (*my*) that possesses alternating zones with both normal and abnormally widened periodicity. Bar = 0.2 μm.

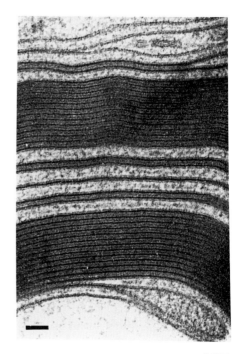

Figure 11—2. Detail of myelin from a patient with an IgM kappa paraproteinemic polyneuropathy. This electron micrograph shows darker areas in which myelin periodicity is normal, separated by zones with abnormally wide spacing. Bar = 0.1 μm.

137

Most cases are sensorimotor but a few are purely sensory; the distribution is distal. A prominent finding in many cases is ataxia in concert with postural tremor in the upper limbs, resembling essential tremor.[53] This can be severe and disabling. The clinical features in neuropathies associated with IgG and IgA paraproteinemias are similar, although tremor and ataxia tend to occur less often than in those related to IgM paraproteins. A predominantly motor neuropathy may be observed in association with IgA paraproteinemia, as may the Crow-Fukase syndrome (POEMS syndrome).[29]

The cerebrospinal fluid (CSF) protein content is usually increased in IgM paraproteinemia. In cases with a demyelinating neuropathy, motor nerve conduction velocity is slow. Sensory nerve action potentials are reduced or absent.

Prognosis and Treatment

The course of IgM paraproteinemic neuropathy is chronic, extending over many years. Slow deterioration may occur or the condition may remain relatively static for prolonged periods. Decisions whether to employ treatments that have potentially deleterious side effects must be tempered accordingly. Improvement in some cases follows repeated plasma exchange combined with immunosuppressive drugs. Corticosteroids are of little use in IgM paraproteinemic neuropathy. In contrast, IgG and IgA paraproteinemias may respond satisfactorily. A recent report suggests that treatment with human immunoglobulin may be beneficial.[9]

Differential Diagnosis

The discovery of a monoclonal serum paraprotein in a patient with neuropathy demands the exclusion of a malignant plasma cell dyscrasia. In older individuals, chance association with a monoclonal serum band is more frequent, and other causes for neuropathy require exclusion. The distinction between paraproteinemic demyelinating neuropathy and chronic inflammatory demyelinating polyneuropathy, even in some cases with IgM paraproteins, is still unresolved, as the clinical picture may be identical. In cases with IgM paraproteins, the deposition of IgM on myelin and the presence of widely spaced myelin may be helpful.

DYSPROTEINEMIC NEUROPATHIES

Waldenström's Macroglobulinemia and Malignant IgM Monoclonal Gammopathies

The paraprotein in Waldenström's macroglobulinemia is usually IgM, rarely IgG or IgA. A monoclonal serum paraprotein may also be present in chronic B-cell lymphomas and leukemias. All may be associated with cranial nerve lesions or other focal and multifocal neuropathies related to cellular infiltration of nerve trunks.[26,30] A generalized polyneuropathy may occur. The IgM paraproteins may be demonstrable on myelin sheaths upon nerve biopsy,[43] and may display anti-MAG activity.

Neuropathy Associated with Anti-G_{M1} Ganglioside Antibodies

There are recent descriptions of motor syndromes associated with high titers of antibodies against G_{M1} and G_{D1b} gangliosides. Such antibodies are detectable in low titer in a proportion of patients with motor neuron disease, acute inflammatory demyelinating polyneuropathy (AIDP), and other disorders affecting the lower motor neurons. Their significance is uncertain, but they most likely represent a nonspecific response. In other patients with multi-focal motor neuropathies, such anti-bodies have been detected in high titer and it has been suggested that they play a causal role in the production of the neuropathy.[1,41,46] Such cases were found refractory to treatment with cor-ticosteroids but responded to a combi-nation of plasma exchange and cyclo-phosphamide.[41,51] The relationship of these cases to chronic inflammatory de-myelinating polyneuropathy is still un-certain. In some but not all of these pa-tients who have responded to therapy, nerve conduction studies have shown evidence of demyelination and multifo-cal conduction block. It is suggested that anti-G_M antibodies may cause mo-tor dysfunction by binding to the nodal region of the peripheral nerve.[48]

Paraproteins to GM_1 and other gangli-osides should be suspected in patients with purely lower motor neuron syn-dromes, increased CSF protein, and electrophysiology showing multifocal conduction block.

Cryoglobulinemia

Both essential and secondary cryo-globulinemia are associated with neu-ropathy in the lower limbs, often pre-cipitated by cold.[35] One study estimates that 7% of all individuals with cryoglob-ulinemia have neuropathy.[57] The pa-thology, pathogenesis, and natural his-tory of this condition are unclear. Endoneurial and intracapillary deposits of cryoglobulin are described with mul-tiple myeloma; it has been proposed that cryoglobulin thrombi render the nerves ischemic. This suggestion ap-pears especially relevant to those cases whose neurologic deficit parallels the occurrence of purpura produced by cold. Vasculitis of the vasa nervorum, probably related to activation of the complement system by intravascular cryoglobulin deposition, has been dem-onstrated in cryoglobulinemic neurop-athy.[7,10] In one report, the cryoprecipi-tate contained monoclonal IgM which had rheumatoid factor activity, to-gether with polyclonal IgG.[11]

MALIGNANCY-ASSOCIATED NEUROPATHIES

Compression and infiltration of pe-ripheral nerve roots and cranial nerves are common peripheral nervous system effects of malignancy. As a remote effect of malignancy, neuropathy is probably rare. Nevertheless, occult malignancy is frequently first on the list of suspected disorders when unexplained polyneur-opathies are encountered; and healthy persons with a mild peripheral neurop-athy may undergo an elaborate, uncom-fortable, expensive, and fruitless evalu-ation for an occult neoplasm. This practice is encouraged by earlier re-ports suggesting that neuropathy or "neuromyopathy" frequently accompa-nies carcinoma.[11,15,39] It seems likely that many of the findings commonly at-tributed to remote effects of cancer may have reflected malnutrition, aging, neu-rotoxic chemotherapeutic agents, and associated or coincidental metabolic and other diseases.

It is difficult to estimate the overall frequency of peripheral neuropathy ac-companying malignancy, because the occurrence of neuropathy will depend upon the type and site of the tumor, the stage and duration of the illness, and the techniques employed in investiga-tion. A current estimate is for clinically evident neuropathy in about 5% of pa-tients with cancer, rising up to 12% if quantitative sensory testing is used, and to 30% to 40% if electrophysiologic studies are undertaken.[36]

Classification

Studies reported over the past few years have indicated that several pat-terns of neuropathy may accompany malignancy and suggest that a classifi-cation based on the neuropathologic scheme in Chapter 2 is appropriate (Table 11–1).

Table 11–1 CLASSIFICATION OF THE PATTERNS OF NEUROPATHY ASSOCIATED WITH MALIGNANCY

CARCINOMA
  Compression and infiltration
  Subacute sensory neuronopathy
  Sensorimotor neuropathy
    Acute, subacute, chronic progressive,
      or relapsing and remitting

LYMPHOMA
  Infiltration
  Sensory neuropathy
  Sensorimotor neuropathy
    Acute, subacute, chronic progressive,
      or relapsing and remitting
  Demyelinating polyneuropathy
  Motor neuronopathy (subacute motor
    neuronopathy)

LEUKEMIA
  Infiltration
  Sensory neuropathy
  Sensorimotor neuropathy
    Acute, subacute, or chronic

POLYCYTHEMIA VERA
  Sensory neuropathy

MYELOMA
  Compression
  Typical myeloma
    Sensory neuropathy
    Sensorimotor neuropathy
  Amyloid polyneuropathy
  Osteosclerotic myeloma
    Sensorimotor neuropathy

Carcinoma

COMPRESSION AND INFILTRATION

Compression of peripheral nerve and roots by adjacent or epidural carcinoma is common. The clinical and pathologic features of chronic and subacute compression are discussed in Chapter 16. Compression syndromes are usually symptomatic, and frequently result in pain or dysfunction and require treatment. Epineural invasion also frequently occurs, is usually asymptomatic, and is especially common in breast and pancreatic adenocarcinoma. The invading tumor spreads into the epineurium and surrounds individual fascicles. The perineurium is relatively resistant to invasion and usually pre-

vents malignant cells from entering the endoneurial space. Asymmetric neuropathies and cranial nerve palsies may result from direct invasion of isolated or multiple nerve roots or cranial nerves. Hematogenous endoneurial metastases are extremely rare.

SUBACUTE SENSORY NEURONOPATHY

This rare syndrome has been recognized since 1948 as a distinct clinicopathologic entity, principally associated with oat-cell carcinoma of the lung.[16] Previously described as "carcinomatous sensory neuropathy," this condition is best understood as a sensory neuronopathy syndrome (see Chapter 2). Its clinical recognition is important, as it frequently produces symptoms when the underlying neoplasm is occult and potentially curable. A wide range of other tumors are occasionally associated with this condition; these include bronchial (squamous) carcinoma, and carcinomas of the breast, gastrointestinal tract, and ovary.[25]

Pathology and Pathogenesis. Primary degeneration of dorsal root ganglion neurons, accompanied by inflammation, is consistently demonstrated both by biopsy and postmortem examination. Salient neuropathologic findings include: (1) neuronal degeneration of the dorsal root ganglia (DRG) accompanied by inflammation and phagocytosis at an early stage, with neuronal loss and fibrosis at later stages; (2) secondary degeneration in dorsal roots, sensory nerve, and dorsal columns of the spinal cord; (3) minimal change in ventral roots and muscle; and (4) occasional evidence of patchy neuronal loss in the brainstem and limbic cortex.[25]

The etiology and pathogenesis have not been established. It appears that an inflammatory assault on the dorsal root ganglion neurons is followed by degeneration of these cells and their processes.[25] Clinical and pathologic evidence of a more diffuse encephalomyelitis is occasionally present.[22] These features have suggested a viral etiology, perhaps accompanied by an immune-mediated cellular response. Viral particles have not been identified and culture of DRG biopsy and postmortem material has been negative.[25] Recently there have been reports of the presence of a polyclonal complement-fixing IgG antibody that has been termed anti-Hu. This is present in serum and CSF of patients with subacute sensory neuronopathy related to small-cell lung carcinoma.[14,17,19,31,42] The antibody reacts with a 35- to 38-kDa brain nuclear protein. It is highly specific for small-cell lung carcinoma and is not associated with other types of malignancy. It may be found in patients with other paraneoplastic syndromes, including encephalomyelitis and subacute cerebellar degeneration. The reason that subacute sensory neuronopathy is the dominant manifestation may be the lack of an effective blood-nerve barrier in the DRG, allowing the antibody preferential access to this component of the nervous system. Central nervous system manifestations may be less frequent because of the presence of the blood-brain barrier.

Clinical Features. This disorder is rare and most commonly affects women in late middle life. The neurologic illness usually precedes the discovery of malignancy, often by a period as long as one year.[25]

The symptoms and signs are characteristic. The onset is subacute, with pain, paresthesias, and dysesthesias commencing in the distal extremities and spreading proximally. The trigeminal nerve is seldom involved. Unsteady gait and clumsy hands are prominent disabling complaints in nearly every case. Weakness is usually absent and wasting may evolve from disuse. Position sense loss in the distal extremities and resulting sensory ataxia are universal signs. Loss of vibration sense parallels the loss of position sense; touch, pin, and temperature sensation are generally decreased as well. Pseudoath-

etoid movements appear in individuals with severe proprioceptive sensory loss. Autonomic neuropathy is not usually a feature of this condition, although urinary bladder dysfunction has been described. Signs of more diffuse nervous system involvement—encephalomyelitis—are frequently present and impaired ocular (nystagmus, anisocoria) and higher cognitive function (memory impairment, dementia) are especially common.[12]

The course of the illness is subacute, initially worsening over a 1- to 3-month interval and then stabilizing. In most cases, the sensory deficit is disabling. The duration and degree of neurologic illness appear unaffected by treatment of the underlying neoplasm. Clearly, successful treatment of the tumor prolongs survival.[25]

The CSF profile is variable. Cells are not present in the majority of patients, but occasionally a mild lymphocytosis occurs and the protein level is frequently elevated. Motor nerve conduction is usually normal; sensory nerve action potentials are often of reduced amplitude or unelicitable in involved limbs.

Differential Diagnosis. *This semiologic constellation represents a well-defined, easily recognized clinical entity that should initiate a thorough search for an underlying malignancy.* Individuals should be re-evaluated after 6 months if the initial investigations proved negative. A history of a febrile illness treated with antibiotics may suggest the acute sensory neuronopathy syndrome,[55] but this condition can only be diagnosed with certainty when neoplasia is eliminated. Cisplatin sensory neuropathy commencing after therapy is completed may present a diagnostic challenge; sparing of small-fiber sensory modalities suggests that the condition represents cisplatin toxicity.

SENSORIMOTOR NEUROPATHY

This is especially frequent in individuals with oat-cell carcinoma of the lung.[20] Primary carcinoma of the breast, stomach, colon, pancreas, uterus, cervix, thyroid, and testis are also associated with distal symmetric neuropathy.

Pathology. Postmortem tissue and nerve biopsies in this disorder predominantly display axonal degeneration and loss of myelinated nerve fibers. These changes are most pronounced in distal limb nerves. Pallor of the dorsal columns and slight loss of anterior horn and dorsal root ganglion cells are also present. There is no systematic morphologic analysis of postmortem tissue in this disorder utilizing current histopathologic techniques, and no animal model.

Pathogenesis. The causation of this presumed distal axonopathy is unknown. The prominent pathologic abnormalities in the dorsal root ganglia in the sensory neuronopathy syndrome suggest that its pathogenesis differs from that in the distal sensorimotor neuropathy. The latter is not associated with anti-Hu antibodies. Possibly the tumor releases a secretory product that leads to the development of a distal axonopathy by the mechanisms described in Chapter 2. Although weight loss usually precedes the onset of this polyneuropathy, patients are neither cachectic nor do they manifest other signs of nutritional disorder.[20] This condition clearly occurs in untreated individuals and the dysfunction cannot be attributed to chemotherapeutic axonal toxins.

Clinical Features. Symmetric distal neuropathy may occur early or late in the course of a known carcinoma. Rarely, it may precede diagnosis of the malignancy. Most experience significant weight loss before neuropathy appears.[20]

Initial symptoms are numbness and weakness of the feet. Progression in most cases is gradual. Eventually the proximal lower limbs and hands are affected. Position, vibration, and touch sense are more affected than pain sense; distal tendon reflexes are lost. Weakness is present in every case; foot drop and atrophy of intrinsic hand muscles occur in some. Autonomic and cranial nerve dysfunction are not usually features of this neuropathy, although trigeminal sensory loss may occasionally develop. The previous natural history was one of slow progression;[11] however, many now experience an illness accelerated by neurotoxic chemotherapeutic agents such as vincristine. It is becoming increasingly difficult to distinguish individuals with this neuropathy from those with drug-induced illness.

In rare instances, an acute neuropathy may develop with clinical features resembling AIDP, but such cases are less well documented than those that occur in relation to lymphoma (see later). Rarely, cases may pursue a relapsing and remitting course.

The CSF in the chronic progressive cases is generally unremarkable. Electromyography usually reveals evidence of denervation in lower limb muscles. Motor and sensory conduction velocities are normal or slightly slowed. Sensory amplitudes are usually diminished.

Electrophysiologic evidence of axonal neuropathy is allegedly common in carcinomatous patients with no symptoms of nervous system dysfunction. Physical examination frequently discloses absent ankle jerks. It is claimed that such individuals have subclinical sensorimotor neuropathy.[11]

Lymphoma

COMPRESSION AND
ENDONEURIAL INVASION

Direct infiltration of cranial nerves and spinal roots is frequent in lymphoma, mainly of non-Hodgkin's type, and is often associated with infiltration of the leptomeninges. Patchy or widespread invasion of the peripheral nerves can also occur and gives rise to focal and multifocal neuropathies that may summate to produce a relatively symmetric syndrome. Nerve biopsy is occasionally required to establish the diagnosis.

SENSORY AND SENSORIMOTOR NEUROPATHY

Paraneoplastic neuropathy from lymphoma is generally considered to be less common than in carcinoma of the lung, although in one prospective series, clinically evident neuropathy was found in 8%.[59] Pure sensory neuropathy is rare, most cases are sensorimotor. These may be subacute, but most are chronic and progressive. The subacute cases may improve with treatment of the underlying lymphoma.[47]

DEMYELINATING POLYNEUROPATHY

Acute neuropathy with the clinical and pathologic features of AIDP[6] is well described, most commonly in patients with Hodgkin's disease.[13] Relapsing and remitting cases similar to chronic inflammatory demyelinating polyneuropathy also occur. It is suggested that the immunosuppression that occurs in the natural course of some malignant reticuloses is a predisposition for AIDP in these individuals.[34]

MOTOR NEURONOPATHY (SUBACUTE MOTOR NEUROPATHY)

This rare syndrome has been recognized since the 1960s as a distinct clinicopathologic entity associated with Hodgkin's disease. Originally described as "subacute poliomyelitis,"[61] it is best understood as a motor neuronopathy syndrome (see Chapter 2). Its clinical recognition is important, as it is a benign syndrome that usually improves and does not require extensive evaluation or indicate further treatment of the underlying neoplasm.[49]

Pathology and Pathogenesis. Primary degeneration of anterior horn cells, accompanied by inflammation, is a consistent feature of postmortem examination. Salient neuropathologic findings[49] include: (1) degeneration of the anterior horn cells at multiple levels of the spinal cord, accompanied by inflammation and phagocytosis at an early stage and neuronal loss and gliosis at late stages; (2) lesser neuronal loss in Clarke's column, the intermediolateral cell column, and the commissural nuclei; (3) varying degrees of axonal loss in the dorsal columns and propriospinal tracts; (4) secondary axonal degeneration in ventral roots and denervation atrophy of skeletal muscles. The overall pathologic pattern suggests a mirror image of the findings in subacute sensory neuronopathy associated with carcinoma, in which DRG cells and dorsal roots are destroyed.

The etiology and pathogenesis are unknown. It appears that an inflammatory assault on the anterior horn cells and, to a lesser extent, other spinal neurons, is followed by phagocytosis of these cells and degeneration of their processes. Many features of this condition resemble poliomyelitis, and it has been suggested that it may represent an infection by an opportunistic virus.[25,61] Patients with Hodgkin's disease are often immunosuppressed and are susceptible to a variety of unusual infections. Radiation therapy may cause selective dysfunction of lower motor neurons and could have a role in some cases, although this syndrome has been described in individuals who have received no radiation.[49]

Clinical Features. This disorder is uncommon, occurs equally in both sexes, at any age, and may begin at any stage in the course of the hematologic malignancy. It has not been shown to precede the discovery of malignancy, in contrast to subacute sensory neuronopathy.[25]

The illness is characterized by the subacute onset of painless asymmetric weakness of the limbs.[49,61] The legs are usually initially involved, although several affected individuals have first experienced weakness in the arms or in neck muscles. Involved limbs are flaccid, muscle atrophy is prominent, tendon reflexes are depressed, and fasciculation may occur. Involvement of bulbar or respiratory muscles is rare, and upper motor neuron signs are not a feature of this condition. Sensation and intellect are spared.

Weakness progresses over a period of months, then stabilizes, and gradual improvement usually occurs. Rarely is this an incapacitating condition. Its course is independent of the activity of the underlying lymphoma. Individuals with this condition are stated to be unusually vulnerable to the neurotoxic effects of vincristine.[49]

The CSF protein content is sometimes increased. Electromyography reveals evidence of denervation in affected limbs. Motor and sensory conduction is normal.

Differential Diagnosis. Diagnosis of this condition is occasionally difficult despite its characteristic clinical profile. Leptomeningeal metastases, plexus compression, and acute and chronic demyelinating neuropathies all occur in Hodgkin's disease and may have similar clinical features. Leptomeningeal metastases frequently produce sensory complaints and cranial nerve findings; CSF cytology may be crucial in distinguishing between these conditions. Plexus compression by lymphoma is usually accompanied by pain and produces a more segmental disturbance of neurologic function. The acute demyelinating neuropathies have a more abrupt onset, weakness is symmetric, and nerve conduction studies demonstrate demyelinating changes.

Leukemia and Polycythemia

Leukemic infiltration of the cranial nerves and spinal roots, rarely of the peripheral nerves, may be encountered both in the acute and chronic leukemias. A noninfiltrative neuropathy is rare in acute leukemia, but subacute

and chronic sensorimotor neuropathies resembling those related to carcinoma are well documented, usually in patients with chronic lymphocytic leukemia.[56,59]

Paresthesias in the extremities occur in 10% to 20% of patients with polycythemia vera.[52] Frank peripheral neuropathy is rare.[62]

Multiple Myeloma

Polyneuropathy is more common in multiple myeloma than in most other malignancies.[29,30,60] Two common (non-amyloid) types are recognized; a sensorimotor (distal axonopathy type) neuropathy seen with typical myeloma[58] and a predominantly motor neuropathy seen with osteosclerotic myeloma. Amyloidosis may coexist, and the neuropathy of primary (AL) amyloidosis,[29,58] described in Chapter 12, also occasionally occurs in individuals with myeloma. Both the subacute sensory neuronopathy syndrome[25] and a polyneuropathy associated with a dermato-endocrine syndrome (Crow-Fukase or POEMS syndrome) may accompany myeloma. The latter condition occurs more commonly in Japan.[27]

ROOT AND NERVE
COMPRESSION

Epidural spinal root and cranial-nerve compression syndromes produced by plasmacytomas, and extending from vertebral bodies or skull, are common. Epineurial involvement with peripheral nerve or plexus compression syndromes are rare. Several reports have demonstrated mild endoneurial infiltration with plasma cells.[3,23] The carpal tunnel syndrome in multiple myeloma suggests systemic amyloidosis.[30]

NONMETASTATIC
POLYNEUROPATHY RELATED
TO TYPICAL MYELOMA

Polyneuropathy related to multiple myeloma is uncommon. In a prospective study, clinical evidence of neuropathy was detected in 13% of patients.[60] Electrophysiologic abnormalities, taken to indicate subclinical polyneuropathy, were detected in about 40% of cases.[60]

Pathology and Pathogenesis. Postmortem and nerve biopsy studies indicate that this condition is a distal axonopathy. Thorough postmortem study has demonstrated axonal degeneration to be more pronounced in distal than in proximal segments of leg nerves.[58] Nerve biopsies consistently reveal profound loss of myelinated axons, and slight evidence of segmental demyelination.[60] Nonspecific axonal changes are described. Amyloid deposition is not a histologic feature of this condition.[60] The pathogenesis is unknown. It is widely held that this neuropathy is somehow related to the sensorimotor neuropathy occurring as a remote complication of carcinoma (vide supra).[30] One unconfirmed study has claimed the production of demyelinating neuropathy in mice by the passive transfer of serum from patients with myeloma.[4]

Clinical Features. The neuropathy may be purely sensory or mixed motor and sensory. It is more common in males in late middle age. Frequently the polyneuropathy antedates discovery of the underlying malignancy. Polyneuropathy is heralded by the insidious onset of numbness and tingling in the feet accompanied by slight weakness. Pain is often prominent. Symptoms intensify over a period of months and the hands are eventually involved. All sensory modalities are impaired, and distal tendon reflexes are lost. Autonomic involvement does not occur. The CSF protein level may be slightly increased. Motor nerve conduction velocities are normal or minimally slowed; sensory nerve action potential amplitudes are low.

The course is variable. Some remain with a mild, nondisabling, stocking-glove sensory loss; others become profoundly weak and bedridden. There is little improvement. Treatment and improvement of the underlying malignancy do not affect the polyneuropathy.

Occasional patients with myeloma develop a subacute sensory neuronopathy resembling the syndrome sometimes associated with bronchial carcinoma. An acute polyneuropathy resembling AIDP or having a relapsing-remitting course similar to that of chronic inflammatory demyelinating polyneuropathy may also occur.

Differential Diagnosis. The clinical features of the more common sensory or sensorimotor distal axonopathy that may be related to myeloma are nonspecific. Because this neuropathy may be painful, myeloma must be considered in the differential diagnosis of painful neuropathies, particularly those developing in middle age or later life in males. Other symptoms of myeloma may or may not be present, especially general malaise and fatigability and bone pain. The erythrocyte sedimentation rate is elevated in most but not all patients; a monoclonal paraprotein band is detectable in 75% of patients,[29] and Bence-Jones protein is usually present in urine. The diagnosis is established by bone marrow biopsy or the finding of a plasmacytoma on skeletal x-ray survey.

SYSTEMIC AMYLOIDOSIS

Systemic amyloidosis may occur in association with multiple myeloma and may be accompanied by neuropathy.[29,30] Formerly it was held that myeloma neuropathy represented a variety of systemic amyloidosis. Abundant postmortem analysis has effectively contradicted this notion, and most individuals with myeloma neuropathy do not have amyloid in the PNS. The peripheral neuropathy of systemic amyloidosis complicating myeloma is similar to that seen in primary (AL) amyloidosis (see Chapter 12).

POLYNEUROPATHY RELATED TO OSTEOSCLEROTIC MYELOMA

Osteosclerotic myeloma occurs in approximately 2% of patients with multiple myeloma. Despite this, polyneuropathy is encountered at least twice as often in patients with this form of myeloma than in osteolytic myeloma.[36] The clinical features of the neuropathy also are more consistent as compared with the variegated syndromes associated with osteolytic myeloma.

Clinical Features. The patients tend to be younger than those with osteolytic myeloma, almost one half being less than 50 years.[29] The symptoms of neuropathy often antedate the diagnosis of myeloma. In some patients a predominantly motor neuropathy occurs. In others, the initial symptoms may be

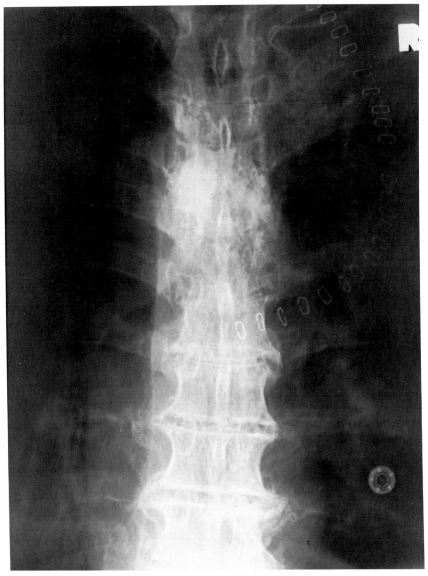

Figure 11–3. An x-ray displaying an osteosclerotic plasmacytoma, at T4, confirmed at biopsy. Skeletal survey in an individual with progressive sensorimotor polyneuropathy of unclear origin detected this lesion. The patient displayed neither systemic signs of myeloma nor paraproteinemia. The polyneuropathy improved following radiation of the plasmacytoma.

sensory with numbness and tingling distally in the extremities. Pain is not a common feature. Distal weakness and wasting may develop later and the tendon reflexes are lost. The course is slowly progressive. Autonomic neuropathy does not occur except for impotence, which may be of endocrine origin. Sensory loss mainly affects large-fiber modalities, and cranial nerves are not usually affected, apart from occasional papilledema.

Nonneurologic features are a frequent accompaniment and include hepatosplenomegaly, cutaneous pigmentation, hypertrichosis, edema, leukonychia, finger clubbing, gynecomastia, and testicular atrophy.[30] These manifestations constitute the Crow-Fukase or POEMS syndrome (P = polyneuropathy, O = organomegaly, E = endocrinopathy, M = M protein, S = skin changes).[2,27,45] Patients with osteosclerotic myeloma may display only some of these features (or none).

Electromyography and nerve conduction studies indicate a combination of axonal loss and demyelination, with moderate slowing of nerve conduction. Nerve biopsies have suggested that the demyelination is secondary to axonal atrophy.[40] A monoclonal paraprotein is detectable in serum in about 75% of cases.[29] The CSF protein is consistently elevated and bony osteosclerotic lesions are seen in all cases (Fig. 11–3). The diagnosis is confirmed by the demonstration of plasma cells on biopsy of such lesions, the cells showing uniform staining for heavy and light Ig chains.

The neuropathy may show slow improvement following appropriate treatment of the plasmacytoma. The mechanism of neuropathy and the nonneurologic features is obscure.

REFERENCES

1. Adams, D, Kuntzer, T, Steck, AJ, et al: Motor neuropathies with antibodies to GM_1 ganglioside. J Neurol (Suppl 1)237:S8, 1990.

2. Bardwick, PA, Zraifler, NJ, Gill, NJ, et al: Plasma cell dyscrasia with polyneuropathy, organomegaly, endocrinopathy, M-protein, and skin changes. The POEMS syndrome. Medicine (Baltimore) 59:311, 1980.

3. Barron, KD, Rowland, LD, and Zimmerman, HM: Neuropathy with malignant tumor metastases. J Nerv Ment Dis 131:10, 1960.

4. Besinger, UA, Toyka, KV, Anzil, AP, et al: Myeloma neuropathy: Passive transfer from man to mouse. Science 213:1027, 1981.

5. Braun, PE, Frail, DE, and Latov, N: Myelin-associated glycoprotein is the antigen for a monoclonal IgM in polyneuropathy. J Neurochem 39: 1261, 1982.

6. Cameron, DG, Howell, DA, and Hutchinson, JL: Acute peripheral neuropathy in Hodgkin's disease. Neurology (Minneapolis) 8:575, 1958.

7. Chad, D, Pariser, K, Bradley, WG, et al: The pathogenesis of cryoglobulinemic neuropathy. Neurology (NY) 32:725, 1982.

8. Chazot, G, Berger, B, Carrier, H, et al: Manifestations neurologiques des gammopathies monoclonales. Rev Neurol (Paris) 132:195, 1976.

9. Cook, D, Dalakas, M, Galdi, A, Biondi, D, and Porter, H: High-dose intravenous immunoglobulin in the treatment of demyelinating neuropathy associated with monoclonal gammopathy. Neurology 40:212, 1990.

10. Cream, JJ, Hern, JEC, Hughes, RAC, et al: Mixed or immune complex cryoglobulinaemia and neuropathy. J Neurol Neurosurg Psychiatry 37:82, 1974.

11. Croft, PB and Wilkinson, M: Carcinomatous neuromyopathy—its incidence in patients with carcinoma of the lung and carcinoma of the breast. Lancet 1:184, 1963.

12. Croft, PB, Henson, RA, Urich, H, et al: Sensory neuropathy with bronchial carcinoma: A study of four cases showing serological abnormalities. Brain 88:501, 1965.

13. Cros, D, Harris, NL, and Hedley-Whyte, ET: Case records. The Massachu-

setts General Hospital. N Engl J Med 323:895, 1990.

14. Dalamau, J, Furneaux, HM, Rosenblum, M, et al: Immunological study of the nervous system and tumor of 3 patients with paraneoplastic encephalomyelitis and small cell lung cancer. Neurology (Suppl)40:166, 1990.

15. Dayan, AD, Crost, PB, and Wilkinson, M: Association of carcinomatous neuromyopathy with different histological types of carcinoma of the lung. Brain 88:435, 1965.

16. Denny-Brown, D: Primary sensory neuropathy with muscular changes associated with carcinoma. J Neurol Neurosurg Psychiatry 11:73, 1948.

17. Dick, DJ, Harris, JB, Falkous, G, et al: Neuronal antinuclear antibody in paraneoplastic sensory neuropathy. J Neurol Sci 85:1, 1988.

18. Freddo, L, Yu, RK, Latov, N, et al: Gangliosides GM_1 and GD_{1b} are antigens for IgM M-protein in a patient with motor neuron disease. Neurology 36:454, 1986.

19. Graus, F, Elkon, KB, Cordon-Cardo, C, et al: Sensory neuropathy and small cell lung cancer. Antineuronal antibody that also reacts with the tumor. Am J Med 80:45, 1986.

20. Hawley, R, Cohen, MH, Saini, M, et al: The carcinomatous neuromyopathy of oat cell lung cancer. Ann Neurol 7:65, 1980.

21. Hays, AP, Latov, N, Takatsu, M, et al: Experimental demyelination of nerve induced by serum of patients with neuropathy and an anti-MAG IgM M-protein. Neurology 37:242, 1987.

22. Henson, RA and Urich, H: Peripheral neuropathy associated with malignant disease. In Vinken, PJ and Bruyn, GW (eds): Handbook of Clinical Neurology, Vol 8. Elsevier/North Holland, New York, 1970, p 131.

23. Hesselvick, M: Neuropathological studies on myelomatosis. Acta Neurol Scand 45:95, 1969.

24. Hodgkinson, SJ, Pollard, JD, and McLeod, JG: Cyclosporin A in the treatment of chronic demyelinating polyradiculoneuropathy. J Neurol Neurosurg Psychiatry 53:327, 1990.

25. Horwich, MS, Cho, L, Porro, RS, et al: Subacute sensory neuropathy: A remote effect of carcinoma. Ann Neurol 2:7, 1977.

26. Iwashita H, Argyrakis, A, Lowitsch, K, et al: Polyneuropathy in Waldenström's macroglobulinaemia. J Neurol Sci 21:341, 1974.

27. Iwashita, H, Ohnishi, A, Asada, M, et al: Polyneuropathy, skin hyperpigmentation, edema and hypertrichosis in localized osteosclerotic myeloma. Neurology (Minneapolis) 27:675, 1977.

28. Kahn, SN, Riches, PG, and Kohn, J: Paraproteinaemia in neurological disease: Incidence, associations and classifications of monoclonal immunoglobulins. J Clin Pathol 33:617, 1980.

29. Kelly, JJ, Kyle, RA, and Latov, N: Polyneuropathies Associated with Plasma Cell Dyscrasias. Martinus Nijhoff, Boston, 1987.

30. Kelly, JJ, Jr, Kyle, RA, and Miles, JM: The spectrum of peripheral neuropathy in myeloma. Neurology (NY) 31:24, 1981.

31. Kimmel, DW, O'Neill, BP, and Lennon, VA: Subacute sensory neuronopathy associated with small cell lung carcinoma: Diagnosis aided by autoimmune serology. Mayo Clin Proc 63:29, 1988.

32. Kyle, RH: Monoclonal gammopathy of undetermined significance: Natural history of 241 cases. Am J Med 64:814, 1978.

33. Latov, N: Plasma cell dyscrasia and peripheral neuropathy: Identification of the myelin antigens that react with human paraproteins. Proc Natl Acad Sci (USA) 78:71, 1981.

34. Lisak, RP, Mitchell, M, Zweiman, B, et al: Guillain-Barré syndrome and Hodgkin's disease: Three cases with immunological studies. Ann Neurol 1:72, 1977.

35. Logothetis, J, Kennedy, WR, Ellington,

A, et al: Cryoglobulinemic neuropathy: A pathological study. Ann Neurol 8:179, 1980.

36. McLeod, JG: Paraneoplastic neuropathies. In Dyck, PJ and Thomas, PK (eds): Peripheral Neuropathy, ed 3. WB Saunders, Philadelphia, 1991.

37. Mendell, JR, Sahenk, Z, Whitaker, JN, et al: Polyneuropathy and IgM monoclonal gammopathy: Studies on the pathogenic role of anti-myelin-associated glycoprotein antibody. Ann Neurol 17:243, 1985.

38. Monoca, S, Bonetti, B, Ferrari, S, et al: Complement-mediated demyelination in patients with IgM monoclonal gammopathy and polyneuropathy. N Engl J Med 332:649, 1990.

39. Morton, DL, Itabashi, HH, and Grimes, OF: Nonmetastatic neurological complications of bronchogenic carcinoma: The carcinomatous neuromyopathies. J Thorac Cardiovasc Surg 51:14, 1966.

40. Ohi, T, Kyle, RA, and Dyck, PJ: Axonal attenuation and secondary segmental demyelination in myeloma neuropathies. Ann Neurol 17:255, 1985.

41. Pestronk, A, Cornblath, DR, Ilyas, AA, et al: A treatable multifocal motor neuropathy with antibodies to GM_1 ganglioside. Ann Neurol 24:73, 1988.

42. Posner, JB, Dalamau, J, Furneaux, HM, et al: Paraneoplastic "Anti Hu" syndrome: A clinical study of 47 patients. Neurology (Suppl)40:165, 1990.

43. Propp, RP, Means, E, Deibel, R, et al: Waldenström's macroglobulinemia and neuropathy: Deposition of M-component on myelin sheaths. Neurology (Minneapolis) 25:980, 1975.

44. Rebai, T, Mhiri, C, Heine, P, et al: Focal myelin thickenings in a peripheral neuropathy associated with IgM monoclonal gammopathy. Acta Neuropathol (Berlin) 79:226, 1989.

45. Reulecke, M, Dumas, M, and Meier, C: Specific antibody activity against neuroendocrine tissue in a case of POEMS syndrome with IgG gammopathy. Neurology 38:614, 1988.

46. Sadig, SA, Thomas, FP, Kilidireas, K, et al: The spectrum of neurologic disease associated with anti GM_1 antibodies. Neurology 40:10678, 1990.

47. Sagar, HJ and Read, DJ: Subacute sensory neuropathy with remission: An association with lymphoma. J Neurol Neurosurg Psychiatry 45:83, 1982.

48. Santoro, M, Thomas, FP, Fink, M, et al: IgM deposits at nodes of Ranvier in a patient with amyotrophic lateral sclerosis, anti GM_1 antibodies and multifocal motor conduction block. Ann Neurol 28:373, 1990.

49. Schold, SC, Cho, ES, Somasunderam, M, et al: Subacute motor neuronopathy: A remote effect of lymphoma. Ann Neurol 5:271, 1979.

50. Sherman, WH, Latov, N, Hays, AP, Takatsu, M: Monoclonal IgM antibody precipitating with chondroitin sulphate C from patients with axonal polyneuropathy and epidermolysis. Neurology 33:192, 1983.

51. Shy, ME, Herman-Patterson, T, Parry, GJ, et al: Lower motor neuron disease in a patient with antibodies against Gal (B1–3) GalNAc in gangliosides GM_1 and GD_{16}: Improvement following immunotherapy. Neurology 40:842, 1990.

52. Silverstein, A, Gilbert, H, and Wasserman, LR: Neurologic complications of polycythemia. Ann Intern Med 57:909, 1962.

53. Smith, IS, Kahn, SM, Lacey, BW, et al: Chronic demyelinating neuropathy associated with benign IgM paraproteinaemia. Brain 106:169, 1983.

54. Steck, AJ, Murray, N, Meier, C, et al: Demyelinating neuropathy and monoclonal IgM antibody to myelin associated glycoprotein. Neurology (NY) 33:19, 1983.

55. Sterman, AB, Schaumburg, HH, and Asbury, AK: The acute sensory neuronopathy syndrome. Ann Neurol 7:354, 1979.

56. Sumi, SM, Farrell, DF, and Knauss, TA:

Lymphoma and leukemia manifested by steroid-responsive polyneuropathy. Arch Neurol 40:577, 1983.

57. Vallat, JM, Desproges-Gotteron, R, Leboutet, M, et al: Cryoglobulinemic neuropathy: A pathological study. Ann Neurol 8:179, 1980.

58. Victor, M, Banker, BQ, and Adams, RD: The neuropathy of multiple myeloma. J Neurol Neurosurg Psychiatry 21:73, 1958.

59. Walsh, JC: Neuropathy associated with lymphoma. J Neurol Neurosurg Psychiatry 34:42, 1971.

60. Walsh, JC: The neuropathy of multiple myeloma, an electrophysiological and histological study. Arch Neurol 25:404, 1971.

61. Walton, JN, Tomlinson, BE, and Pearce, GW: Subacute "poliomyelitis" and Hodgkin's disease. J Neurol Sci 6:435, 1968.

62. Yiannikas, C, McLeod, JG, and Walsh, JC: Peripheral neuropathy associated with polycythemia vera. Neurology (NY) 33:139, 1983.

Chapter 12

OTHER SYSTEMICALLY RELATED DISORDERS

AMYLOID NEUROPATHY
UREMIC NEUROPATHY

AMYLOID NEUROPATHY

Extracellular deposition of the fibrous protein amyloid is associated with peripheral neuropathy in two distinct classes of neuropathic disorders, non-hereditary (immunoglobulin-derived) amyloidosis and hereditary (non-immunoglobulin-derived) amyloidosis. Although diverse etiologies underlie the accumulation of amyloid in these conditions, they share common neuropathologic and pathophysiologic features.*

Pathology and Pathogenesis

Nerve biopsy studies indicate that three patterns of amyloid deposition occur in the peripheral nervous system (PNS), and all may contribute to neuropathy. One is amyloid deposition in connective tissue surrounding peripheral nerves, leading to nerve compression at potential entrapment sites[38,42] such as the median nerve at the wrist, giving rise to a carpal tunnel syndrome. The second pattern is widespread endoneurial deposition of amyloid[16,41,42] (Fig. 12–1). The third is amyloid deposition within the walls of the vasa nervorum of both epineurium and endoneurium.[16,32] Two recent postmortem studies of hereditary amyloid neuropathy demonstrate that amyloid is deposited in either a diffuse or multifocal pattern,

most often in the endoneurium and within endoneurial vessels, but also in the perineurium and epineurium.[32,55] Endoneurial edema may be evident. In the early stages there is a predominant depletion of unmyelinated and small-diameter myelinated fibers, which correlates closely with clinical findings of pain and temperature sensory loss and autonomic dysfunction.[22,41] Although axonal degeneration is the most conspicuous feature, a substantial degree of segmental demyelination is also described.[32] Much of the nerve fiber degeneration in peripheral nerves could result from lesions affecting dorsal root and autonomic ganglion cells, where amyloid deposits also occur.[1,3,17,31,34,35,55] Autopsy studies in patients with familial amyloid polyneuropathy (FAP) show greater deposition of amyloid protein in the vagus nerve and celiac ganglia than in patients with acquired amyloidosis. This may account for the greater incidence of bowel symptoms in FAP than in the acquired form.[34]

The pathogenesis is unknown. There are three hypotheses: one states that nerve fiber degeneration results from a generalized metabolic disorder and that amyloid deposits are an epiphenomenon;[12,45] the second is that the peripheral nerve changes result from ischemia induced by amyloid compromising endoneurial vessels;[32] and the third holds that the amyloid deposits exert a direct mechanical effect,[41] or lead to the loss of sensory and autonomic ganglion neurons and peripheral axons by a mechanism that is not yet understood.[3,41] Recent evidence has raised the possibility that endoneurial and

---

*3,12,13,16,22,36,39,41,42,47,63,66

151

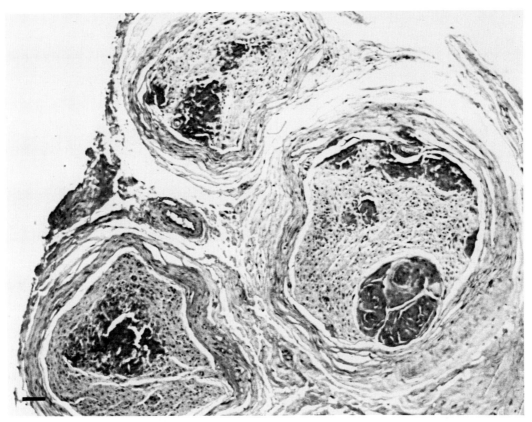

Figure 12–1. Transverse section through sural nerve biopsy specimen from a patient with type I (Portuguese) familial amyloid polyneuropathy. The specimen has been immunostained for transthyretin (TTR). The amyloid deposits, derived from TTR, stand out as the dark areas. Bar = 50 μm.

blood vessel deposition of amyloid results in altered vascular permeability resulting in endoneurial edema. The space-occupying effect of edema, combined with the mass of focal amyloid deposits, may compress perineurial blood vessels to produce ischemic nerve fiber injury.[32] It is also suggested that Schwann and satellite cells may be directly affected by amyloid deposits.[55]

Clinical Features

NONHEREDITARY AMYLOIDOSIS
(PRIMARY AND
DYSPROTEINEMIC TYPES)

Systemic organ involvement by amyloidosis is apparent in about half of the cases of primary amyloidosis with neu-ropathy, and the disorder is most common in middle-aged males.[16] The amyloid neuropathy of dysproteinemic states, especially myeloma, appears to be identical to that of primary amyloidosis, and should be distinguished from the other neuropathies associated with the dysproteinemias[15,37] (see also Chapter 11). In both primary amyloidosis and multiple myeloma, the amyloid fibril appears homologous to the terminal region of an immunoglobulin light chain, lambda more often than kappa.[9,29,30,56] Peripheral neuropathy occurs in 20% of patients with immunologic light-chain-type amyloid and is the presenting symptom in about 10%.[18]

Symptoms, Signs, and Course of Neuropathy. Sensory symptoms are the most prominent feature in all but the few patients who present with autonomic or carpal tunnel involvement.[16,63] Symmetric numbness of hands and feet heralds this disorder; dysesthesias and spontaneous aching pains are especially frequent and often predate systemic organ involvement.[30] Symptoms of autonomic dysfunction (postural hypotension, bladder atony, impotence, diarrhea, and impaired sweating) eventually appear in most affected individuals and are often an early feature.[27,50,71] Distal weakness appears later. The severity of systemic involvement is independent from that of the neuropathy.

Signs of distal limb sensory impairment are a constant feature and, characteristically, pain and temperature sensation are more affected than position and vibration sense. Distal weakness and atrophy appear, usually mild at first. Signs of autonomic dysfunction include orthostatic hypotension, hypoactive pupils, diminished sweating, and abnormal cystometrograms.[27,50,71] The peripheral nerves may be enlarged.

The course is one of gradual incapacitation by sensory, motor, and autonomic dysfunction.[1,63] The distal limbs become numb and atrophic, and sensorimotor impairment progressively moves proximally. Most patients eventually require orthopedic appliances or become unable to walk. Primary amyloidosis is fatal, usually secondary to cardiac or renal decompensation. Survival of individuals with neuropathy ranges from 2 to 10 years following the initial sensory symptoms. Patients exhibiting dysproteinemia with amyloid neuropathy usually die because of the underlying disorder, but occasionally succumb from systemic amyloidosis. There is no specific treatment for systemic amyloidosis.

Laboratory Studies

Clinical Laboratory Findings. Abnormal routine laboratory values are common in primary amyloidosis, often reflecting renal and cardiac involvement. Serum protein electrophoretic patterns are abnormal in half the patients and urine patterns in 70%. Monoclonal proteins are present in 90% of patients when both urine and blood are studied. Amyloid may be demonstrated by histologic examination of gingival or rectal biopsy tissue, or peripheral nerve if neuropathy is present. Myeloma patients display additional abnormalities, including an excessive number of plasma cells in the bone marrow, large amounts of monoclonal protein in urine and serum, lytic bone lesions, and severe anemia.[15,16,37]

Cerebrospinal Fluid. Protein levels are usually moderately elevated, and electrophoresis may reveal elevation of the IgG fraction.

Electrodiagnostic Studies. Motor conduction velocities usually show evidence of an axonal neuropathy with either absent or reduced amplitudes of sensory and motor potentials. Conduction velocities are either normal or slowed—appropriate for the degree of axon loss. Electromyography reveals evidence of active and chronic denervation in distal limb muscles.

Nerve Biopsy. Amyloid deposits are usually present in sural nerve biopsy specimens on light microscope examination of sections stained with Congo red or thioflavin T.

Differential Diagnosis. Sensorimotor-autonomic neuropathy appearing in a nondiabetic middle-aged male, especially if painful, should strongly suggest the diagnosis of amyloid neuropathy. Demonstration of amyloid on nerve, rectal, or gingival biopsy confirms its presence. Diabetic autonomic and sensorimotor neuropathy resembles this disorder, although distal motor involvement is less severe and dissociated pain and temperature sensory loss less apparent. Tabes dorsalis is also featured by pupillary abnormalities, bladder dysfunction, and pains in the extremities; abnormal proprioception and preservation of strength help distinguish tabes from amyloid neuropathy.

Case History: Primary Amyloid Neuropathy*

A 61-year-old patient developed a feeling of coldness in his fingers and feet which slowly spread proximally as far as the wrists and knees. About 2 years later, he became aware that he was unable to perceive the temperature of bath water with his feet. At about the same time, he began to experience occasional stabbing pains in the legs and, a little later, a persistent aching in both lower legs. He became impotent and constipated and on a few occasions had felt faint on standing up abruptly. He had no definite urinary symptoms. He also slowly became aware of wasting of his legs and of unsteadiness in walking. Apart from a myocardial infarct at the age of 53 years, his previous health had been good and there was no family history of similar disorder.

He was admitted for investigation in the following year. Examination showed small, unequal pupils; the response to light was absent on the left and reduced on the right; they constricted normally on convergence. There was slight, generalized wasting in the arms, with focal wasting of the small hand muscles. The small hand muscles were weak bilaterally, as were the finger and wrist extensors. There was generalized wasting and weakness in the legs, maximal distally. The right triceps and both ankle jerks were absent; the other tendon reflexes were obtainable only on reinforcement. The plantar responses were absent. The appreciation of light touch was lost in the hands and below both knees. Pain and temperature sensation was lost over a similar distribution. Joint position sense was impaired in the toes but preserved in the fingers. Vibration sense was lost in the feet but remained in the hands. The peripheral nerves were not thickened. A sweating test showed absent sweating over the trunk and limbs except in the axillae. Intradermal histamine failed to produce a flare in the legs, but did so on the forearm and on the abdominal wall. His blood pressure decreased from 140/90 when lying to 90/70 on standing. The liver and spleen were not enlarged.

A full blood count was normal and the

ESR was 8 mm at one hour. A bone-marrow biopsy was normal, as was a glucose tolerance test. The blood urea concentration was 23 mg per dl and the creatinine clearance was 109 ml per min. A 3-day fecal fat collection yielded 3 to 5 gm per day. Chest x-ray was normal. The CSF showed no pleocytosis but the protein content was 2.4 g/L. Motor nerve conduction velocity in the right median and peroneal nerves was 46 and 31 m/sec, respectively. No sensory action potentials were detectable with percutaneous recordings at the wrist following stimulation of the digital nerves of the index and fifth fingers.

Sural nerve biopsy revealed gross depletion of myelinated nerve fibers; remaining fibers showed evidence both of degeneration and regeneration. The number of unmyelinated axons was grossly reduced. Congo red staining revealed the presence of amyloid deposits around endoneurial blood vessels and free in the endoneurium.

Comment. This case clearly illustrates the cardinal clinical features of primary amyloid neuropathy: autonomic involvement, onset in late middle age, insidious progression, and distal limb loss of pain, temperature, and tactile sense. No evidence of multiple myeloma was found and nerve biopsy was diagnostic.

HEREDITARY AMYLOIDOSIS (FAMILIAL AMYLOID POLYNEUROPATHY)

Seven different forms of familial amyloid polyneuropathy (FAP) are currently recognized (Table 12–1), characterized by their clinical features and, except in type IV, by the specific mutation that has given rise to the amyloid. All show prominent PNS involvement and are inherited in an autosomal dominant manner. Sporadic cases are encountered, identified either by immunohistochemical staining of amyloid in tissue sections[44] or by the use of cDNA probes.[14,33] The pathologic features of peripheral nerve involvement appear identical to those of the nonhereditary types (see text preceding). Abnormal immunoglobulins are not present in in-

---

*From Thomas and King,[63] p 401, with permission.

Table 12–1 FAMILIAL AMYLOID POLYNEUROPATHIES

TYPE	SOURCE OF AMYLOID*
I (Portuguese)	met/val 30 TTR substitution
II (Indiana)	ser/ile 84 TTR substitution
III (Van Allen)	variant apolipoprotein A1
IV (Finnish)	not yet established
OTHER FORMS	
Jewish	ile/phe 30 TTR substitution
Appalachian	ala/thr 60 TTR substitution
German	tyr/ser 77 TTR substitution

*met = methionine; val = valine; ser = serine; ile = isoleucine; phe = phenylalanine; ala = alanine; thr = threonine; tyr = tyrosine

dividuals with hereditary amyloidosis. Except in two instances, the amyloid associated with the FAP is known to be the result of a genetic mutation in the transthyretin (prealbumin) gene, which produces a variant transthyretin (TTR) protein (Table 12–1). In type I FAP, the variant TTR results from substitution of methionine for valine at position 30 in the molecule.[19,48,59,60] Other forms of FAP have different amino acid substitutions accounting for the variant TTR, such as a serine to isoleucine substitution at position 84 in type II FAP.[48] A variant apolipoprotein A1 has been shown to be the source of the amyloid in type III FAP.[45] The presence of abnormal TTR[45] or detection of the abnormal gene using cDNA probes[14,33] is possible before clinical manifestations develop. Although a marker for FAP, the presence of variant TTR is not predictive of the clinical expression of neuropathy: asymptomatic gene carriers may be detected in late life. Direct analysis of the TTR gene now allows prenatal detection of FAP without the need for the examination of extended pedigrees.

Type I (Portuguese). This form was originally recognized in Portugal,[3] Japan,[2] Sweden,[28] and elsewhere.[14] The onset is usually in the third decade, with initial symptoms of numb paresthetic feet or gastrointestinal or other evidence of autonomic dysfunction. The disorder progresses slowly, leading to the development of a severe sensory and autonomic neuropathy, featured by dissociated sensory loss with pain and temperature sense most affected. Attacks of stabbing pain, trophic changes, and ulcerated skin are common. Signs of autonomic dysfunction include impotence, constipation, diarrhea, postural hypotension, and pupillary inequality. Asymptomatic cardiac conduction deficits are common and include left bundle branch block, atrioventricular block, and the loss of normal beat-to-beat variation. Vitreous opacities occasionally occur. Distal limb weakness appears later with muscle atrophy and sometimes fasciculation. Amyloid deposits in the heart, kidneys, and liver commonly occur in this disorder. Except for rare renal disease, signs of organ dysfunction are not prominent features of the condition. The degree of renal compromise varies among different kinships. Median nerve compression (carpal tunnel syndrome) by extraneurial deposits of amyloid is not a feature of this type of FAP. The CSF protein is commonly increased due to hyperalbuminosis. The diagnosis of amyloidosis is confirmed by skin, rectal, abdominal fat pad, or nerve biopsy, and FAP I by cDNA studies. Most patients die before the age of 50 years, although in some the disorder does not become symptomatic until late life.

Type II (Indiana). This variant is a milder disorder featured by the carpal tunnel syndrome and vitreous opacities, and later a generalized neuropathy. It has been described in families of Swiss origin in Indiana[39] and German origin in Maryland.[66] The condition begins in middle life with progressive bilateral median nerve compression at the wrist, sometimes accompanied by vitreous opacities. In time, there is more widespread motor and sensory loss, but peripheral neuropathy is rarely disabling and autonomic involvement is not a prominent feature of the disorder. Older men appear more likely to manifest diffuse limb involvement, while older females may have only the carpal tunnel syndrome.[20] Although systemic involvement is not usual, some Indiana kinships have had hepatosplenomegaly and congestive heart failure. Cerebrospinal fluid examination is normal. The carpal tunnel syndrome is usually relieved by decompression of the flexor retinaculum. Individuals may survive as long as 35 years with little disability, although vitreous opacities may lead to serious impairment of vision. Postmortem examination in several instances has revealed widespread, clinically inapparent, systemic amyloid deposition. The Indiana families have been determined to have variant transthyretin due to substitution of serine for isoleucine at position 84 in the TTR molecule.[43]

Type III (Van Allen). This disorder was described in an Iowa family whose ancestors originated in Britain.[12] Both upper and lower extremities are severely affected, but the carpal tunnel syndrome does not appear. The onset is in the fourth decade and death from renal involvement occurs within 20 years. Initial symptoms of pain, dysesthesias, and weakness of the distal lower limbs are followed by progression and subsequent proximal upper limb involvement. Although autonomic dysfunction is not usually prominent, diarrhea, constipation, vomiting, and epigastric distress may be present. Peptic ulcers may occur. Most individuals are disabled by the peripheral neuropathy within 10 years. Hypertension and uremia are additional debilitating features of the illness. Postmortem examination has revealed abundant amyloid deposition in the kidneys, liver, adrenal glands, testes, peripheral nerve, and sympathetic ganglia. As already stated, the amyloid is derived from a variant apolipoprotein A1.[48]

Type IV (Finnish). A further distinctive form was first described in Finland; it is characterized by the combination of lattice corneal dystrophy and multiple cranial nerve involvement.[11] Onset is usually in the third decade with bulbar signs, especially facial palsies. A mild generalized neuropathy may develop later.[46]

Other Forms. Other kinships have been described with FAP that are so far incompletely characterized clinically; a Jewish form with an onset at 20 to 30 years of age of vitreous opacities followed by a rapidly progressive sensory-autonomic neuropathy and visceral involvement related to an isoleucine to phenylalanine TTR substitution at position 30;[68] an Appalachian form with cardiomegaly and the carpal tunnel syndrome and an alanine to threonine TTR substitution at position 60;[67] and a German form with the onset at 50 to 60 years of age of neuropathy, autonomic failure and renal disease and a tyrosine to serine TTR substitution at position 77.[67] Frederiksen[26] described a Danish family in which the features were not primarily neurologic and comprised vitreous opacities and progressive cardiomegaly. Paresthesias occurred in the hands. This kinship had a methionine for leucine substitution, accounting for the variant transthyretin.[51]

UREMIC NEUROPATHY

Definition and Etiology

Uremic neuropathy is best defined as the distal symmetric sensorimotor polyneuropathy associated with chronic renal insufficiency. The nature of the underlying renal disease appears immaterial, since virtually all types of kidney disorders that can lead to uremia have now been associated with this neuropathy.[25]

The etiology is elusive. It is widely held that uremic neuropathy is secondary to retained, dialyzable toxins or metabolites normally excreted by the kidneys. Two features support this notion: one is the close clinicopathologic resemblance of uremic and the toxic neuropathies;[54] the other is improvement following dialysis and, more dramatically, after renal transplantation.[10,61] The responsible agent clearly has a molecular weight exceeding that of urea or creatinine. Initial reports of improvement of neuropathy following long-duration dialysis suggested that this agent had a molecular weight between 1350 and 5000 daltons, the so-called middle molecule.[8] The middle molecule hypothesis has not been strongly supported by subsequent experience with dialysis.[53] Elevation of myoinositol, parathyroid hormone, or magnesium, and the presence of vitamin deficiencies or transketolase inhibition, have also been proposed as candidates, but none has been established as being responsible.[7,23,49] It has recently been suggested that retained ethylene oxide, an axonal neurotoxin used to sterilize dialysis equipment, may contribute to the progressive neuropathy observed in some patients on long-term hemodialysis.[70] Focal ischemic neuropathies may develop in extremities undergoing access procedures for dialysis; presumably these represent vascular steal syndromes.

Pathology and Pathogenesis

Axonal degeneration is characteristic of this disorder, and it can be classified as a distal axonopathy. This conclusion is supported both by comprehensive postmortem examination of the peripheral nervous system from several advanced cases, and by detailed ultrastructural study of nerve biopsy material from early and advanced cases.

Postmortem studies indicate the following common distribution of changes in all cases: striking loss of nerve fibers in distal nerve trunks of the legs, intense fiber breakdown in distal nerves with less active changes proximally, normal spinal roots, and degeneration in the cervical portion of the gracile fasciculi. Anterior horn cells remain intact but show chromatolysis.[6,24] Nerve biopsy study of early and advanced cases indicates the nature of PNS change to be nonspecific axonal shrinkage, secondary myelin breakdown, and eventually, fiber loss. One clinicopathologic study, that utilized nerve biopsies, has suggested that there are three distinct patterns of uremic polyneuropathy: an acute axonal type, a slowly progressive

axonal neuropathy with secondary demyelination, and a slowly progressive, predominantly demyelinating neuropathy.[53]

The distribution and nature of the postmortem pathologic changes, taken in concert with clinical findings, constitute overwhelming evidence that uremic polyneuropathy is a distal axonopathy. Presumably, its pathogenesis is similar to that proposed for other metabolic or toxic axonopathies (see Chapter 2), and its elucidation awaits identification of the underlying metabolic disturbance. There is no animal model of uremic neuropathy.

Clinical Features

INCIDENCE

Evidence of uremic neuropathy is present in approximately half of all patients in hemodialysis programs, and the disorder is more common in males. Since the proliferation of hemodialysis and renal transplant centers, advanced disabling cases of neuropathy are less frequent. Many mild or subclinical instances probably go undetected. This notion is supported by the observation of nerve conduction or nerve biopsy abnormalities in individuals with chronic renal insufficiency and no clinical evidence of PNS dysfunction.[52,65]

SYMPTOMS, SIGNS, COURSE

The cardinal clinical features of fully developed uremic neuropathy are similar to those described for most distal axonopathies and include distal-to-proximal progression of signs, legs affected more than arms, symmetric loss of both motor and sensory function, and slow recovery.[64]

Initially, sensory symptoms often predominate, and tingling paresthesias of the legs are especially frequent. Occasionally, the "burning feet" syndrome, similar in every respect to that encountered in nutritional deficiency, occurs. Weakness of foot dorsiflexion is the usual first motor complaint.

The *restless legs* syndrome may accompany sensory symptoms in the early stages of uremic neuropathy. Leg movements are especially frequent at night. Occasionally, this syndrome appears to herald the development of neuropathy.[65a] Muscle cramps in distal extremities are common.

Loss of the Achilles reflex is usually an early sign in uremic neuropathy, often accompanied by diminished vibration sense in the toes.[62] Advanced cases almost always display distal diminution of vibration, touch, and position sense. Pain and temperature senses are less frequently involved. Weakness and atrophy of distal muscles are also common in advanced uremic neuropathy. Although the overwhelming majority of cases develop a mixed sensorimotor neuropathy, rare instances of pure motor and pure sensory patterns are well documented. Autonomic dysfunction, especially parasympathetic, occurs in the more severe cases; prominent features include postural hypotension, impaired sweating, and abnormal Valsalva maneuver.[57] Abnormalities of eighth nerve function, both auditory and vestibular, may be present; their pathogenesis is obscure and may reflect, in part, CNS dysfunction or the effects of ototoxic antibiotics.

Isolated mononeuropathy syndromes, especially carpal tunnel syndrome, may occur in individuals with chronic renal failure, and are generally attributed to an abnormal susceptibility to pressure palsies or to vascular steal syndromes from forearm access shunts.[58]

Gradual onset of a progressively disabling sensorimotor neuropathy is usual. Most progress over several months to reach a plateau despite worsening of the renal state. However, there may be considerable variation in tempo, and cases with a near-apoplectic development of an advanced neuropathy occur.[53]

The advent of long-term hemodialysis and aggressive renal transplant services have significantly altered the natural history of uremic neuropathy.[6] These are discussed below.

Laboratory Studies

CLINICAL LABORATORY FINDINGS

Abnormal values of routine clinical laboratory tests are common and reflect the effects of chronic renal dysfunction such as anemia and electrolyte abnormalities.

CSF

The CSF is acellular; protein may be moderately elevated, perhaps reflecting coexistent uremic encephalopathy.

ELECTRODIAGNOSTIC STUDIES

Abnormalities of motor nerve conduction may be present in the legs of asymptomatic individuals with chronic renal disease, and may indicate subclinical neuropathy.[65] In more advanced cases, nerve conduction velocity and muscle and nerve action potential amplitudes generally parallel the degree of clinical and pathologic impairment, and probably reflect changes in large, rapidly conducting myelinated axons.[62] Following successful renal transplantation, nerve conduction gradually returns to normal levels, *pari passu* with clinical improvement.[25] Hemodialysis has much less effect on nerve conduction.[61]

Electromyographic signs of denervation in distal leg muscles are a consistent early feature of uremic polyneuropathy, and frequently disappear following renal transplantation.

Prognosis and Treatment

The prognosis of untreated uremic neuropathy is usually poor, in the past it was among the disabling complications of chronic renal failure.[6]

Successful renal transplantation is unquestionably effective in the prevention and reversal of uremic neuropathy. Mild cases display prompt relief of paresthesias and a steady return of strength and sensibility. Recovery is more prolonged in advanced cases and not always complete.[25,53]

Repeated hemodialysis is considerably less effective in ameliorating neuropathy. While it is generally held that some patients improve or become stable following repeated hemodialysis, its effectiveness in treating uremic neuropathy is still debated.[25,61] One report claims that neuropathy improves more with ambulatory peritoneal dialysis.[40] Paresthesias frequently disappear soon after beginning dialysis. Features of uremic neuropathy most resistant to dialysis are sensory loss for vibration, touch, and position. The *restless legs* syndrome also responds poorly (if at all) to dialysis. It is suggested that administration of methylcobalamin to patients receiving dialysis ameliorates autonomic neuropathy.[58]

It has been proposed that close monitoring of peripheral nerve function (electrophysiology, quantitative sensory testing) in individuals with chronic renal disease might prove helpful in indicating the need for dialysis. One clinical study, using quantitative tests of sensation, sensitive electrophysiologic techniques, and the services of an active nephrology unit, challenges this assumption and indicates a limited role for the neurologist in the decisions that dictate the need for dialysis.[21]

Differential Diagnosis

The diagnosis of uremic neuropathy is not difficult in an individual with chronic renal disease who develops signs of a progressive, distal, symmetric sensorimotor polyneuropathy.

Uremia is a common condition, and its associated polyneuropathy is clinically indistinguishable from many other distal axonopathies. Therefore, the diagnosis of uremic polyneuropathy should be entertained with caution in individuals with coexistent metabolic disorders, or in those who may also be exposed to neurotoxic agents. This is an especially common dilemma in uremic patients who may be taking neurotoxic antibiotics such as nitrofurantoin.

Case History: Uremic Neuropathy*

A 38-year-old printer was admitted to the hospital in February, 1967. He had been admitted to another hospital 1 year earlier because of headaches and occasional vomiting which had been present for 3 years; these symptoms became worse 1 week before admission. He was found to have malignant hypertension (220/150) and chronic pyelonephritis with bilateral contracted kidneys.

Hb was 8.4 g per dl; plasma sodium 137 mEq per liter, potassium 5.3 mEq per liter; blood urea 330 mg per dl. His urine showed proteinuria with a few granular casts and polymorphonuclear leukocytes and an IVP revealed bilateral small contracted kidneys.

While in the hospital he developed weakness of both legs. He was treated initially by a 20-g-protein, low-salt diet, and peritoneal dialysis, with no improvement in his neuropathy and was therefore transferred to the Royal Free Hospital, London, for hemodialysis. On admission, no cranial nerve abnormality or abnormalities in the upper limbs were found. In the legs, there was bilateral wasting of the anterior tibial and small foot muscles, and weakness of all muscle groups below the knees, particularly dorsiflexion and eversion at the ankles and extension of the toes. Knee jerks were present and symmetric, but neither ankle jerk was obtainable. Plantar responses were flexor. No sensory loss was detectable.

Following hemodialysis, his neuropathy deteriorated, with increased weakness in his legs and the development of numbness and tingling paresthesias in his legs and hands. He became unable to walk. Examination showed mild weakness and wasting of the small hand muscles and more severe weakness distally in his legs, loss of both knee and ankle jerks, mild distal sensory impairment for all modalities in the legs, and slight cutaneous sensory loss over the hands.

Slow recovery occurred, but the patient remained confined to a wheelchair for 6 months. Nine months after starting hemodialysis, he was able to walk and climb stairs. When examined in July 1970, he was

*From Thomas,[64] et al, p 769, with permission.

still aware of some persisting weakness and numbness of his feet and had "restless legs" in bed at night. There was moderate distal weakness in his legs, mainly for dorsiflexion and eversion at the ankles. The knee jerks had returned, but the ankle jerks remained absent. Sensory testing still revealed distal impairment for all modalities in the legs. Successful renal transplantation was performed in November 1970, after which he was maintained on prednisolone and azathioprine. The sensory symptoms from his neuropathy improved substantially after the operation, although the weakness in his legs improved only slightly. When last reviewed in April 1971, he still showed distal weakness in his legs and his ankle jerks remained absent. The impairment of cutaneous and vibration sensibility was less extensive and joint position sense in his toes was no longer inaccurate.

Motor nerve conduction velocity in the median nerve shortly after his transfer to the Royal Free Hospital was slightly reduced (47 m/sec). A median sensory nerve action potential of reduced amplitude (4 μV), but with a normal velocity (55 m/sec), was obtained. Following the deterioration after the commencement of hemodialysis, motor nerve conduction velocity in the median nerve was 45 m/sec, but could not be estimated in the peroneal nerve as both extensor digitorum brevis muscles became completely denervated and had not become reinnervated at the time of his examination in July 1970. Motor-nerve conduction velocity in the median nerve was then 50 m/sec, and a median-sensory nerve action potential 6 μV in amplitude with a velocity of 56 m/sec was obtained. In April 1971, 5 months after renal transplantation, motor nerve conduction velocity in the median nerve was 55 m/sec. A median sensory nerve action potential 12 μV in amplitude was recorded with a velocity of 59 m/sec. The extensor digitorum brevis muscles had become reinnervated, and a motor conduction velocity of 27 m/sec was obtained in the common peroneal nerve.

Comment. This case displays many of the cardinal features of uremic neuropathy. PNS dysfunction appeared years following symptoms related to re-

nal failure, and although initially motor signs predominated, an eventual sensorimotor neuropathy developed. Hemodialysis was associated with some improvement of neurologic dysfunction, but did not prevent the appearance of the "restless legs" syndrome. Renal transplantation was eventually necessary and resulted in substantial relief from sensory symptoms.

REFERENCES

1. Andersson, R and Bjerle, P: Peripheral circulation, particularly heat regulation in patients with amyloidosis and polyneuropathy. Acta Med Scand 199:191, 1976.
2. Andersson, R: Hereditary amyloidosis with polyneuropathy. Acta Med Scand 188:85, 1970.
3. Andrade, C: A peculiar form of peripheral neuropathy: Familial atypical generalized amyloidosis with special involvement of the peripheral nerves. Brain 75:408, 1952.
4. Araki, S, Mawatari, S, Ohta, M, et al: Polyneuritic amyloidosis in a Japanese family. Arch Neurol 18:593, 1968.
6. Asbury, AK, Victor, M, and Adams, RD: Uremic polyneuropathy. Arch Neurol 8:413, 1963.
7. Avram, MM, Feinfeld, DA, and Huatuco, AH: Search for the uremic toxin: Decreased motor conduction velocity and elevated parathyroid hormone in uremia. N Engl J Med 298:1000, 1978.
8. Babb, AL, Ahmad, S, Berstrom, J, et al: The middle molecule hypothesis in perspective. Am J Kid Dis 1:46, 1981.
9. Benson, MD: Partial amino acid sequence homology between an heredofamilial amyloid protein and human plasma prealbumin. J Clin Invest 67:1035, 1981.
10. Bolton, CF, Baltzan, MA, and Baltaan, RG: Effects of renal transplantation on uremic neuropathy. A clinical and electrophysiologic study. N Engl J Med 284:1170, 1971.

11. Boysen, G, Galassi, G, Kamieniecka, Z, Schlaeger, J, and Trojaborg, W: Familial amyloid neuropathy and corneal lattice dystrophy. J Neurol Neurosurg Psychiatry 42:1020, 1979.

12. Coimbra, A and Andrade, C: Familial amyloid polyneuropathy: An electron microscope study of the peripheral nerve in five cases. I. Interstitial changes. Brain 94:199, 1971.

13. Coimbra, A and Andrade, C: Familial amyloid polyneuropathy: An electron microscope study of the peripheral nerve in five cases. II. Nerve fiber changes. Brain 94:207, 1971.

14. Costa, PP, Figueira, AS, and Bravo, FR: Amyloid fibril protein related to prealbumin in familial amyloidotic polyneuropathy. Proc Natl Acad Sci (USA) 75:4499, 1978.

15. Davies-Jones, GAB and Esiri, MM: Neuropathy due to amyloid in myelomatosis. Br Med J 2:444, 1971.

16. Dayan, AD, Urich, H, and Gardner-Thorpe, C: Peripheral neuropathy and myeloma. J Neurol Sci 14:2, 1971.

17. De Navasquez, S and Treble, HA: A case of primary generalized amyloid disease with involvement of the nerves. Brain 61:116, 1938.

18. Duston, MA, Skinner, J, Anderson, J, and Cohen, AS: Peripheral neuropathy as an early marker of AL amyloidosis. Arch Intern Med 149:358, 1989.

19. Dwulet, JE and Benson, MD: Characterization of a transthyretin (prealbumin) variant associated with familial amyloidotic polyneuropathy type II. J Clin Invest 78:880, 1986.

20. Dwulet, FE and Benson, MD: Primary structure of an amyloid prealbumin and its plasma precursor in a heredofamilial polyneuropathy type II. J Clin Invest 78:880, 1984.

21. Dyck, PJ, Johnson, WI, Lambert, EH, et al: Comparison of symptoms, chemistry and nerve function to assess adequacy of hemodialysis. Neurology (Minneapolis) 29:1361, 1979.

22. Dyck, PJ and Lambert, EH: Dissociated sensation in amyloidosis: Compound action potential, quantitative histologic and teased-fiber, and electron microscopic studies of sural nerve biopsies. Arch Neurol 20:490, 1969.

23. Egan, JD and Wells, IC: Transketolase inhibition and uremic peripheral sensory neuropathy. J Neurol Sci 41:379, 1979.

24. Forno, L and Alston, W: Uremic polyneuropathy. Acta Neurol Scand 43:640, 1967.

25. Fraser, C and Arieff, A: Nervous system complications in uremia. Ann Intern Med 109:143, 1988.

26. Frederiksen, T, Gotzsche, H, Harboe, N, Kiaer, W, and Mellengaard, K: Familial primary amyloidosis with severe amyloid heart disease. Am J Med 33:328, 1963.

27. French, JM, Hall, G, Parish, DJ, et al: Peripheral and autonomic nerve involvement in primary amyloidosis associated with uncontrollable diarrhea and steatorrhea. Am J Med 39:277, 1965.

28. Glenner, GG and Murphy, MA: Amyloidosis of the nervous system. J Neurol Sci 94:1, 1989.

29. Glenner, GG, Ein, D, and Terry, WD: The immunoglobulin origin of amyloid. Am J Med 52:141, 1972.

30. Glenner, GG: Amyloid deposits and amyloidosis: The B-fibrillosis. N Engl J Med 302:1283, 1980.

31. Gotze, W and Krücke, W: Über Paramyloidose mit besonderer Beteiligung der peripheren Nerven und granular Atrophie des Gehirns, und über ihre Beziehungen zu den intracerbralen Gefassverkalkungen. Arch Psychiat Nerve Krank 114:182, 1941.

32. Hanyu, N, Ikeda, S, Nakadai, A, et al: Peripheral nerve pathologic findings in familial amyloid polyneuropathy: A correlative study of proximal sciatic nerve and sural nerve lesions. Ann Neurol 25:340, 1989.

33. Holt, IJ, Harding, AE, Middleton, L, et al: Molecular genetics of amyloid neuropathy in Europe. Lancet 1: 524, 1989.

34. Ikeda, S, Yanagiawa, N, Hongo, M, and Ito, N: Vagus nerve and celiac ganglion lesions in generalized amyloidosis. A correlative study of familial amyloidotic polyneuropathy and AL-amyloidosis. J Neurol Sci 79:129, 1987.

35. Jedrzejowska, H. Some historic aspects of amyloid polyneuropathy. Acta Neuropathol 37:119, 1977.

36. Kelley, JJ, Kyle, RA, O'Brien, PC, et al: The natural history of peripheral neuropathy in primary systemic amyloidosis. Ann Neurol 6:1, 1979.

37. Kelly, JJ, Kyle, RA, Miles, JM, et al: The spectrum of peripheral neuropathy in myeloma. Neurology (Minneap) 31:24, 1981.

38. Kernohan, JW and Woltman, HW: Amyloid neuritis. Arch Neurol Psychiatry 47:132, 1942.

39. Krücke, W: Zur pathologischen anatomie der paramyloidose. Acta Neuropathol (Suppl)11:74, 1963.

40. Kurz, SB, Wong, VH, Anderson, CF, et al: Continuous ambulatory peritoneal dialysis. Three years experience at the Mayo Clinic. Mayo Clin Proc 58:633, 1983.

41. Kyle, RA and Bayrd, ED: Amyloidosis: Review of 236 cases. Medicine 54:271, 1975.

42. Mahloudji, M, Teasdall, RD, Adamkiewicz, JJ, et al: The genetic amyloidoses with particular reference to hereditary neuropathic amyloidosis, type II (Indiana or Rukavina type). Medicine 48:1, 1969.

43. Meretoja, J: Familial systemic paramyloidosis with lattice dystrophy of the cornea, progressive cranial neuropathy, skin changes, and various internal symptoms. Ann Clin Res 1:314, 1969.

44. Mita, S, Maeda, S, Ide, M, Tsuzuki, T, et al: Familial amyloidotic polyneuropathy diagnosed by cloned human prealbumin cDNA. Neurology 36:298, 1986.

45. Nakazato, M, Tanaka, M, Yamamura, Y, et al: Abnormal transthyretin in asymptomatic relatives in familial amyloidotic polyneuropathy. Arch Neurol 44:1275, 1987.

46. Nakazoto, M, Kangawa, K, Minamino, N, Tawara, S, Matsuo, H, and Araki, S: Revised analysis of amino acid replacement in a prealbumin variant (SKO-III) associated with familial amyloidotic polyneuropathy of Jewish origin. Biochem Biophys Res Comm 123:921, 1984.

47. Neundörfer, B, Meyer, JG, and Volk, B: Amyloid neuropathy due to monoclonal gammopathy: A case report. J Neurol 216:207, 1977.

48. Nichols, WC, Dwulet, FE, Liepnieks, J, and Benson, MD: Variant apolipoprotein A1 as a major constituent of a human hereditary amyloid. Biochem Biophys Res Comm 156:762, 1988.

49. Niwa, T, Asada, H, Maeda, K, et al: Profiling of organic acids and polyols in nerves of uraemic and non-uraemic patients. J Chromatog 377:150, 1986.

50. Nordberg, C, Kristensson, K, Olsson, Y, et al: Involvement of the autonomous nervous system in primary and secondary amyloidosis. Acta Neurol Scan 49:31, 1973.

51. Nordlie, M, Sletten, K, and Ranolov, PJ: A new prealbumin variant in familial amyloid cardiomyopathy of Danish origin. Scand J Immunol 27:119, 1988.

52. Rosales, RL, Navarro, J, Isumo, S, et al: Sural nerve morphology in asymptomatic uremia. Eur Neurol 28:156, 1988.

53. Said, G, Boudier, L, Selva, J, et al: Different patterns of uremic polyneuropathy: Clinicopathologic study. Neurology 33:567, 1983.

54. Schaumburg, HH and Spencer, PS: Chemical neurotoxocity. In Asbury, AS, McKann, GM, and McDonald, WI (eds): Diseases of the Nervous System, Vol 2. WB Saunders, Philadelphia, 1986, p 1303.

55. Sobue, G, Nakao, N, Murakami, K, et al: Type 1 familial amyloid polyneuropathy. Brain 113:903, 1990.

56. Soji, S, Kameko, M, Tsukada, N, et al: Immunologic and immunohistochemical study of familial amyloid polyneuropathy. Neurology (Minneapolis) 31:1493, 1981.

57. Solders, G: Autonomic function tests in

healthy controls and in terminal uremia. Acta Neurol Scand 73:638, 1986.

58. Taniguchi, H, Ejire, K, and Baba, S: Improvement of autonomic neuropathy after mecobalamin treatment in uremic patients on hemodialysis. Clin Therap 9:607, 1987.

59. Tawara, S, Nakazato, M, Kangawa, K, et al: Identification of amyloid prealbumin variant in familial amyloidotic polyneuropathy (Japanese type). Biochem Biophys Res Commun 116:880, 1983.

60. Tawara, S, Araki, S, Toshimori, K, et al: Amyloid fibril protein in type I familial amyloidotic polyneuropathy in Japan. J Lab Clin Med 98:811, 1981.

61. Tegner, R and Lindblom, B: Uremic polyneuropathy: Different effects of hemodialysis and continuous ambulatory peritoneal dialysis. Acta Med Scand 218:409, 1985.

62. Tegner, R and Lindblom, U: Vibratory perception threshold compared with nerve conduction velocity in the evaluation of uremic neuropathy. Acta Neurol Scand 71:284, 1985.

63. Thomas, PK and King, RHM: Peripheral nerve changes in amyloid neuropathy. Brain 97:395, 1974.

64. Thomas, PK, Hollinrake, K, Lascelles, R, et al: The polyneuropathy of chronic renal failure. Brain 94:761, 1971.

65. Thomas, PK: Screening for peripheral neuropathy in patients treated by chronic hemodialysis. Muscle Nerve 1:396, 1979.

65a. Tyler, HR and Tyler, KL: Neurologic complications. In Eknoyan, G and Knochel, JP (eds): The Systemic Consequences of Renal Failure. Grune and Stratton, Orlando, 1984, p 302.

66. Van Allen, MW, Frohlich, JA, and Davis, K, Jr: Inherited predisposition to generalized amyloidosis. Clinical and pathological study of a family with neuropathy, nephropathy and peptic ulcer. Neurology (Minneapolis) 19:10, 1969.

67. Wallace, MR, Dwulet, FE, Williams, EC, et al: Identification of a new hereditary amyloidosis prealbumin variant, Tyr-77, and detection of the gene by DNA analysis. J Clin Invest 81:189, 1988.

68. Wallace, MR, Dwulet, FE, Conneally, PM, and Benson, MD: Biochemical and molecular genetic characterization of a new variant prealbumin associated with hereditary amyloidosis. J Clin Invest 78:6, 1986.

69. Wilborn, AJ, Furlan, AJ, Hulley, W, et al: Ischemic monomelic neuropathy. Neurology 33:447, 1983.

70. Windebank, A and Blexrud, M: Ethylene oxide and dialysis-associated neuropathy. Ann Neurol 26:63, 1989.

71. Yamada, M, Tsukagoshi, H, Satoh, J, et al: Sporadic prealbumin-related amyloid polyneuropathy: Report of two cases. J Neurol 235:69, 1987.

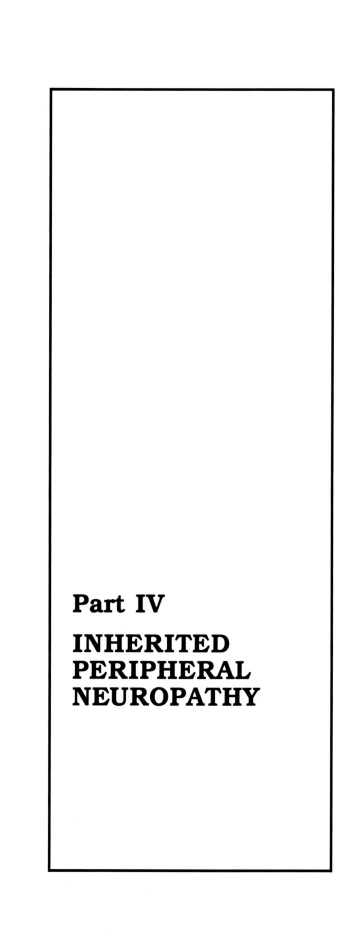

Part IV

INHERITED PERIPHERAL NEUROPATHY

Chapter 13

METABOLIC NEUROPATHY: THE PORPHYRIAS

DEFINITION AND ETIOLOGY
PATHOLOGY
PATHOGENESIS
CLINICAL FEATURES OF ACUTE
  INTERMITTENT PORPHYRIA

DEFINITION AND ETIOLOGY

The porphyrias are a group of seven rare hereditary disorders, five hepatic and two erythropoetic, characterized by disturbances in heme biosynthesis.[6] Four of the hepatic disorders are associated with peripheral neuropathy and mental disturbance: variegate porphyria (VP), acute intermittent porphyria (AIP), hereditary coproporphyria (HCP), and delta-aminolevulinic aciduria.[7] Porphyric neuropathy is a serious, life-threatening illness and occurs in acute episodes, often induced by drugs, hormones, or nutritional factors[8] (Table 13–1).

Each of the porphyrias is characterized by a unique pattern of overproduction, accumulation, and excretion of intermediates of heme biosynthesis. They reflect genetically determined deficiencies of specific enzymes in the heme synthetic pathway (Fig. 13–1). The primary enzymatic defects in the porphyrias associated with neuropathy are known: porphobilinogen deaminase deficiency in AIP, coproporphyrin oxidase deficiency in HCP, protoporphyrinogen oxidase in VP, and porphyrobilinogen synthetase deficiency in delta-aminolevulinic aciduria.

PATHOLOGY

Detailed postmortem and sural nerve biopsy studies of the peripheral nervous system (PNS) in AIP, HCP, and VP suggest that distal axonopathy is the pathologic pattern in this neuropathy. Curiously, these studies suggest that in some cases, short motor axons appear preferentially affected, in striking contrast to the early changes in long axons that characterize most metabolic distal axonopathies.[3,9] Early weakness is a prominent clinical phenomenon in AIP; frequently it is proximal or asymmetric, and may in part reflect selective involvement of short motor fibers. One study also demonstrates involvement of large-diameter axons supplying muscle spindles.[3] Selective involvement of short motor axons is not an inevitable finding, and there are also well-documented instances with predominant dysfunction of long motor and sensory fibers. This variability remains unexplained. Severe cases may exhibit axonal degeneration in the ventral roots, and fiber loss in the gracile columns occasionally occurs.[5] Nerve biopsies from cases of porphyric neuropathy display paranodal demyelination, probably secondary to axonal alterations.[1,10] The nature of

Table 13–1 DRUGS THAT INDUCE PORPHYRIC ATTACKS

Barbiturates	Chlorpropamide
Chlordiazepoxide	Hydantoins
Meprobamate	Glutethimide
Sulfonamides	Griseofulvin
Estrogens	Rifampin
Oral contraceptives	Chloroquine
Ergot preparations	Dichloralphenazone
Ethanol	Imipramine
	Methyldopa

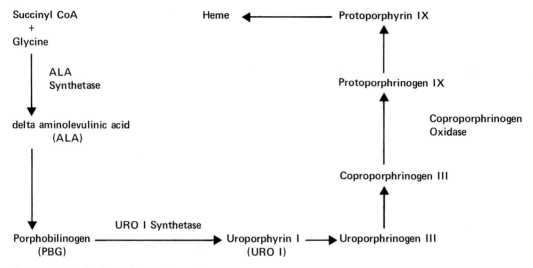

Figure 13–1. Outline of heme biosynthesis.

the axonal change has not been carefully studied by modern morphologic techniques, and despite considerable effort, there is no satisfactory animal model of human porphyric neuropathy. In sum, the mechanisms underlying this unusual distal axonopathy have not been elucidated by pathologic studies.

There are no characteristic postmortem CNS changes that can account for the prominent psychiatric phenomena in the porphyrias. Scattered areas of ischemic necrosis, present in some cases,[5] are scarcely an adequate substrate for the major behavioral aberrations characteristic of these conditions.

PATHOGENESIS

The pathogenetic link between peripheral neuropathy and biochemical abnormalities in heme synthesis is unknown.[6] AIP is the most studied of the porphyrias, and several hypotheses have been proposed to explain both its neurologic and psychiatric manifestations. None is firmly supported by experimental evidence. Three suggestions are currently considered: (1) porphobilinogen deaminase is deficient in neurons, leading to local cytotoxic accumulations of porphobilinogen (PBG) or aminolevulinic acid (ALA);[1,7] (2) excess circulating delta-ALA or PBG formed in the liver gains access to the nervous system and is toxic (an attractive notion since there exists a rough correlation between plasma levels of ALA, PBG, and neuropsychiatric symptoms);[1,6] and (3) another hereditary metabolic defect, not directly related to heme synthesis, is present in the nervous system.[2,6]

Additionally, there is no satisfactory explanation for the precipitation of acute clinical attacks by such unrelated events as fasting, fever, drugs, toxins, and hormones.[1] The mechanism of drug idiosyncrasy is especially clinically relevant, since certain drugs (see Table 13–1) appear to be precipitating factors in most attacks in asymptomatic individuals with the genetic trait. It is suggested that these drugs induce production of hepatic ALA synthetase,[6] the rate-limiting enzyme for heme biosynthesis in the liver.

CLINICAL FEATURES OF ACUTE INTERMITTENT PORPHYRIA

The neurologic manifestations of AIP, HCP, and VP are identical, and presumably have a common pathogenesis.[2,6,7] These three conditions are distinguished principally by their different enzymatic defects, variations in urinary and fecal excretion of heme precursors, and the occurrence of photocutaneous lesions in VP. In other respects they are similar, and discussion of minor clinical variations is beyond the scope of this volume. Since AIP is the most commonly encountered and thoroughly delineated of the three conditions, this section is confined to its description.

Incidence and Heredity

AIP is rare, occurs worldwide and its incidence range is between 1:5000 and 1.5:100,000. In Lapland, the incidence is 1:1000, largely because of one kindred of 137 in this sparsely populated region. Diagnostic screening of psychiatric patients yields, not surprisingly, a higher incidence than in the general population. AIP is of autosomal dominant inheritance. It seems likely that homozygosity for the hereditary defect is not compatible with life.

Signs and Symptoms

SYMPTOMS AND SIGNS OF GENERAL DISEASE

This condition is often clinically latent; affected individuals may be asymptomatic or only manifest vague feelings of anxiety, malaise, and so forth. Latency is the rule in childhood, and attacks before puberty are extremely rare. Variability in phenotypic expression is common in autosomal dominant disorders; in AIP this genetic variability appears further compounded by the exquisite sensitivity of the disorder to environmental factors (drugs, fever, fasting).[6]

An acute attack of colicky abdominal pain, often combined with constipation, vomiting, fever, and leukocytosis is frequently the initial presentation. The duration is variable (days to months), and intermittency is the rule. Exploratory laparotomy is frequent in unsuspected cases. The pathogenesis of the abdominal crises is obscure. Possibly they are related to autonomic neuropathy that results in gastrointestinal dysfunction with alternate periods of spasm and hypomotility. Mental disturbance may accompany or occasionally precede abdominal attacks. Many patients have long histories of emotional instability, agitation, and occasionally, sociopathic behavior. Severe psychiatric episodes are frequently drug-provoked, follow abdominal crises, and may be characterized by psychotic behavior, visual hallucinations, and delirium.

PNS SYMPTOMS

Symptoms of PNS dysfunction are rarely present in individuals without previous abdominal or mental disturbances. Most commonly, the peripheral neuropathy develops acutely or subacutely following the general manifestations of AIP. Indeed, the sequence of consultation may (in retrospect) be diagnostic: surgeon, psychiatrist, neurologist.[2,6,7]

Weakness is the cardinal symptom in the neuropathy of AIP. Pain in affected limbs or back may precede or be coincident.[7] Weakness may either begin in proximal or distal portions of any extremity, and both symmetric and asymmetric patterns occur. Sudden onset of asymmetric, painful, shoulder-girdle weakness may initially suggest an erroneous diagnosis of brachial neuropathy. Sensory symptoms occasionally are prominent and may herald neuropathy. Paresthesias, often distressingly severe, may occur in proximal or distal extremities, and in the face and neck.[8]

PNS SIGNS

Motor signs are predominant.[7] The pattern of weakness in early stages is extremely variable, upper limbs are usually most affected, and proximal muscles may be weakest. This contrasts strikingly with the pattern in other metabolic distal axonopathies, and correlates well with predominance of distal axonal degeneration in short nerves found at postmortem examination.[3] Some cases present with distal weakness of upper and lower extremities, and rare individuals display a monoplegic or hemiplegic pattern of flaccid paralysis. Most eventuate with severe, flaccid weakness of all extremities, also involving respiratory muscles. Muscle tenderness and cramps usually are present, and eventually profound wasting may occur.

Weakness of muscles innervated by cranial nerves, especially the facial and vagus, is frequent at later stages. Extraocular muscle palsy, ptosis, and anisocoria are also described.[7]

Tendon reflexes are unpredictable. Usually they are absent or diminished; however, profound weakness may occur with relative preservation of tendon jerks, especially the Achilles reflex. Sensory loss parallels the distribution of weakness. Diminished pain, thermal, and touch sensation is usual; vibration is less frequently affected and joint position sense is often normal.

Autonomic dysfunction is common. Persistent tachycardia is present in almost every case and is widely held to be characteristic of porphyric neuropathy. Other signs attributed to autonomic neuropathy are fluctuations in blood pressure, urinary and anal sphincter dysfunction, and fever. Many of the prominent gastrointestinal features of AIP may reflect autonomic involvement (*vide supra*).

Laboratory Studies

CLINICAL LABORATORY FINDINGS

Abnormal values for routine clinical laboratory tests are common and usually reflect associated metabolic problems: inappropriate secretion of antidiuretic hormone, salt-losing renal lesions, and mild hepatic dysfunction.

The urine may be brown if porphobilinogen is present in high concentrations. During attacks, urine porphobilinogen and ALA are usually readily detected in the urine; however, they may be absent between attacks and are variably present in asymptomatic carriers. Assay of erythrocyte porphobilinogen deaminase is the most accurate means of detecting AIP.[6]

The cerebrospinal fluid is usually acellular and the protein content modestly elevated, rarely exceeding 1 g/liter.

ELECTRODIAGNOSTIC STUDIES AND NERVE BIOPSY

Electromyographic signs of denervation appear in most involved muscles. Motor nerve conduction velocities are generally normal or slightly decreased. There are no reports of sensory conduction in AIP neuropathy.[7]

One report describes paranodal demyelination with slight segmental loss of myelin, and attributes this to a primary axonal alteration;[1] another describes axonal degeneration.[10] It is likely that the disturbance is a primary axonal degeneration.

Course

The progression of limb weakness or sensory signs is usually continuous and, in most cases, the neurologic deficit peaks within 6 weeks.[7] Variation exists, and stepwise progression, plateaus, mild fluctuations, and prolonged courses are all documented.[7] It should be emphasized that *continuous* progression does not always mean *contiguous* progression in AIP. Proximal weakness in the arm may be followed by distal weakness in the same extremity, or by symmetric weakness of the feet. Ascending paralysis or sensory loss is unusual in this disorder. Most attacks eventuate in weakness and sensory loss in all extremities and trunk and cranial nerves. Many patients are desperately ill, require both ventilatory and alimentary support, and can die from either respiratory or cardiac (sympathetic) complications. In the past, death was not unusual in severe cases. One study documented 10 deaths in 29 attacks of neuropathy that occurred in 25 individuals—7 died during the first attack and 3 in subsequent episodes.[8]

Recovery usually commences within 2 months of the peak of neurologic disability. Reinnervation of proximal muscles and dermatomes occurs initially, and eventually strength and sensation are restored in the distal extremities. Thus, the pattern of recovery further supports the notion of distal axonopathy in this condition. Complete or near-complete recovery is the rule, although it may take years to restore function to distal extremities.[7]

Repeated attacks of AIP at widely ranging intervals may be accompanied by neuropathy, and are documented to occur within weeks or decades.[7]

Treatment

The most important measure in the management of AIP is prevention of acute attacks by instructing carriers about the provocative factors shown in Table 13–1. A dedicated search for carriers and genetic counseling are crucial in AIP.

Hematin and glucose both prevent induction of ALA synthetase in experimental animals and appear to reverse the human biochemical abnormalities in AIP, usually resulting in clinical improvement of abdominal and psychiatric signs within 48 hours. Glucose should be administered intravenously at a rate of 10 to 20 g every hour in amounts approaching 400 g/day; if there is no improvement within 48 hours, intravenous hematin, 2 mg/kg, should be infused every 24 hours.[4] Hematin infusion clearly can arrest a progressing neuropathy but does not reverse established lesions.

Hyponatremia, hypomagnesemia, and azotemia require skilled manipulation of fluid and electrolyte therapy. Agitation, combativeness, and delirium usu-

ally respond well to phenothiazines.

Severe autonomic and somatic complications of AIP neuropathy frequently occur rapidly, may be life-threatening, and require sophisticated management. These patients, like individuals with AIDP, are best treated in respiratory care facilities. Tachycardia and hypertension usually respond to beta-adrenergic blocking drugs; propranolol is widely used and should be introduced at very low doses. Respiratory failure is frequent in severe cases, and usually requires tracheostomy and ventilatory assistance.[7]

Prognosis

Although AIP remains a serious illness, its prognosis has improved considerably in recent years, and fatalities are now rare. Many are able to avoid attacks by not taking drugs and avoiding provocative activities. Most cases of neuropathy recover and gradually regain strength and sensation.

Differential Diagnosis

The diagnosis of porphyric neuropathy is not difficult in an individual known to carry the genetic trait. Unsuspected AIP can constitute a formidable diagnostic problem for the surgeon and psychiatrist, since its clinical manifestations are protean. Differential diagnosis for the neurologist usually involves two conditions, AIDP and toxic neuropathy.

1. AIDP may be readily misdiagnosed as AIP neuropathy since both may be asymmetric, proximal, subacute, dominated by motor signs, accompanied by tachycardia and respiratory failure, and involve cranial nerves. AIDP is not usually accompanied by abdominal crises or mental disturbances, and the urine does not contain porphobilinogen or ALA. Electrophysiologic evaluation may be of use in differentiating the two.

2. Toxic neuropathies with a subacute onset may be caused by thallium,

organophosphate, dapsone, nitrofuran-toin, and lead. While the pharmaceutical agents usually are readily ruled out by history, lead presents a more difficult diagnostic problem. Plumbism may be accompanied by major gastrointestinal symptoms, porphyrin metabolism is deranged, and weakness often dominates the neurologic syndrome. There is a clear biochemical distinction between the two disorders: urine lead levels are not elevated in AIP, and erythrocyte porphobilinogen deaminase is normal in lead neuropathy.

3. Hereditary neuropathy with liability to pressure palsies may have an episodic course with abdominal colic resembling AIP.[11] A careful history, nerve biopsy, and appropriate biochemical tests usually establish the correct diagnosis.

Case History: AIP Neuropathy

A 20-year-old male experienced intermittent abdominal cramping. Radiographic studies of the abdomen were negative. Two months later he noticed weakness and numbness in his fingers, which progressively involved hands, arms, and distal lower extremities. Within 6 days he was unable to walk, had difficulty swallowing, and was admitted to a hospital. General physical examination was unremarkable. Mental state appeared normal except for an inappropriate affect. There was severe proximal weakness and slight distal weakness of the arms, mild diffuse weakness of the legs, and bilateral facial weakness. Tendon reflexes were absent, except for brisk Achilles reflexes. Sensation was normal. Routine clinical laboratory tests were normal, as was examination of the CSF. Electromyography revealed denervation potentials in most muscles tested, and motor nerve conduction velocities were normal in arms and legs. He began to improve 1 week following admission, and was ambulatory at discharge 5 weeks later. Within 3 months there were no detectable neurologic signs, and he was told that he had recovered from an *atypical AIDP*.

Two months later he developed fever, sore throat, and cramping abdominal pain. He was given erythromycin and aspirin and soon afterward he was admitted to a hospital. Examination revealed a temperature of 34°C, purulent exudate over the tonsils, and despite severe abdominal pain, a soft abdomen. Neurologic examination was unremarkable. He was treated with intravenous fluids and erythromycin. On the 9th day of hospitalization, his urine was noted to become red when standing exposed to room air. A Watson-Schwartz test was positive, and subsequently elevated urine levels of ALA and PBG were obtained. He gradually improved and was discharged without signs or symptoms of neuropathy. He denied any family history of a similar condition.

Comment. This case demonstrates many of the cardinal findings of AIP neuropathy—rapid onset, proximal weakness, and diffuse motor involvement—yet the diagnosis was only established following the serendipitous observation of red discoloration of the urine during a subsequent hospitalization. It seems likely that the febrile illness triggered the abdominal pain of the latter admission, and it is fortunate that he was not given a drug that might have precipitated another episode of neuropathy.

REFERENCES

1. Asbury, AK, Sidman, RL, and Wolf, MK: Drug induced porphyrin accumulation in the nervous system. Neurology (Minneapolis) 16:320, 1966.
2. Bissell, DM: Haem metabolism and the porphyria. In Wright, R, Milward-Sadler, GH, Alberti, KG, and Karran, MM (eds): Liver and Biliary Disease, ed 2. Balliere, Tindall and Cox, London, 1985, p 387.
3. Cavanagh, JB and Mellick, RS: On the nature of the peripheral nerve lesions associated with acute intermittent porphyria. J Neurol Neurosurg Psychiatry 28:320, 1965.
4. Goetsch, CA and Bissell, DM: Instability of hematin used in the treatment of acute hepatic porphyria. N Engl J Med 315:235, 1986.
5. Hierons, R: Change in the nervous sys-

tem in acute porphyria. Brain 80: 176, 1957.

6. Keppes, A, Saua, S, and Andersen, KG: The porphyrias. In Stanbury, JB, Wyngaarden, JB, and Fredrickson, DS (eds): The Metabolic Basis of Inherited Disease, ed 5. McGraw-Hill, New York, 1983, p 1301.

7. Ridley, A: Porphyric neuropathy. In Dyck, PJ, Thomas, PK, and Lambert, EH (eds): Peripheral Neuropathy, ed 2, Vol 2. WB Saunders, Philadelphia, 1984, p 1704.

8. Ridley, A: The neuropathy of acute intermittent porphyrias. Q J Med 38: 307, 1969.

9. Sweeney, VP, Pathak, MH, and Asbury, AK: Acute intermittent porphyria. Increased ALA-synthetase activity during an acute attack. Brain 93:369, 1970.

10. Trapani, G, Casal, C, Tonal, P, et al: Peripheral nerve findings in hereditary coproporphyria. Light and ultrastructural studies in the sural nerve biopsies. Acta Neuropathol 63:96, 1984.

11. Trockel, U, Schroeder, JM, Reiners, KH, et al: Multiple exercise-related mononeuropathy with abdominal colic. J Neurol Sci 60:431, 1983.

12. Yamada, M, Kondo, M, Tanaka, M, et al: An autopsy case of acute porphyria with decrease of both uroporphyrinogen 1 synthetase and ferrochelatase activities. Acta Neuropathol 64:6, 1984.

Chapter 14

METABOLIC NEUROPATHY: HEREDITARY DISORDERS OF LIPID METABOLISM

**SULFATIDE LIPIDOSIS
(METACHROMATIC
LEUKODYSTROPHY)
GALACTOSYLCERAMIDE LIPIDOSIS
(GLOBOID CELL
LEUKODYSTROPHY, KRABBE
DISEASE)
LIPOPROTEIN DEFICIENCIES
PHYTANIC ACID STORAGE DISEASE
(REFSUM DISEASE)
ALPHA-GALACTOSIDASE A
DEFICIENCY (FABRY'S DISEASE,
ANGIOKERATOMA CORPORIS
DIFFUSUM)**

This is a heterogeneous group of disorders characterized by accumulation of lipids in tissues, but with widely different clinical and morphologic manifestations. These are rare disorders, and in most, peripheral neuropathy is not the predominant feature.

SULFATIDE LIPIDOSIS (METACHROMATIC LEUKODYSTROPHY)

Biochemical Abnormality

Accumulation of galactosyl-3-sulfate and lipids containing the galactosyl-3-sulfate moiety characterizes this disorder. The enzyme arylsulfatase A is deficient in the more common forms, and its assay in blood leukocytes and cultured skin fibroblasts is used both as a standard diagnostic test and a means of heterozygote detection.[27]

General Features

Sulfatide lipidosis includes several autosomal recessive conditions. The three most common types, all associated with arylsulfatase A deficiency, are the late infantile, juvenile, and adult forms.[27] The late infantile form is by far the most common, and its clinical features develop in four stages. **Stage 1** manifests in the second year by weakness and hypotonia; a spastic gait occasionally appears in this stage. **Stage 2** follows within 18 months; the child is unable to stand, speech is slurred, and intellect dulled. **Stage 3** follows within 6 months, featured by quadriparesis, abnormal posture, and further deterioration in speech and intellect. In the **final stage,** patients are blind, unable to move, and fed by nasogastric tube. Life span is usually 5 to 6 years after the first signs. The juvenile form is clinically similar to the infantile type. The adult-onset variety is characterized by slowly progressing dementia, spasticity, and a protracted course. Most are labeled as having schizophrenia or multiple sclerosis, unless other family members are known to be affected. Occasional cases present with neuropathy; in others, signs of neuropathy such as depressed tendon reflexes in a patient with a psychiatric disorder may suggest this diagnosis. The sulfatide lipidoses are uniformly fatal and there is no specific therapy.

Diagnosis is usually established by the determination of leukocyte or fibroblast arylsulfatase A levels. This test is not fully reliable for heterozygote detection. Prenatal diagnosis of sulfatide lipidosis is possible, allowing selective termination of pregnancy. However, there is an overlap between the enzyme levels for homozygotes and heterozygotes.[27]

The most striking pathologic change is widespread degeneration of white matter of the CNS, accompanied by the accumulation of masses of sulfatide (metachromatic-staining) material in macrophages and in certain groups of neurons.[43] Deposits of metachromatic material are also present in Schwann cells, renal epithelium, gall bladder, pancreas, and anterior pituitary. Ultrastructurally, the sulfatide deposits display a characteristic lamellated pattern.[41]

Pathology and Pathogenesis of Neuropathy

Segmental demyelination with little alteration in axons is characteristic of this neuropathy. Metachromatic granular inclusions in the perinuclear region of Schwann cells of myelinated and unmyelinated nerve fibers is pathognomonic.[44] This material is composed of sulfatide, and on ultrastructural examination displays a lamellated pattern identical to the CNS inclusions. It is likely that these inclusions indicate abnormal storage within Schwann cells and do not merely represent myelin breakdown products.[5]

The pathogenesis of this demyelinative neuropathy presumably is related to the generalized disorder of lipid metabolism. Myelin breakdown probably results from a metabolic disturbance that affects the ability of the Schwann cell to maintain its myelin segment. The mechanism underlying myelin loss is unclear, as is its relationship to sulfatide storage. The degree of demyelination correlates poorly with the amount of inclusion material present.[5]

Clinical Features of Neuropathy

Diffuse limb weakness, hypotonia, and diminished tendon reflexes are early findings in the infantile and juvenile conditions. Muscle cramps and extremity pain are occasional complaints in the initial stages. Signs of peripheral neuropathy are always present from the outset in these forms, and occasionally dominate the early clinical profile. As CNS dysfunction progressively develops, a curious mixture of hypertonicity, ataxia, and areflexia appears. In the adult form, signs and symptoms of neuropathy are usually few, and often are obscured by severe dementia and spasticity.[5,44]

The cerebrospinal fluid (CSF) protein level is usually elevated to 100 to 200 mg/dl. Motor nerve conduction velocity is frequently profoundly slowed, characteristic of a demyelinating neuropathy. Nerve biopsy is almost always diagnostic for sulfatide lipidosis, and prior to the development of leukocyte and serum enzyme assays, sural nerve biopsy was considered one of the most reliable tests.

GALACTOSYLCERAMIDE LIPIDOSIS (GLOBOID CELL LEUKODYSTROPHY, KRABBE'S DISEASE)

Biochemical Abnormality

The enzyme galactosylceramide beta-galactosidase is deficient in all cases, leading to the accumulation of galactocerebroside in CNS and PNS. This diagnosis can now be established by enzymatic assay in affected individuals, fetuses, and heterozygous carriers.[42]

General Features

This is an autosomal recessive condition and occurs equally in both sexes. Onset is usually in the first year of life, rarely juvenile. Typically, the affected child, apparently normal during the first months of life, becomes irritable, hypersensitive to sounds, and may have a seizure. Within a few months motor and mental deterioration become apparent and hypertonicity appears. Eventually the child becomes blind, mute, and quadriplegic. The duration of the illness is usually 2 years, and most clinical phenomena reflect CNS involvement. Visceral enlargement or skin changes are not part of this illness. The sole indications of peripheral nerve dysfunction are hyporeflexia and abnormal nerve conduction.[22]

The most striking pathologic change is widespread degeneration of CNS white matter, accompanied by profuse clusters of characteristic cells (globoid cells) with abundant cytoplasm and one or more nuclei.

Globoid cells are macrophages, contain galactocerebroside, and occur throughout CNS white matter. Ultrastructurally, they are characterized by abnormal, intracellular tubular profiles that have a crystalloid appearance in cross section. Injection of galactocerebroside into a rat's brain produces typical globoid cells with identical ultrastructural features.[42]

The PNS is affected in nearly every case.[5,14] Segmental demyelination may be prominent, presumably related to a primary metabolic disorder in Schwann cells. Globoid cells do not occur in the PNS, but similar ultrastructural inclusions are present in Schwann cells and in perivascular histiocytes.[5] Usually there is a mild or moderate degree of nerve fiber loss, and nonspecific axonal changes occasionally occur.[40] Nerve biopsy may be diagnostic if enzyme assay is not readily available.

LIPOPROTEIN DEFICIENCIES

Two hereditary lipoprotein-deficiency states are associated with nervous system dysfunction. One is deficiency of

beta-lipoprotein, with principal involvement of the CNS together with moderate PNS dysfunction; the other is deficiency of high-density lipoprotein (HDL), prominently featured by peripheral neuropathy.

Abetalipoproteinemia (Bassen-Kornzweig Disease)

BIOCHEMICAL ABNORMALITY

The specific metabolic deficit in abetalipoproteinemia is unknown. Reduced synthesis of very low density lipoprotein, low-density lipoprotein, and chylomicrons is characteristic of this disorder. Consequently, triglycerides accumulate in intestinal mucosa, fat-soluble vitamins (A and E) are not absorbed, and plasma cholesterol levels are reduced.[21]

PATHOLOGY AND PATHOGENESIS

Although it has been suggested that peripheral neuropathy of the distal axonopathy type is a feature of this disorder, this has not been adequately documented morphologically. Autopsy and nerve biopsy studies have revealed segmental demyelination and predominant large-fiber axonal loss.[13,28,35,46] Electrophysiologic studies do not support the notion of primary demyelination. One postmortem study depicts diffuse fiber loss in gracile and cuneate tracts and in the ascending spinocerebellar systems.[35] These changes are compatible with primary neuronal dysfunction in these systems (neuronopathy). Unfortunately, the changes in the dorsal root ganglia and Clarke's column neurons were not adequately described.

The pathogenesis of the nervous system dysfunction is unknown. Two hypotheses are current.[21] One is that malabsorption of fat-soluble vitamins at an early age results in this widespread disorder. The other holds that absent low-density lipoprotein impairs cholesterol synthesis and affects the content of sterols in membranes of many cell types, leading to early dysfunction and eventual disappearance of these cells. The prevention of its development or progression by the administration of vitamin E favors the former explanation and is discussed later.

CLINICAL FEATURES

This is an autosomal recessive disorder that occurs equally in males and females. Onset is in infancy and the clinical picture is dominated by intestinal malabsorption. Growth retardation is soon apparent, accompanied by clumsiness and limb ataxia. Subsequently, tremor and dysarthria indicate cerebellar system abnormality, while hyporeflexia and diffuse proprioceptive dysfunction imply large-fiber sensory malfunction.[35,46] The clinical profile is thus similar in some aspects to that of Friedreich's ataxia.[28] Stocking-glove sensorimotor loss, distal muscle atrophy, and occasional spasticity may occur. Motor conduction velocity is moderately reduced and sensory action potentials are reduced or absent. Pigmentary retinopathy appears in adolescence, accompanied by night blindness and diminished visual acuity. Myocardial fibrosis associated with congestive heart failure, ventricular enlargement, and arrhythmias has been described.[13] Individuals may live until middle age, but are usually incapacitated by neurologic involvement. Clinical laboratory evaluation is usually diagnostic and is featured by low plasma cholesterol, absent chylomicrons, abnormal levels of low-density lipoprotein, and bizarre erythrocytes with spiny processes (acanthocytes).[21] Reports have implicated vitamin E deficiency secondary to intestinal malabsorption in the causation of the retinal and neurologic changes, indicating that dietary fat restriction and high doses of vitamins A and E may prevent or arrest these complications.[2,20,23,29]

High-Density Lipoprotein Deficiency (Tangier Disease)

BIOCHEMICAL ABNORMALITY

The nature of the fundamental biochemical defect is unknown; presumably there is either decreased synthesis or increased catabolism of HDL.[21] The cardinal biochemical features of Tangier disease are the virtual absence of serum HDL, a low serum cholesterol, and massive deposition of cholesterol esters within tissues. The latter accounts for some of the most striking clinical signs in this illness. The biochemical disturbance underlying cholesterol ester deposition remains obscure, but may relate to instability of chylomicrons leading to the uptake of their contents by reticuloendothelial cells throughout the body.

GENERAL FEATURES

HDL deficiency is autosomal recessive and occurs equally in both sexes. There is a considerable spectrum both in degree and variety of clinical phenomena in this condition. Some experience only mild enlargement of the tonsils throughout life, while others become incapacitated with neuropathy, have severe hepatosplenomegaly, and corneal opacities. In general, it is not a life-threatening condition. Tonsilar enlargement is dramatic and the condition may be first suspected on discovering large, yellow multilobular glands on oropharyngeal examination. Evidence of PNS dysfunction is present in most cases, and peripheral neuropathy may be the initial manifestation of the condition (vide infra). Splenomegaly secondary to cholesterol-ester deposits occurs in two thirds of the cases and may produce anemia and thrombocytopenia. Lymph nodes, cornea, and rectal mucosa are also frequently distorted by cholesterol ester deposition.[21]

PATHOLOGY AND PATHOGENESIS

The CNS is normal.[3] PNS changes are reported from several nerve biopsies and one postmortem study.[3,15,16,18,19,] [25,26] Diminished numbers of small myelinated and unmyelinated fibers occur in both radial and sural nerves, without evidence of segmental demyelination or active axonal degeneration. A variable number of small lipid droplets are present in Schwann cells. A postmortem study demonstrated abnormal foamy histiocytes in the distal segments of cutaneous nerves, but did not document such changes in large nerve trunks.[3] In sum, the morphologic studies of the PNS in HDL deficiency provide little insight into the pathogenesis of this condition. Both clinical and pathologic data are consistent with a primary degeneration of motor and sensory axons. The lipid droplets within Schwann cells are probably derived from nerve-fiber breakdown, but the reason for the failure of these cells to dispose of the lipid, as occurs during wallerian degeneration, for example, is unknown. Presumably, it is related to the HDL deficiency in some way.[45]

CLINICAL FEATURES

A curious variety of symptoms and signs has been reported in cases of HDL deficiency. Some resemble a multiple mononeuropathy[26] and others a progressive loss of small myelinated and unmyelinated fibers.[15] A distal mixed motor and sensory neuropathy affecting fibers of all diameters may also occur.

The multiple mononeuropathy may affect cranial or limb nerves, be permanent or transient, occur several times in the same individual, and occasionally give rise to changes so subtle as to escape detection by the patient.[36] Nerve biopsies or detailed electrophysiologic studies are not available in these cases.

The second pattern, suggestive of small myelinated and unmyelinated fiber loss, has been carefully documented in three individuals. The initial clinical profile may superficially resemble syringomyelia, with severe impairment of pain and temperature sense over the face and the proximal portions of the

limbs, and wasting of the small hand muscles.[15,25,26] Gradually, over 10 or more years, disability progresses, other muscles become weak, and the tendon reflexes diminish. Nerve biopsies from such cases reveal predominant loss of small myelinated and nonmyelinated fibers, corresponding to the pattern of sensory loss.

PHYTANIC ACID STORAGE DISEASE (REFSUM'S DISEASE)

Two forms exist: classic Refsum's disease (heredopathia atactica polyneuritiformis) and infantile Refsum's disease.

Classic Refsum's Disease

BIOCHEMICAL ABNORMALITY

The biochemical abnormality is defective oxidative catabolism of phytanic acid derived from ingested phytol. The initial alpha oxidation of phytanic acid to yield alpha-hydroxyphytanic acid does not occur, and it has been proposed that a deficiency of phytanic acid alpha-hydroxylase underlies this disorder. Biochemical investigations now permit identification of presymptomatic cases in families. Treatment of such cases with a diet low in phytol[37] to prevent the development of neurologic damage and for the treatment of established cases is available.

GENERAL FEATURES

This is an extremely rare autosomal recessive condition, clearly related to an inability to degrade ingested phytanic acid. The diet of affected individuals is not responsible, nor is absorption of phytanic acid. Dairy products, fish oils, and ruminant fats appear to be the major source of dietary phytanic acid. Serum levels of phytanic acid are elevated in every case, with excessive storage throughout the body. Dietary restriction of phytol can reduce plasma levels of phytanic acid to normal, and may improve neurologic signs and symptoms. Attempts to produce the disease in experimental animals by feeding large amounts of phytanic acid have not been successful, and it remains to be established just how tissue accumulation of phytanic acid relates to clinical manifestations.[38]

The illness is usually first manifest before age 20. Cardinal clinical features are atypical retinitis pigmentosa, peripheral neuropathy, and ataxia.[32] Most individuals initially experience failing night vision, followed by gradual deterioration of visual function, and eventually near-blindness. Pigmentary degeneration of the retina is of the "salt-and-pepper" variety. Cataracts appear in one third of cases, further compromising visual function. Evidence of PNS dysfunction is present at some time in most cases (vide infra). Anosmia, deafness, and disordered cardiac function are common; cardiomyopathy is the most frequent cause of death.[33] Dry, scaly skin or diffuse ichthyosis is frequent, especially among children.

Few postmortem studies are available, none has utilized modern histopathologic techniques and, in general, the findings do not satisfactorily account for the clinical phenomena.[10] Liver and kidneys contain increased lipid, and myocardial fibrosis is common. Atrophy of the inferior olive, the olivocerebellar fibers, and medial lemniscus is prominent. In the spinal cord, axonal reaction of anterior horn cells and degeneration of the gracile fasciculi occur.[10]

PATHOLOGY AND PATHOGENESIS OF NEUROPATHY

The salient gross morphologic feature of this condition is enlargement of peripheral nerves and proximal nerve trunks, related to the predominance of hypertrophic changes in these areas.[10] Microscopically, fibers are reduced in number, and surviving myelinated axons are often surrounded by concentric layers of Schwann cell processes. Such "onion bulbs" may completely enclose demyelinated axons. Segmental demyelination and remyelination are evident in teased fibers. Nonspecific crystalline inclusions are present in some Schwann cells and may originate within mitochondria.[17,34]

In sum, the few pathologic descriptions of the PNS in phytanic acid storage disease provide little insight into either the basic pathologic process or its pathogenesis. Onion-bulb formation is a nonspecific process associated with chronic myelinopathy or axonopathy.[31] Changes in the gracile fasciculi suggest the presence of a distal axonopathy (see Chapter 2), but further postmortem studies are needed to verify this notion.

CLINICAL FEATURES OF PERIPHERAL NEUROPATHY

Initial manifestations of peripheral neuropathy are symmetric distal weakness and reflex loss in the lower limbs.[32] This is accompanied by diminished position and vibration sense, and paresthesias. Atrophy of the lower leg and intrinsic foot muscles is common, and pes cavus may occur. Generally, weakness and sensory loss appear in the hands as well. With time, all tendon reflexes are lost, all sensory modalities become impaired, and proximal weakness develops. This is accompanied by electrophysiologic evidence of denervation in affected muscles and profoundly slowed nerve conduction. The CSF is acellular and the protein concentration markedly elevated. Initially, most patients experience a gradual progression of neurologic deficit.[33] In over half, remission of signs and a fluctuating course is seen, probably related to dietary factors. Most individuals are disabled and eventually develop moderate to severe peripheral nervous system dysfunction, in combination with ataxia, impaired vision, and poor hearing.[33]

TREATMENT

Treatment with a diet low in phytanic acid[37] leads to stabilization of the disease and may result in improvement of neurologic signs. The diet is unpalatable and compliance may be a problem. Acute exacerbations can be treated by plasma exchange to reduce the plasma phytanic acid level.

Infantile Refsum's Disease

A further type of phytanic acid storage disease, the manifestations of which appear in childhood, has been recognized.[1,9] The clinical aspects include dysmorphic facial features, hepatomegaly, single palmar creases, mental retardation, self-mutilation, pigmentary retinopathy, severe sensorineural deafness, ataxia, and hypotonia. Peripheral neuropathy is present in some. Seizures, cardiomyopathy, steatorrhea, and osteopenia are inconstant features.

This peroxisomal disorder is presumed to be of autosomal recessive inheritance. As in classic Refsum's disease, phytanic acid oxidase activity is absent.[1,9] Hepatic peroxisomes are lacking. Additional biochemical changes are the accumulation of very long chain fatty acids in the serum and pipecolic acidemia.[1] Restriction of dietary phytanic acid intake has produced improvement in behavior but not in neurologic signs.[9]

ALPHA-GALACTOSIDASE A DEFICIENCY (FABRY'S DISEASE, ANGIOKERATOMA CORPORIS DIFFUSUM)

Biochemical Abnormality

Defective activity of the lysosomal enzyme alpha-galactosidase characterizes this disorder, resulting in progressive deposition of neutral glycosphingolipids in most visceral tissues and fluids.[7] Globotriosylceramide is the predominant lipid stored in this condition. Elucidation of the specific enzymatic defect has permitted accurate diagnosis of affected individuals and asymptomatic carriers of unborn fetuses, and allowed therapeutic trials of enzyme replacement, fetal liver transplants, and plasmapheresis.[8,11]

General Features

This disorder is inherited in an X-linked recessive fashion, and usually first appears in childhood or adolescence. The course is steadily progressive and the mean age at death is 41 years. In general, there is an excellent correlation between the clinical manifestations and progressive deposition of glycosphingolipids.[11] In addition to signs of neurologic dysfunction, dermatologic, ocular, cardiovascular, and renal disease are prominent.[8]

The characteristic skin lesion is a cluster of punctate, red angiectases usually located between the umbilicus and the knees. Corneal opacities and tortuous conjunctival and retinal vessels are common.[11] Cardiovascular and renal failure account for most deaths in Fabry's disease. Renal failure is an especially disabling feature; many individuals become uremic, require hemodialysis, and transplantation is occasionally necessary.[8]

In homozygous males, failure of each organ system can be directly linked with deposition of lipid within the malfunctioning tissue: angiokeratomas reflect lipid storage in vessel walls resulting in aneurysm formation; ocular findings are secondary to lipid infiltrates in the cornea and conjunctival vessels; cardiac failure results from deposits within the myocardium and valves; and glomerular and distal tubular epithelial lipid infiltration produce renal failure. Lipid deposits in endothelial cells may occlude the lumen of arterioles in any tissue, including the CNS, thereby producing a variety of signs. This phenomenon probably accounts for the dissimilarity among cases in the same sibship. Heterozygous females may show less obtrusive manifestations, most frequently corneal opacification.[11]

Treatment of Fabry's disease has centered around relief of pain and management of renal failure. Circulating enzyme replacement has been attempted in affected individuals,[6,12] but, unfortunately, without benefit on the neuropathy. It is hoped that a method of delivering exogenous alpha-galactosidase A to affected target organs can be devised.

Pathology and Pathogenesis

This condition most probably represents a sensory and autonomic neuronopathy. The primary morphologic event is lipid storage in neurons, presumably reflecting a local metabolic disorder. This process results in cell loss and disappearance of their processes. Study of postmortem and nerve biopsy material reveals selective loss of small neurons in lumbar dorsal root ganglia, and lipid inclusions in others.[30] Correlating with the neuronal changes are moderate fiber loss in the gracile fasciculi of the spinal cord, and selective loss of small myelinated fibers and unmyelinated fibers in peripheral nerves. Segmental demyelination and distal axonal degeneration are not prominent features.[24] Abnormal lipid inclusions are described in perineurial cells and vascular endothelial cells, but not in Schwann cells. There are no studies of the peripheral autonomic ganglia, but lipid-distended neurons are present both in the intermediolateral columns of the thoracic spinal cord and in Onuf's (parasympathetic) nucleus in the sacral cord.[39] Neuronal changes, apparently asymptomatic, are present in scattered loci in the CNS, including the substantia nigra, amygdala, hypothalamus, and nucleus ambiguus.

Clinical Features of Neuropathy

Pain is probably the single most debilitating symptom of Fabry's disease. Two types of pain occur: episodic bouts and constant acral discomfort. The episodic bouts (Fabry's crises) consist of periods of excruciating burning pain over hands and feet, often spreading to other areas. Crises may occur spontaneously, are sometimes triggered by fatigue, temperature change, and exercise; increase in frequency with age; and may be severe enough to cause the afflicted individual to contemplate suicide.[11] Most patients also experience constant, low-level burning discomfort in the hands and feet between crises. Curiously, there are usually no objective signs of sensory impairment, reflex loss, or weakness in this condition. Motor and sensory conduction velocities are normal, as are sensory amplitudes.

Episodic diarrhea, nausea, vomiting, abdominal pain, and diminished sweating in the legs are generally attributed to involvement of autonomic ganglia.[4]

REFERENCES

1. Asbury, AK and Johnson, PC: Pathology of Peripheral Nerve. WB Saunders, Philadelphia, 1978, p 136.
2. Azizi, E, Zaidman, JL, Eshchar, J, et al: Abetalipoproteinemia treated with parenteral and oral vitamins A and E, and with medium chain triglycerides. Acta Pediatr Scand 67: 797, 1978.
3. Bale, PM, Clifton-Bligh, P, Benjamin, BN, et al: Pathology of Tangier disease. J Clin Pathol 24:609, 1971.
4. Bannister, R and Oppenheimer, DR: Degenerative diseases of the nervous system associated with autonomic failure. Brain 95:457, 1972.
5. Bischoff, A: Neuropathy in leukodystrophies. In Dyck, PJ, Thomas, PK, and Lambert, EH (eds): Peripheral Neuropathies, Vol 2. WB Saunders, Philadelphia, 1975, p 891.
6. Brady, RO, Tallman, JF, Johnson, WG, et al: Replacement therapy for inherited enzyme deficiency. Use of purified ceramidetrihexosidase in Fabry's disease. N Engl J Med 289: 9, 1973.
7. Brady, RO, Gal, AE, Bradley, RM, et al:

Enzymatic defect in Fabry's disease: Ceramidetrihexosidase deficiency. N Engl J Med 276:1163, 1967.

8. Brady, RO and King, FM: Fabry's disease. In Dyck, PJ, Thomas, PK, and Lambert, EH (eds): Peripheral Neuropathy, Vol 2. WB Saunders, Philadelphia, 1975, p 914.

9. Budden, SS, Kennaway, NG, Buist, NRM, et al: Dysmorphic syndrome with phytanic acid oxidase deficiency, abnormal very long chain fatty acids, a pipecolic acidemia: Studies in four children. J Pediatr 108:33, 1986.

10. Cammermeyer, J: Pathology of Refsum's disease. In Vinken, PJ and Bruyn, GW (eds): Handbook of Clinical Neurology, Vol 21. Elsevier/North Holland, New York, 1975, p 234.

11. Desnick, RJ, Klionsky, B, and Sweeley, CC: Fabry's disease (X-linked galactosidase A deficiency). In Stanbury, JB, Wyngaarden, JB, and Fredrickson, DS (eds): The Metabolic Basis of Inherited Disease, ed 4. McGraw-Hill, New York, 1978, p 810.

12. Desnick, RJ, Thorpe, SR, and Fiddler, MB: Towards enzyme therapy for lysosomal storage disease. Physiol Rev 56:57, 1976.

13. Dische, MR and Porro, RS: The cardiac lesions in Bassen-Kornzweig syndrome. Am J Med 49:568, 1970.

14. Dunn, HG, Lake, BD, Dolman, CL, et al: The neuropathy of Krabbe's infantile cerebral sclerosis (globoid cell leukodystrophy). Brain 92:329, 1969.

15. Dyck, PJ, Ellefson, RD, Yao, JK, et al: Adult-onset of Tangier disease. I. Morphometric and pathologic studies suggesting delayed degradation of neutral lipids after fiber degeneration. J Neuropathol Exp Neurol 37:119, 1978.

16. Engel, WK, Dorman, JD, Levy, RI, et al: Neuropathy in Tangier disease: Alpha lipoprotein deficiency manifesting as familial recurrent neuropathy and intestinal lipid storage. Arch Neurol 17:1, 1967.

17. Fardeau, M and Engel, WK: Ultrastructural study of a peripheral nerve biopsy in Refsum's disease. J Neuropathol Exp Neurol 28:278, 1969.

18. Ferrans, VJ and Fredrickson, DS: The pathology of Tangier disease. A light and electron microscopic study. Am J Pathol 78:101, 1975.

19. Haas, LF, Austad, WI, and Bergin, JD: Tangier disease. Brain 97:351, 1974.

20. Harding, AE, Matthews, S, Jones, S, et al: Spinocerebellar degeneration associated with a selective defect of Vitamin E absorption. N Engl J Med 313:32, 1985.

21. Herbert, PN, Gotto, AM, and Fredrickson, DS: Familial lipoprotein deficiency. In Stanbury, JB, Wyngaarden, JB, and Fredrickson, DS (eds): The Metabolic Basis of Inherited Disease, ed 4. McGraw-Hill, New York, 1978, p 544.

22. Hogan, GR, Gutmann, L, and Chou, SM: The peripheral neuropathy of Krabbe's (globoid) leukodystrophy. Neurology (Minneapolis) 19:1094, 1969.

23. Illingworth, DR, Connor, WE, and Miller, RG: Abetalipoproteinemia. Report of two cases and review of therapy. Arch Neurol 37:659, 1980.

24. Kocen, RS and Thomas, PK: Peripheral nerve involvement in Fabry's disease. Arch Neurol 22:81, 1970.

25. Kocen, RS, King, RHM, Thomas, PK, et al: Nerve biopsy findings in two cases of Tangier disease. Acta Neuropathol 26:317, 1973.

26. Kocen, RS, Lloyd, JK, Lascelles, PT, et al: Familial alpha-lipoprotein deficiency (Tangier disease) with neurological abnormalities. Lancet 1: 1341, 1967.

27. Kolodny, EH and Moser, HW: Sulfatide lipidosis: Metachromatic leukodystrophy. In Stanbury, JB, Wyngaarden, JB, and Fredrickson, DS (eds): The Metabolic Basis of Inherited Disease, ed 5. McGraw-Hill, New York, 1983, p 881.

28. Miller, R, Davis, JF, Illingworth, R, et al: The neurology of abetalipoproteinemia. Neurology (Minneapolis) 30:1286, 1980.

29. Muller, DPR and Lloyd, JK: Effect of large oral doses of Vitamin E on the neurological sequelae of abetalipoproteinemia. Ann NY Acad Sci 393:133, 1982.

30. Ohnishi, A and Dyck, PJ: Loss of small peripheral sensory neurons in Fabry disease. Arch Neurol 31:120, 1974.

31. Poulos, A, Pollard, AC, Mitchell, JD, et al: Patterns of Refsum's disease. Phytanic acid oxidase deficiency. Arch Dis Child 59:222, 1984.

32. Refsum, S: Heredopathia atactica polyneuritiformis: A familial syndrome not hitherto described. A contribution to the clinical study of the hereditary disease of the nervous system. Acta Psychiatr Scand (Suppl)38:1, 1964.

33. Refsum, S: Heredopathia atactica polyneuritiformis. In Vinken, PJ and Bruyn, GE (eds): Handbook of Clinical Neurology, Vol 21. Elsevier/North Holland, New York, 1975, p 181.

34. Savettieri, G, Camarda, R, Galatioto, S, et al: Refsum disease. Clinical and morphological report on a case. Ital J Neurol Sci 3:241, 1982.

35. Sobrevilla, LA, Goodman, ML, and Kane, CA: Demyelinating central nervous system disease, macular atrophy and acanthocytosis (Bassen-Kornzweig syndrome). Am J Med 37:821, 1964.

36. Spiess, H, Ludin, HP, and Kummer, H: Polyneuropathie bei familiare Analphalipoproteinamie (Tangier disease). Nervenarzt 40:191, 1969.

37. Steinberg, D: Phytanic acid storage disease: Refsum's syndrome. In Stanbury, JB, Wyngaarden, JB, and Fredrickson, DS (eds): The Metabolic Basis of Inherited Disease, ed 5. McGraw-Hill, New York, 1983, p 731.

38. Stokke, O: Alpha oxidation of fatty acids in various mammals, and a phytanic acid feeding experiment in an animal with low alpha oxidation capacity. Scand J Clin Lab Invest 20:305, 1967.

39. Sung, JH: Autonomic neurons affected by lipid storage in the spinal cord in Fabry's disease: Distribution of autonomic neurons in the sacral cord. J Neuropathol Exp Neurol 38:87, 1979.

40. Suzuki, K and Grover, WD: Krabbe's leukodystrophy (globoid cell leukodystrophy): An ultrastructural study. Arch Neurol 22:385, 1970.

41. Suzuki, K, Suzuki, Y, and Chen, G: Metachromatic leukodystrophy: Isolation and chemical analysis of metachromatic granules. Science 151:1231, 1966.

42. Suzuki, K and Suzuki, Y: Galactosylceramide lipidosis: Globoid cell leukodystrophy. In Stanbury, JB, Wyngaarden, JB, and Fredrickson, DS (eds): The Metabolic Basis of Inherited Disease, ed 4. McGraw-Hill, New York, 1978, p 787.

43. Terry, RD, Suzuki, K, and Weiss, M: Biopsy study in 3 cases of metachromatic leukodystrophy. J Neuropathol Exp Neurol 25:141, 1966.

44. Webster, H deF: Schwann cell alterations in metachromatic leukodystrophy. Preliminary phase and electron microscopic observations. J Neuropathol Exp Neurol 21:534, 1962.

45. Yao, JK, Herbert, PN, Fredrickson, DS, et al: Biochemical studies in a patient with a Tangier syndrome. J Neuropathol Exp Neurol 37:138, 1978.

46. Yuill, GM, Scholz, C, and Lascelles, RG: Abetalipoproteinemia: A case report with pathological studies. Postgrad Med J 52:713, 1976.

Chapter 15

HEREDITARY NEUROPATHIES WITHOUT AN ESTABLISHED METABOLIC BASIS

There is a wide range of inherited neuropathies in which a metabolic basis has not been identified. Most are rare. The commonest consist of the hereditary motor and sensory neuropathies (HMSN). Studies in recent years have indicated that cases diagnosed clinically as peroneal muscular atrophy, Charcot-Marie-Tooth disease and Dejerine-Sottas disease are genetically heterogeneous. Some, without sensory involvement, have been shown to consist of a distal spinal muscular atrophy (the spinal form of Charcot-Marie-Tooth dis-

ease) and will not be considered here. The remainder, designated HMSN, display both motor and sensory involvement and are subdivided into types I, II, and III to replace the previous confusing eponyms.* Their clinical and pathologic features are summarized in Table 15–1. More complex forms also exist.

A second group consists of neuropathies in which the salient manifestations are either sensory or autonomic, or both. They have been linked together as the hereditary sensory and autonomic neuropathies (HSAN).[15] Additional conditions, without a known metabolic basis in which neuropathy is the main or an important aspect have identifying or pathologic features. Those to be considered here are hereditary pressure-sensitive neuropathy (HPSN), pressure palsies, X-linked bulbospinal neuronopathy, and giant axonal neuropathy.

HMSN TYPE I

Definition

This is a slowly progressive, relatively benign motor and sensory polyneuropathy. Most cases display autosomal dominant inheritance, but kinships with probable autosomal recessive inheritance are reported.[33] HMSN I includes the majority of individuals previously labeled as either having peroneal mus-

---

*3,6,13,14,17,29,30,33,56,57

184

cular atrophy, Charcot-Marie-Tooth disease, or Roussy-Lévy syndrome.[29]

Pathology

Abnormalities are confined to the spinal cord and peripheral nerves. The most prominent spinal cord alteration is loss of myelinated axons in the gracile fasciculi at upper cervical levels. Shrunken anterior horn cells in the lumbar cord are occasionally present. PNS changes are prominent in this condition and usually include: (1) enlargement of nerves, "hypertrophic neuropathy"; (2) abundant segmental demyelination, remyelination, and "onion-bulb" formation in distal nerves (Fig. 15–1); and (3) reduced numbers of axons in distal nerves. It has been sug- gested that many of the surviving distal axons are atrophic. Ultrastructural examination of advanced cases reveals onion bulbs frequently devoid of axons, occasional vacuolar endoneurial fibroblasts, and an abundant fibrillar material in the endoneurial spaces.[19]

Pathogenesis

The pathogenesis of HMSN I is unknown. Xenograft studies have yielded conflicting results. One study found no evidence of a Schwann cell abnormality,[18] whereas another did.[1] Slowed axonal transport of dopamine beta-hydroxylase has been demonstrated.[5] The pathologic changes, in general, suggest that a distal axonal change is the primary event, and that segmental demye-

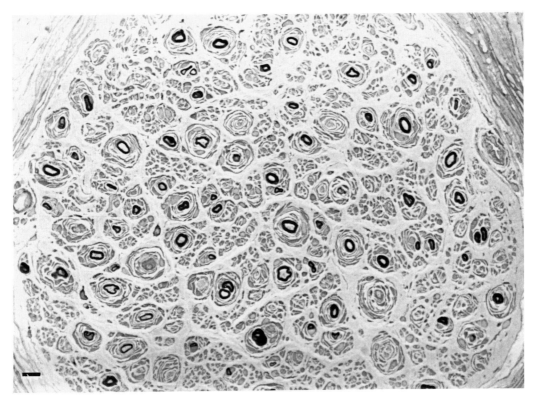

Figure 15–1. A low-power photomicrograph of myelinated nerve fibers from a sural nerve biopsy in a case of HMSN I. The myelinated nerve fiber population is reduced. Many of the surviving fibers are surrounded by an onion-bulb proliferation of Schwann cells. Some of the onion bulbs do not contain myelinated fibers or contain a demyelinated axon. These appearances constitute "hypertrophic" neuropathy and indicate a chronic process with repeated demyelination and remyelination. Bar = 10 μm.

**Table 15–1 THE PRINCIPAL HEREDITARY MOTOR AND
SENSORY NEUROPATHIES (HMSN)**

Type	Alternative Names	Inheritance	Clinical & Electrophysiologic Features
HMSN Type I	Hypertrophic form of peroneal muscular atrophy (PMA). Hypertrophic form of the disease. Roussy-Lévy syndrome (some cases).	Usually autosomal dominant, rarely autosomal recessive. Recessive cases are severely affected. Marriage of two affected individuals has produced children resembling recessively inherited HMSN.	Not an uncommon condition. Many mild, asymptomatic cases. Onset in childhood, adolescence, or later. Slowly progressive distal atrophy and weakness. Little sensory loss. Nerves often enlarged. Pes cavus common, scoliosis unusual. Essential tremor in some individuals. Sensory and motor conduction diffusely affected, motor may be extremely slow. Abnormal visual and auditory evoked potentials indicate optic and acoustic nerve involvement in some cases. Normal active life span common.
HMSN Type II	Neuronal form of PMA. Neuronal form of Charcot-Marie-Tooth disease.	Usually autosomal dominant, rarely autosomal recessive; recessive cases more severely affected.	Less common than HMSN I. Onset most often in second decade. Progressive distal weakness and atrophy similar to HMSN I. Sensory and motor nerve conduction only mildly abnormal.
HMSN Type III	Dejerine-Sottas disease. Hypertrophic neuropathy of infancy. Congenital hypomyelination neuropathy.	Autosomal recessive, probably genetically heterogeneous.	Rare. Onset in infancy, or from birth. Slowly progressive motor and sensory loss, and ataxia. Scoliosis and pes cavus frequent. Enlarged nerves. Short stature. Patients often severely disabled in adult life. Occasional pupillary abnormality. Motor conduction velocity severely reduced, sensory action potentials unrecordable.
X-linked HMSN	X-linked CMT disease	X-linked dominant locus in pericentromeric region.	Clinical features in males resemble those of autosomal dominant HMSN I. Onset usually at 5–15 yrs, slowly progressive. Obligate female carriers may be mildly affected. Motor nerve conduction velocity moderately reduced in affected males, normal or mildly reduced in female carriers.

Table 15–1 *Continued*

Pathology	Pathogenetic Hypothesis	Comments
Distal segmental demyelination, remyelination, and onion bulbs. Fewer myelinated axons of large diameter in distal nerves. CNS normal except for dorsal columns.	Pathology and morphometry suggest primary axonal disorder (distal axonopathy). Xenograft studies and abnormal axonal transport of dopamine beta-hydroxylase support this hypothesis.	The "classic" form of CMT disease or PMA. Mild cases widely misdiagnosed as orthopedic foot disorders.
Nerves not enlarged. Fewer myelinated axons of large diameter in distal nerves. Rare demyelination, few onion bulbs. Loss of anterior horn and dorsal root ganglion cells.	Generally held to represent disease of motor and sensory neurons. Not a variant of spinal muscular atrophy (SMA), although a distal form of SMA with features resembling other types of CMT disease but without sensory involvement exists.	Often clinically indistinguishable from HMSN I. Nerve conduction studies usually essential for differential diagnosis.
Enlarged nerves. Hypomyelination. Long demyelinated axon segments and many onion bulbs.	One case with decreased nerve cerebroside and increased liver ceramide monohexoside sulfate. Primary Schwann-cell disorder possible.	Need for confirmation of disordered lipid metabolism.
Axonal loss, segmental demyelination, moderate hypertrophic onion-bulb changes.	Probably a disease of motor and sensory neurons with an axonopathy.	

lination and remyelination are secondary. It is possible that this disorder results from an inherited abnormality of neuronal or axonal metabolism, but no human study or analysis of an appropriate experimental animal has indicated the nature or existence of a metabolic disorder.

Clinical Features

INCIDENCE AND INHERITANCE

HMSN I remains undiagnosed in many instances, and therefore most studies of prevalence are inaccurate. It appears to be a fairly common disorder with a world-wide occurrence.[19] The inheritance pattern is usually autosomal dominant, but families with a probable autosomal recessive inheritance have also been identified. In some dominantly inherited families the abnormal gene is linked to the Duffy locus near the centromere on chromosome 1,[54] but it is now clear that in the majority of families the abnormal gene is located on chromosome 17.[58] They have been designated HMSN types Ia and Ib, respectively. Families with X-linked inheritance also exist.[27,49] The abnormal gene is on the proximal part of the long arm. Variable expression of this genetic disorder probably accounts for the presence of individuals with slightly deformed feet as the sole manifestation of disease in kinships with advanced cases.[30] Slightly affected individuals often first become aware of their disease when it appears, in a more serious form, in their children.

SYMPTOMS AND SIGNS

HMSN I may be most simply described as a chronic polyneuropathy featured by the presence of weakness and wasting of distal limb muscle and foot deformity.

Common initial complaints are weakness of distal lower extremities beginning in the first or second decade. The disorder may be present from birth.

Subsequently, a gait disorder is noted, usually a nonspecific "unsteady ankle" sensation while walking; eventually foot-drop and steppage gait may occur. Weakness and wasting of the hands usually begin some years after the lower extremities are involved. Paresthesias are uncommon and, if prominent, challenge the diagnosis. Frequently, the only complaint is progressive deformity of the feet accompanied by atrophy of the legs (Fig. 15–2). Such individuals may seek orthopedic consultation early in the illness. Kyphoscoliosis occurs in some cases, particularly in severe cases with an early onset. In some families, distal sensory loss may be prominent and may lead to persistent foot ulceration.

The progression of neurologic signs in a fully developed case is stereotyped.[19,29] Individuals display initial

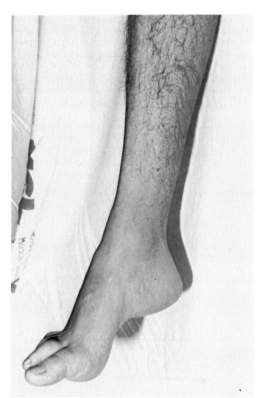

Figure 15–2. Atrophy of distal leg and intrinsic foot muscles with hammer toes and high arch in an individual with hereditary peripheral neuropathy.

weakness in the extensor hallucis and digitorum longus and the anterior tibial and peroneal muscles before calf muscles are affected. Fasciculations may occur in weakened muscles. Footdrop and contracture of calf muscles may result from the unopposed flexor action of posterior compartment muscles; eventually they too undergo atrophy, sometimes resulting in a storklegged or "inverted champagne bottle" appearance of the lower limbs. This pattern of weakness is occasionally accompanied by progressive pes cavus and clawed toes. The typical pes cavus deformity is a foreshortened, high-arched foot attributable to weakness of the intrinsic foot muscles and the unequal action of long toe flexors and extensors. Atrophy seldom extends above the mid-thigh, and weakness only very rarely involves the girdle muscles. Absence of the Achilles tendon reflex is an early sign and generalized tendon areflexia is usual in established cases. Sensory involvement, although present in all individuals, is generally not a very prominent feature of this condition, and may be difficult to elicit on routine clinical examination. In advanced cases, careful analysis of pain, touch, and thermal sense will usually demonstrate some disorder, but in mild cases may only be detectable from sensory nerve conduction or quantitative sensory testing studies. Discolored skin, edema, and cold extremities probably result from inactivity, diminished blood flow, and muscle loss, rather than autonomic dysfunction, which is not a feature of HMSN I.

The hands usually do not become weak until some years after the condition has become advanced in the lower extremities.[19] Occasionally, the intrinsic hand muscles may atrophy to an extreme degree, resulting in a "claw hand" deformity, analogous to pes cavus. Weakness of forearm muscles occurs in the more severe cases. Mild involvement of upper arm or shoulder girdle muscles is detectable in severe cases. Mild sensory loss may accompany the hand weakness. In severe cases, diaphragmatic weakness may develop, resulting in dyspnea when the subject is recumbent.[28] Firm, thickened nerves can be palpated in 50% of cases. There is considerable variation in atrophy, weakness, and distal skeletal deformity. Individuals may frequently be profoundly weak with only slight atrophy and minimal pes cavus. Contrary to popular belief, the majority of cases do not display the triad of pes cavus, "champagne bottle" legs, and claw hands, but rather manifest a mild degree of weakness, atrophy, or foot deformity. In children, slowed nerve conduction may be the sole manifestation of disease. Such cases are usually only detected in genetic screening of relatives of severe cases.

Essential tremor accompanies the neuropathy in members of some kinships. Since the features of the illness in these instances are otherwise identical to HMSN I, there seems little justification for not including them under this classification.[29] Previously, such cases were designated as the Roussy-Lévy syndrome.

Pupillary abnormalities,[50] optic atrophy, and deafness[51] are rare associations. In the absence of optic atrophy, optic nerve involvement may be demonstrable from visual evoked potential studies.

Laboratory Studies

CLINICAL LABORATORY FINDINGS AND CEREBROSPINAL FLUID

Cerebrospinal Fluid. This is usually normal, although the CSF protein content may be mildly elevated.

ELECTRODIAGNOSTIC STUDIES

Profoundly slow motor nerve conduction is a hallmark of this condition; velocities as low as 10 to 20 m/sec are common. Sensory nerve conduction in digital and sural nerves is also slowed, and the potential amplitude decreased. Frequently, sensory action potentials are not detectable by routine recording techniques.

Slowed motor nerve conduction may be abnormal in cases with few neurologic findings; for this reason, electrodiagnosis has been a useful tool in genetic counseling.[19]

NERVE BIOPSY

Nerve biopsy in most cases reveals fiber loss, segmental demyelination, and sometimes onion-bulb formation (see Fig. 15–1). These findings are characteristic but not pathognomonic of HMSN I, since they may occur in other chronic neuropathies characterized by repeated episodes of segmental demyelination and remyelination.

Course, Prognosis, and Treatment

Progression is gradual and individuals may appear to plateau for many years before worsening again. Rapid deterioration raises the possibility of a superimposed inflammatory demyelinating polyneuropathy.[16] HMSN I is frequently a benign condition, and need not seriously handicap affected individuals. Even in severe cases there is no evidence that it shortens life expectancy, and most working individuals remain employed until retirement age.

Preservation of proximal strength often enables even seriously involved cases to remain ambulatory if aided by braces, special shoes, and surgery to correct deformity, and occasionally by tibialis posterior tenodesis to reduce foot-drop.

There is no specific treatment; genetic counseling is usually simple.

Differential Diagnosis

The principal conditions misdiagnosed as HMSN I are:

• *Friedreich's ataxia:* Excluded by severely reduced nerve conduction in HMSN I.

• *HMSN II:* Excluded by severely abnormal nerve conduction tests and usually earlier onset in HMSN I (although motor conduction velocities may overlap between the two forms).

• *Hereditary distal spinal muscular atrophy:* Excluded by abnormal motor and sensory nerve conduction studies and presence of sensory loss in HMSN I.

• *Refsum's disease:* Excluded by absence of pigmentary retinal degeneration and elevated serum phytanic acid levels in HMSN I.

• *Chronic inflammatory demyelinating polyneuropathy:* A markedly fluctuating course, absence of family history and foot deformity (except with childhood onset), and a substantially elevated CSF protein level will suggest CIDP. Differential diagnosis in chronic progressive cases may be difficult. The presence of focal conduction block in nerve conduction studies and inflammatory infiltrates in nerve biopsies from some cases of chronic inflammatory idiopathic polyneuropathy may be helpful.

• *Paraproteinemic neuropathy:* Demyelinating polyneuropathy related to a benign monoclonal paraproteinemia, usually IgM kappa in type, does not commonly enter into the differential diagnosis, as the age of onset is much later. It occasionally commences in the third decade and thus requires exclusion.

Case History: HMSN TYPE I

A 50-year-old accountant was admitted to the hospital because of urinary frequency and hesitancy. He underwent surgery for prostatic hyperplasia and made an uneventful recovery. During his hospitalization he slipped on a wet floor and sustained a mild concussion. The neurologist was called and noted, in addition to a postconcussive state, mild pes cavus, moderate atrophy of the calf and anterior compartment muscles, 4/5 (MRC scale) strength, and fasciculations in the extensors and flexors of toes and feet, 4/5 strength of the intrinsic hand muscles, tendon areflexia, and slightly diminished pinprick and position sense in toes and fingers. Walking barefoot was difficult because of ankle instability. Walking was normal while wearing high-topped shoes. The remainder of the examination was unremarkable. There was no palpable nerve enlargement. Motor nerve conduction velocity in the peroneal nerves was 10 to 15 m/sec, and in the ulnar, 20 to 25 m/sec. Median nerve sensory conduction was 30 to 40 m/sec.

Comment. This man had a mild hereditary polyneuropathy of early onset with deformed feet and marked slowing of nerve conduction. The diagnosis of HMSN I was apparent to the neurologist, but had been overlooked by other physicians. The patient was either unaware of his illness or wished to deny it, and was distressed that his 18-year-old son also was found to be affected. This case emphasizes the insidious course of the condition and the need for genetic counseling.

HMSN TYPE II

Definition

This is less common than HMSN type I and displays similar clinical features. Inheritance is usually autosomal dominant. Factors that separate this disorder from type I are normal or only moderately reduced nerve conduction velocity, later onset, and absent or minimal hypertrophic changes on nerve biopsy.[13,14,57]

Pathology

Nerve biopsies show a decreased number of myelinated nerve fibers with, in contrast to HMSN I, little segmental demyelination or "onion-bulb" formation.[19] A postmorten study has indicated that the underlying pathology is a primary neuronopathy affecting motor and sensory neurons that gives rise to a distally accentuated loss of axons and also loss of anterior horn and dorsal root ganglion cells.[4] The disorder is sometimes referred to as the "neuronal form" of peroneal muscular atrophy.[19,29]

Clinical Features

The onset of symptoms in this disorder is most often in the second decade, but many cases begin in the first decade and onset may be delayed until middle or even late life. Clinical features, including signs, symptoms, course, and prognosis, are similar to those of HMSN type I, except that the upper extremities are less severely affected, sensory loss and tremor are less prominent,[29] widespread tendon areflexia is less common, foot deformity tends to be less severe, and nerve thickening does not occur. Occasional cases are encountered with an onset in early childhood that show a much more aggressive course.[43]

Motor and sensory nerve conduction velocities are within normal limits or only moderately reduced, in contrast to the usually profoundly slowed conduction of type I. Indeed, clinical electrophysiology and nerve biopsy are the cardinal means of distinguishing between these disorders (see Table 15–1). There is some overlap for motor conduction velocity between the two forms, but not for sensory conduction.[6]

HMSN TYPE III

Definition

The status of this condition is confusing. It is a rare, probably genetically heterogeneous, slowly progressive, debilitating peripheral neuropathy, beginning in infancy or present from birth. It includes examples of progressive childhood-onset hypertrophic neuropathy of autosomal recessive inheritance that have been categorized as Dejerine-Sottas disease,[19] and others, also probably of autosomal inheritance with hypotonic weakness present from birth.[38] Both are characterized by severe hypomyelination and demyelination in peripheral nerves and by severely reduced nerve conduction velocity.

Pathology and Pathogenesis

Pathologic changes are confined to the peripheral nervous system and the dorsal columns of the spinal cord. Salient features are gross enlargement of distal peripheral nerves, loss of myelinated axons, segmental demyelination and remyelination, and "onion-bulb" formation. Myelin sheaths of normal thickness are not present (hypomyelination),[19,42] and many fibers lack myelin sheaths (amyelination). Particularly in the congenital cases, the onion bulbs may be composed mainly of concentric layers of basal lamina with few Schwann cell processes.[26]

The pathogenesis is unknown. One patient has demonstrated a decrease in peripheral nerve cerebrosides and an increase in liver ceramide monohexoside sulfate.[19] The relationship of this biochemical abnormality to the peripheral neuropathy is unknown.

Clinical Features

INCIDENCE AND HEREDITY

This is a rare condition with a worldwide incidence. Most cases are isolated or, if multiple, confined to a single sibship, suggesting autosomal recessive inheritance.

SIGNS AND SYMPTOMS

Cases identifiable as HMSN III begin in infancy. Early motor landmarks, such as walking, are often delayed. Initially, weakness appears in the distal lower extremities. Soon afterward, the upper extremities become involved. Proximal extremity muscles are eventually affected in most cases, resulting in severe disability. Areflexia is the rule and, in almost all cases, enlarged peripheral nerves can be easily palpated. Sensory loss may be severe, usually affecting touch, position, and vibration sense more than pain. Pseudoathetosis and sensory ataxia may occur. Autonomic function is spared. Hearing loss is common and occasional patients are nearly deaf. Abnormal pupillary responses to light occasionally occur. The condition may stabilize once adult life is reached.

Other cases (congenital hypomyelination neuropathy) present with hypotonic weakness from birth and display the features described above in more severe degree.[26,38]

A distinguishing feature of HMSN type III is the frequency of skeletal abnormality. Many affected individuals are of short stature, kyphoscoliotic, and have severe deformities of hands and feet.

Laboratory Studies

CEREBROSPINAL FLUID

The CSF protein content is often elevated, presumably because lumbar roots are involved.

ELECTRODIAGNOSTIC STUDIES

Motor conduction velocity is extremely slow, often less than 10 m/sec. Sensory nerve action potentials are unrecordable by standard techniques.

NERVE BIOPSY

The nerve biopsy findings are characteristic and have been detailed above.

Course and Prognosis

The illness progresses more rapidly and is generally more disabling than HMSN type I; many patients are confined to a wheelchair in adulthood. There is no study of longevity in this illness.

Differential Diagnosis

Diagnosis of this disorder is rarely difficult. The congenital or infantile onset, recessive inheritance, and skeletal deformities distinguish it from most cases of HMSN types I and II, the usual absence of CNS signs and the severely reduced nerve conduction velocity from Friedreich's ataxia, and normal serum phytanic acid levels from Refsum's disease. The rare recessively inherited instances of HMSN type I are distinguishable on nerve biopsy, where some of the myelinated fibers possess myelin sheaths of normal thickness.

X-LINKED HMSN

Families have been described with HMSN in which the disorder shows an X-linked pattern of inheritance.[24,25,27,49] Affected males develop symptoms between the ages of 5 and 15 years with clinical features that are similar to those of autosomal dominant HMSN I. Some obligate female heterozygotes have neurologic symptoms or are affected subclinically.[24,27,49] Motor nerve conduction velocity in affected males is moderately reduced (26 to 38 m/sec in upper limbs) and normal or mildly reduced in affected females.[27,49] Nerve biopsies show segmental demyelination, hypertrophic changes, and fiber loss, the demyelination probably being secondary to an axonopathy.[27] The disorder has been mapped to the centromeric region of the X chromosome.[23]

COMPLEX FORMS OF HMSN

The clinical syndrome of peroneal muscular atrophy is occasionally associated with other features, including pyramidal signs[32] and optic atrophy and deafness.[48] These variants are not yet well characterized.

HSAN TYPE I (DOMINANTLY INHERITED SENSORY NEUROPATHY)

This is a rare, dominantly inherited sensory neuropathy primarily affecting the distal lower extremities. The onset is in the second decade and progression is slow. Loss of unmyelinated fibers and mutilated feet usually occur. This disorder was previously labeled hereditary sensory radicular neuropathy of Denny-Brown and, before that, lumbosacral syringomyelia or maladie de Thévenard (Table 15–2).[8,9,12]

Pathology and Pathogenesis

There is a marked loss of unmyelinated axons in distal nerves and a moderate loss of small myelinated fibers. Histometric studies of sensory nerves have demonstrated a progressive depletion of small myelinated fibers at distal levels. Large myelinated axons are usually spared, and vacuolated fibroblasts are prominent in the endoneurium.[52] Postmortem examination has revealed reduced numbers of lumbar dorsal root ganglion neurons. Nerves are not enlarged, segmental demyelination and remyelination are not present, and onion-bulb formation is not a feature of this illness.[12] There are no CNS changes other than loss of axons in the gracile fasciculi.

The pathogenesis is unknown. It is generally held that this disorder results from slowly progressive distal axonal atrophy and degeneration, and that dorsal root ganglion cells degenerate after years of centripetal spread of axonal disease. The histopathology is suggestive of a distal axonopathy.

Clinical Features

This is an uncommon condition and occurs throughout the world. Its inheritance is autosomal dominant.[12]

The onset is usually in the second decade or later. Individuals initially notice painless injuries in the sole of the foot. Calluses often form on the base of the toes, become discolored, and then ulcerate. Frequently, the ulcers become infected and cellulitis develops or purulent drainage occurs. Repeated trauma to the anesthetic feet may result in joint destruction, pathologic fractures, and

Table 15–2 THE PRINCIPAL HEREDITARY SENSORY AND AUTONOMIC NEUROPATHIES (HSAN)

Type	Alternative Names	Inheritance	Clinical & Electrophysiologic Features
HSAN Type I	Dominantly inherited sensory neuropathy. Hereditary sensory neuropathy of Denny-Brown. Maladie de Thévenard.	Autosomal dominant.	Rare. Onset in second decade. Progressive distal extremity sensory loss. Mutilation of feet. Pain and temperature sense more affected than touch-pressure. Occasional lancinating pain. Sweating impaired in distal extremities. Motor nerve conduction normal. Preserved sensory action potentials (A-alpha component) in earlier stages with abnormal A-delta and C-fiber potentials. Proximal tendon reflexes and autonomic function largely spared (except sweating). Life expectancy normal with good foot care.
HSAN Type II	Congenital sensory neuropathy. Recessive hereditary sensory neuropathy. Morvan's disease. Infantile syringomyelia.	Autosomal recessive.	Rare. Onset in early childhood or at birth. May be substantially nonprogressive. Hands and feet mutilated, pathologic fractures common. Distal touch-pressure may be affected earlier; eventually all modalities involved. Sensory loss not confined to extremities. All tendon reflexes lost. Distal sensory conduction profoundly affected. Motor nerve conduction near normal. Prognosis not known.
HSAN Type III	Riley-Day syndrome. Familial dysautonomia.	Autosomal recessive. Predominantly in Jewish families.	Rare. Onset in infancy. Autonomic dysfunction prominent: absent lacrimation, labile sweating, blood pressure, and temperature. Loss of taste. Generalized diminution of pain-temperature sensation. Preserved touch sensation. Short stature, scoliosis. Decreased amplitude of sensory action potentials, mild slowing of motor conduction. Mutilation unusual. Decreased life expectancy.
HSAN Type IV	Congenital sensory neuropathy with anhidrosis.	Autosomal recessive.	Very rare. Onset in infancy. Widespread absence of pain-temperature sensation. Strength normal. Episodic fever. Absent sweating. Mental retardation. Mutilation usual. Short stature.

Table 15–2 *Continued*

Pathology	Pathogenetic Hypothesis	Comments
Proximal to distal gradient of fiber loss. Unmyelinated and small myelinated fibers more depleted than large myelinated fibers.	Pathology and clinical data support hypothesis of slowly progressive distal axonopathy.	Firm correlation of sensory deficit with fiber-type loss on morphologic and electrophysiologic studies. Sparing of proximal autonomic function helpful in differentiation from amyloid neuropathy. Increased synthesis of immunoglobulin A in one kinship.
Mild proximal to distal gradient of fiber loss. Myelinated fibers severely depleted, unmyelinated fibers less so. Occasional degenerating fibers present on biopsy. Some distal segmental demyelination and remyelination.	Morphologic evidence somewhat supports hypothesis similar to HSAN I. Xenograft study indicates no disorder of Schwann cells.	Morphology supports notion of a progressive degenerative condition; clinical state often seems static.
Sural nerve has near total absence of unmyelinated axons and reduced numbers of myelinated axons. Slow progression with age. Reduced numbers of neurons in sympathetic, dorsal root, gasserian, and sphenopalatine ganglia. Ciliary ganglia normal. No CNS change aside from progressive dorsal column degeneration and depletion of preganglionic sympathetic neurons.	Congenital absence of autonomic and sensory ganglia and peripheral processes indicates disorder of embryogenesis, with mild progressive, degenerative disease of neurons.	Relationship of abnormal levels of catecholamine metabolites and low serum dopamine beta-hydroxylase to the labile autonomic clinical phenomena unclear. Diagnosis usually established shortly after birth; may be initially misdiagnosed as HSAN IV.
PNS incompletely studied. Reduced number of smaller neurons in dorsal root ganglia.	Pathogenesis may resemble HSAN III, but insufficient data to support this notion firmly.	Mental retardation and lack of sweating help to distinguish from HSAN III. More clinical, pathologic, and basic studies needed.

grossly deformed feet. Mild denervation of distal muscles may be present and pes cavus may occur in rare cases. It is always accompanied by severe sensory loss in the feet. Episodes of lancinating pain of the feet or arms are an intermittent phenomenon in about one half of the cases, and burning pain in the feet can be troublesome. Occasional kinships with features similar to those of HSAN type I have shown hearing loss, and others, spastic paraparesis.[7]

Neurologic examination in the early stages of disease usually reveals a striking dissociated pattern of sensory change, featured by a profound loss of pain and temperature sense with only slight loss of touch, pressure, and position sense. Such cases display diminished ankle jerks and intact strength. This pattern of sensory loss correlates with histometric and physiologic studies that have demonstrated a loss of myelinated fibers and greatly diminished C and A-delta action potentials. With time, the entire foot and lower leg become anesthetic and ankle deformities develop. The ankle jerks disappear and the knee jerks become depressed. Atrophy of the small muscles of the feet is common, and mild weakness of toe extension and dorsiflexion and eversion at the ankles may occur, but sensory loss always dominates the clinical profile.[12]

Upper extremity sensory loss is usually mild and, when present, confined to the fingers. Infection of the fingers or mutilation of the hand is rare in this disorder. Many individuals are unaware of sensory changes in the hands, and strength and tendon reflexes are usually preserved in the upper limbs.

Autosomal dysfunction is not an important feature of type I HSAN, bladder dysfunction being the only autonomic disturbance that has been described.

Laboratory Studies

The CSF is usually normal. Increased serum levels of immunoglobulin A may occur.

ELECTRODIAGNOSTIC STUDIES

Sensory nerve conduction studies of the sural nerve show diminished amplitude but near-normal conduction velocities in early cases. Motor nerve conduction velocities are normal. Advanced cases may display slightly prolonged motor and sensory latencies in the lower extremities. Sensory and motor conduction studies in the upper extremities are usually normal.

NERVE BIOPSY

Sural nerve biopsy in early cases reveals a grossly normal specimen which, on light microscope examination, has a normal complement of large diameter myelinated fibers and a moderate reduction of small myelinated fibers. Electron microscope examination will usually display a striking depletion in the number of unmyelinated fibers.[52]

Course, Prognosis, and Treatment

The course is a gradual, proximal progression of the sensory loss in the feet, usually to a level below the knee. The proximal lower extremities are not affected, and the hands are usually spared or sustain only minor sensory loss. Life expectancy is normal in this condition with proper treatment, and many patients remain gainfully employed.

There are two concerns for the physician. One is genetic counseling on the probability and nature of the illness. Treatment revolves around proper foot care. The chief disability in this illness stems from bone and joint deformity and repeated infected ulcerations of the feet. Occasionally, amputation is necessary. These complications are probably largely preventable by scrupulous maintenance of foot hygiene and skilled podiatric management.

Differential Diagnosis

Conditions that are occasionally misdiagnosed as HSAN type I include:

• *Hereditary Amyloidosis.* In contrast to HSAN type I, this condition is usually featured by prominent autonomic dysfunction and usually has a later age of onset.

• *Syringomyelia.* In contrast to HSAN type I, this condition usually begins in the cervical segments, is not hereditary, and involvement of the lower extremities is extremely rare.

• *Diabetic Polyneuropathy.* In contrast to HSAN type I, diabetes mellitus is not dominantly inherited, the onset is usually later in life, the upper extremities are commonly also involved, and impaired glucose tolerance is demonstrable.

• *HSAN Type II.* In contrast to HSAN type I, this condition has an early onset, is recessively inherited, the upper extremities are usually involved, and the sensory loss affects all modalities. The differential diagnosis between HSAN types I and II may be impossible, on occasion, since the clinical features appear to overlap in some kinships.

• *Leprosy.* In areas in which leprosy is endemic, type I HSAN may be mistaken for lepromatous leprosy. The distinction may be made clinically by the presence of nerve enlargement and skin lesions and confirmed by the finding of leprosy bacilli on skin or nerve biopsy.

Case History: HSAN Type I

A 35-year-old clerk was admitted to the hospital with septicemia from an infection of the left foot. Ten years earlier, she first noticed painless bruises on the metatarsal pads and recurrent calluses of the dorsum of the great toe. A podiatrist informed her that she had hammer toes and peripheral neuropathy. She declined to see a neurologist and continued to work as a filing clerk. In the ensuing ten years, repeated infections of the soles of both feet occurred, and a nonhealing ulcer developed on the first metatarsal pad on the left. Her gait had become uneven and bony deformities gradually developed in both feet.

The sepsis responded to a three-week course of antibiotics. Neurologic consultation revealed bilateral hammer toe deformi-

ties and grossly deformed feet with limitation of motion at the ankles. There was a partially healed ulcer at the base of the left metatarsal. The toes of both feet were shortened, with only remnants of toenails. Sensory examination revealed anesthesia to pinprick and markedly diminished thermal sensation over the plantar surfaces bilaterally. There was no weakness of dorsiflexion or plantar flexion of the ankle. The intrinsic muscles of the feet and long toe flexors and extensors could not be tested because of limitation of movement. The ankle jerks were depressed while other reflexes were brisk. There were no abnormalities of mental status and the cranial nerves were intact. Urinary bladder function was normal and there was no orthostatic hypotension.

Routine laboratory evaluation, including serum proteins and a glucose tolerance test, were normal. X-rays of the feet showed healed bilateral osteomyelitis of the tarsal and metatarsal bones, with fusion of the tarsal bones. Electromyography of the muscles of upper and lower extremities was normal. Sensory conduction in the sural and median nerves was normal. Nerve biopsy was declined.

The patient is married and has two adolescent daughters who are neurologically normal. The patient's mother had developed recurrent foot ulcers in the sixth decade and wore orthopedic shoes until her death from a myocardial infarct at age 60.

Comment. This case demonstrates many of the salient features of HSAN type I: probable dominant inheritance, late onset, dissociated sensory loss confined to the feet, and recurrent foot infections with bony deformity; of the only serious alternative diagnoses, diabetes mellitus was excluded by tests of carbohydrate metabolism; amyloid neuropathy was unlikely, as tests of autonomic function were normal.

HSAN TYPE II (RECESSIVELY INHERITED SENSORY NEUROPATHY)

This is a rare, recessively inherited disorder, characterized by a sensory neuropathy present at birth or with an onset in infancy. Absence of myelinated sensory fibers is correlated with anesthesia and mutilation of the extremities. This disorder was previously labeled Morvan's disease, infantile syringomyelia, or congenital sensory neuropathy. The clinical criteria for HSAN type II are based on detailed studies of relatively few patients.[11,12]

Pathology and Pathogenesis

There are no autopsy reports of this condition. The following features are prominent in nerve biopsies: a striking decrease in myelinated axons of all diameters, more pronounced at distal levels; a moderate decrease in the number of unmyelinated fibers; some evidence of segmental demyelination and remyelination; occasional degenerative change in myelinated axons; and vacuolated endoneurial fibroblasts.[11,12]

The pathogenesis is unknown. It is generally held that this disorder, like HSAN type I, is associated with distal axonal changes. Xenograft studies have suggested that the Schwann cell does not have a primary role.[11] The evidence that this disorder is degenerative stems solely from observation of scattered fiber breakdown in nerve biopsies; the clinical illness appears static, and it has been suggested that HSAN II represents a failure in differentiation of primary sensory neurons or their prenatal degeneration. A study of an animal model of sensory neuropathy suggests that there is progressive loss of sensory neurons during late prenatal development.[35]

Clinical Features

HSAN type II is a very rare condition and has a world-wide distribution. It is inherited in an autosomal recessive manner.

Almost all cases are recognized in infancy and, in some, the disorder appears to have been present at birth. Ulcerating wounds develop on the fingers and feet at an early age, and painless fractures of the fingers may occur in falls. At an age when a reliable sensory examination can be performed, there is usually a total absence of position, touch, and pressure sensation over the hands and feet, moderate to severe loss of pain and thermal sense in these areas, absence of all tendon reflexes, and normal muscle strength. Commonly, there are similar sensory abnormalities over the proximal extremities, trunk, and forehead. Mental status, cranial nerves, and autonomic function are usually preserved, but in some individuals there may be disturbances of bladder function and impotence.[11]

Laboratory Studies

Routine laboratory tests are usually unremarkable.

ELECTRODIAGNOSTIC STUDIES

Electromyography is usually normal, although occasionally fibrillation potentials are present in the intrinsic foot muscles. Motor nerve conduction velocities are generally normal or in the low normal range. Unobtainable sensory action potentials from either digital or sural nerves are characteristic of this condition.

NERVE BIOPSY

See *Pathology,* above. Sensory nerves are extremely atrophic and may be very difficult to locate at biopsy.

Course, Prognosis, and Treatment

Unequivocal evidence of progression is lacking in most cases. Children grow into adolescence and accommodate to severe sensory loss of the extremities. Most have deformed, shortened, ulcerated fingers and toes. The problems and therapies for foot care are identical to those outlined for HSAN type I, compounded by severe limitations imposed by deformed and deafferented hands. Unless exceptional parental care is afforded to these children at a young age, and self-care instruction given during adolescence, they may become severely handicapped. There are no data on longevity in this condition.

Differential Diagnosis

Conditions occasionally misdiagnosed as HSAN type II include:

• *HSAN Type I.* This condition is dominantly inherited, usually confined to the lower extremities, and has a later onset than HSAN type II.

• *HSAN Type III.* This condition can also be present in infants, but its prominent autonomic phenomena readily distinguish it from HSAN type II.

• *HSAN Type IV.* This condition is characterized by high fevers and mental retardation, but initially may be confused with HSAN type II since both are present in infancy and featured by mutilation of the extremities.

HSAN TYPE III (FAMILIAL DYSAUTONOMIA; RILEY-DAY SYNDROME)

This rare, autosomal recessive condition is probably a disorder of embryogenesis. Congenital absence of autonomic and sensory ganglion cells results in a characteristic clinical syndrome.

Pathology and Pathogenesis

There are diminished numbers of sympathetic ganglion cells throughout, with a corresponding decrease in peripheral sympathetic terminals. Degenerative changes have not been described, but older individuals have fewer neurons.[44] The preganglionic sympathetic neurons in the intermediolateral columns of the thoracic spinal cord are reduced in number. Parasympathetic neurons are strikingly depleted in some ganglia (sphenopalatine), but normal in others (ciliary).[46]

Sensory neurons in dorsal root ganglia are fewer than normal in young individuals, and depletion continues with aging. There is a corresponding progressive loss of axons in the dorsal columns and sensory nerves.[45]

The pathogenesis is unknown. Since the neuronal deficits are present at birth and degeneration slowly continues, an antenatal deficit of some trophic mechanism has been proposed.[45] Presumably this trophic mechanism is inadequate during life to sustain the remaining neurons; thus, degeneration continues.

Clinical Features

INCIDENCE AND HEREDITY

This is a rare disorder. It is inherited in an autosomal recessive fashion and the majority of patients have been Jewish.

SIGNS AND SYMPTOMS

Many of the clinical manifestations of HSAN type III correlate well with the congenital absence of autonomic and sensory neurons.

In infancy, poor sucking, crying without tears (alacrima), vomiting crises, blotchy skin, and unexplained fluctuating body temperatures are diagnostic hallmarks. Sweating is usually normal. Older children are noted to have postural hypotension and absence of taste. Responses to hypoxia or hypercarbia may be abnormal.[21] Reduced ambient oxygen tension, as at high altitudes, does not stimulate an increased ventilation rate. Hypoxemia may result and even lead to coma. Patients can hold their breath for a prolonged period without discomfort.

Sensory dysfunction is prominent, featured by absent corneal reflexes, diminished sensation of pain and temperature, and areflexia. Touch-pressure sensation is relatively spared and strength is often normal. Mutilation of the extremities is rare. Seizures may occur. Intelligence is usually normal. Short stature and scoliosis are common; the tongue is smooth and fungiform papillae are lacking.[45]

Laboratory Studies

SPECIAL BIOCHEMICAL STUDIES

Norepinephrine and epinephrine catabolite excretion is diminished; dopamine products are excreted in normal amounts. A block in epinephrine synthesis has not been demonstrated.

ELECTRODIAGNOSTIC STUDIES

Electromyography is usually normal. Motor nerve conduction may be slightly slowed. Sensory nerve action potentials are strikingly diminished in amplitude.

NERVE BIOPSY

Sural nerve biopsies in young patients have demonstrated a markedly decreased number of unmyelinated fibers and a lesser depletion of larger myelinated fibers. Segmental demyelination and onion-bulb formation are not features in this condition.

Course, Prognosis, and Treatment

This is a serious illness associated with a diminished life expectancy. Many patients die in infancy or childhood from multiple causes related to autonomic dysfunction: aspiration during incoordinated sucking, severe postural hypotension, abnormal responses to hypoxia, or dehydration following vomiting. Occasional individuals survive to middle age, but have limited activity because of autonomic lability and sensory dysfunction. There is no treatment aside from symptomatic care dur-

ing crises. Pharmacologic manipulation of autonomic dysfunction is inadvisable because of the unpredictable response of denervated receptors.[45]

Differential Diagnosis

The correct diagnosis is usually rapidly established in infancy. Few syndromes in pediatric neurology are as dramatic and precise as HSAN type III. On occasion, a diagnosis of HSAN type IV may be entertained; however, mental retardation and absence of sweating (prominent in type IV) are not features of HSAN type III.

HSAN TYPE IV (CONGENITAL SENSORY NEUROPATHY WITH ANHIDROSIS)

This extremely rare, autosomal recessive disorder is manifested in infancy by bouts of high fever. Older children display absence of pain sensation, absent sweating, and mental retardation.[55] The salient features of this condition are presented in Table 15–2.

"CONGENITAL INDIFFERENCE TO PAIN"

Reports in the earlier literature described cases of "congenital indifference to pain" assumed to be related to central disturbances such as asymbolia for pain. Contemporary morphometric techniques have demonstrated that most are related to selective small-fiber neuropathies.[9,15]

HEREDITARY PRESSURE-SENSITIVE NEUROPATHY (HPSN)

Recurrent pressure palsies precipitated by trivial trauma are the hallmark of this condition.[20,36] It is inherited in an autosomal dominant fashion. Onset of symptoms is usually in the second or third decade. Initially, the patient may sleep on a limb and the resultant palsy clears after some weeks, instead of within minutes. The resultant mononeuropathies are both sensory and motor. HPSN is accompanied by a chronic asymptomatic generalized polyneuropathy in most cases; occasionally pes cavus and hammer toes are present.

Electrodiagnostic studies disclose evidence of a mild diffuse neuropathy in all limbs, superimposed upon the expected conduction changes in the affected limb. Histopathologic findings in sural nerve biopsies from unaffected limbs have revealed sausage-shaped myelin thickenings ("tomaculous neuropathy") that are characteristic of this condition[39,40] (Fig. 15–3). There is no specific treatment; most patients learn to prevent compressions and have normal longevity.

X-LINKED BULBOSPINAL NEURONOPATHY

This disorder, sometimes referred to as Kennedy's syndrome,[37] was initially considered to be a "spinal muscular atrophy." It is now recognized to involve both the lower motor and primary sensory neurons.[31,53] The affected individuals usually present in the third or fourth decade with proximal limb weakness followed by weakness of the bulbar muscles giving rise to dysarthria and dysphagia. There is often a history of muscle cramps and of upper limb postural tremor dating back to adolescence or early adult life. Approximately 50% of male patients have gynecomastia and some develop diabetes mellitus. The disorder is slowly progressive.

Neurologic examination shows facial weakness with contraction and fascicu-

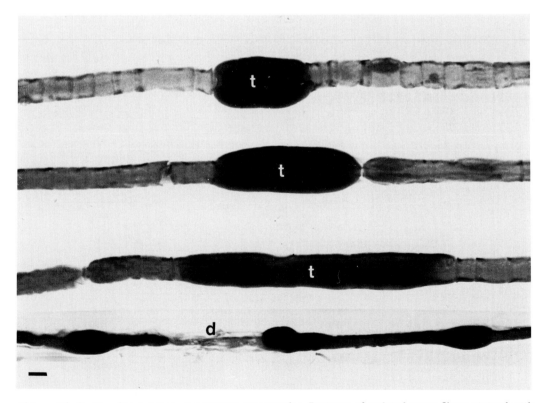

Figure 15–3. Hereditary pressure-sensitive neuropathy. Portions of isolated nerve fibers stained with osmium tetroxide showing focal sausage-shaped expansions (*tomaculi, t*). The lowermost fiber possesses a short region of demyelination (*d*).

lation, dysarthria, wasting, and fasciculation of the tongue. Limb weakness and wasting, often with fasciculation, that is mainly proximal in distribution, is also present. The tendon reflexes are depressed or absent. There may be mild sensory impairment distally in the lower extremities.

Electromyography of affected muscles shows evidence of chronic partial denervation. Motor nerve conduction velocity is normal or mildly reduced. Sensory nerve action potentials are depressed or absent. Sensory nerve biopsy demonstrates axonal loss. Autopsy has shown distally accentuated axonal degeneration and loss of spinal and bulbar motoneurons and dorsal root ganglion cells.[53]

The gene has been mapped to the proximal long arm of the X chromosome,[22] close to that for X-linked HMSN.[25] Although affected males have reduced fertility,[31] they are able to reproduce. All their daughters will be obligate carriers and have a 1:2 risk of transmitting the disorder to male offspring. Obligate female carriers show no manifestations of neurologic disease.

GIANT AXONAL NEUROPATHY

The major feature of this rare, autosomal recessive disorder is a slowly progressive peripheral neuropathy with an onset in childhood.[2] There is also variable CNS involvement with cerebellar ataxia, mental retardation, optic atrophy, nystagmus, and corticospinal tract dysfunction. Most of the reported children have had abnormal, tight curly hair. Motor nerve conduction is only moderately reduced and sensory nerve action potentials are depressed or absent. Nerve biopsies show axonal loss. Surviving axons display large focal swellings that contain accumulations of neurofilaments (Fig. 15–4). These

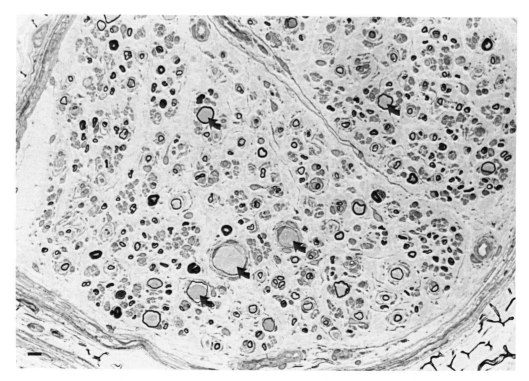

Figure 15–4. Hereditary giant axonal neuropathy. This light micrograph of a transverse section through a sural nerve biopsy specimen shows multiple enlarged "giant" axons (arrows), either lacking myelin sheaths or surrounded by attenuated sheaths. Bar = 10 μm.

accumulations are also evident in a variety of other cell types including Schwann cells, fibroblasts, and vascular endothelial cells, indicating a widespread disturbance of intermediate filaments,[10,47] providing an explanation for the curly hair. The disorder is steadily progressive, death usually occurring before adolescence.

HEREDITARY ATAXIAS

Dorsal root ganglion cell degeneration, resulting in loss of the larger myelinated nerve fibers in peripheral nerves, is a feature of a number of spinocerebellar degenerations including Friedreich's ataxia, and in disorders such as ataxia telangiectasia and xeroderma pigmentosum, in which a spinocerebellar degeneration is a conspicuous part of the clinical picture.[34] A minor loss of larger sensory axons may also occur in the late-onset hereditary ataxias.[58] In all of these disorders some loss of anterior horn cells may occur, resulting in mild distal amyotrophy in the limbs.

REFERENCES

1. Aguayo, AJ, Perkins, S, Bray, G, et al: Transplantation of nerves from patients with Charcot-Marie-Tooth (CMT) disease into immune-suppressed mice. J Neuropathol Exp Neurol 37:582, 1978.
2. Asbury, AK, Gale, MK, Cox, SC, et al: Giant axonal neuropathy: A unique case with segmental neurofilamentous masses. Acta Neuropathol (Berl) 20:237, 1972.
3. Behse, F and Buchthal, F: Peroneal muscular atrophy and related disorders. 2. Histological finding in sural nerves. Brain 100:67, 1977.
4. Berciano, J, Combames, O, Figols, J, et al: Hereditary motor and sensory neuropathy, type II. Clinicopathological study of a family. Brain 198:897, 1986.
5. Brimijoin, S, Capek, P, and Dyck, PJ: Axonal transport of dopamine-beta-hydroxylase by human sural nerves in vitro. Science 180:1295, 1973.
6. Buchthal, F and Behse, F: Peroneal muscular atrophy and related disorders. 1. Clinical manifestations are related to biopsy findings, nerve conduction and electromyography. Brain 100:41, 1977.
7. Cavanagh, NPC, Eames, RA, Galvin RJ, et al: Hereditary sensory neuropathy with spastic paraplegia. Brain 102:79, 1979.
8. Denny-Brown, D: Hereditary sensory radicular neuropathy. J Neurol Neurosurg Psychiatry 14:237, 1951.
9. Donaghy, M, Hakin, RN, Bamford, JM, et al: Hereditary neuropathy with neurotrophic keratitis. Description of an autosomal recessive disorder with a selective reduction of small myelinated nerve fibres and a discussion of the classification of the hereditary sensory neuropathies. Brain 110:563, 1987.
10. Donaghy, M, King, RH, Thomas, PK, et al: Abnormalities of the axonal cytoskeleton in giant axonal neuropathy. J Neurocytol 17:197, 1988.
11. Dyck, PJ, Lais, AC, Sparks, MF, et al: Nerve xenografts to apportion the role of axonal and Schwann cell in myelinated fiber absence in hereditary sensory neuropathy, type II. Neurology (Minneap) 29:1215, 1979.
12. Dyck, PJ and Ohta, M: Neuronal atrophy and degeneration predominantly affecting peripheral sensory neurons. In Dyck, PJ, Thomas, PK, and Lambert, EH (eds): Peripheral Neuropathy, Vol 2. WB Saunders, Philadelphia, 1975, p 791.
13. Dyck, PJ and Lambert, EH: Lower motor and primary sensory neuron disease with peroneal muscular atrophy. II. Neurologic, genetic and electrophysiologic findings in various neuronal degenerations. Arch Neurol (Chicago) 18:619, 1968.
14. Dyck, PJ and Lambert, EH: Lower motor and primary sensory neuron

disease with peroneal muscular atrophy. I. Neurologic, genetic and electrophysiologic findings in hereditary polyneuropathies. Arch Neurol (Chicago) 18:603, 1968.

15. Dyck, PJ, Mellinger, JF, Reagan, TJ, et al: Not 'indifference to pain' but varieties of hereditary sensory and autonomic neuropathy. Brain 106: 373, 1983.

16. Dyck, PJ, Low, PA, Bartleson, JD, et al: Prednisone responsive hereditary motor and sensory neuropathy. Mayo Clin Proc 57:239, 1982.

17. Dyck, PJ: Definition and basis of classification of hereditary neuropathy with neuronal atrophy and degeneration. In Dyck, PJ, Thomas, PK, and Lambert, EH (eds): Peripheral Neuropathy, Vol 2. WB Saunders, Philadelphia, 1975, p 755.

18. Dyck, PJ, Lais, AC and Low, PA: Nerve xenographs to assess cellular expression of the abnormality of myelination in inherited neuropathy and Friedreich's ataxia. Neurology (Minneapolis) 28:261, 1978.

19. Dyck, PJ: Inherited neuronal degeneration and atrophy affecting peripheral motor, sensory and autonomic neurons. In Dyck, PJ, Thomas, PK, Lambert, EH, and Bunge, R (eds): Peripheral Neuropathy, ed 2, Vol 2. WB Saunders, Philadelphia, 1984, p 1600.

20. Earl, CJ, Fullerton, PM, Wakefield, GS, et al: Hereditary neuropathy with liability to pressure palsies. Q J Med 33:481, 1964.

21. Edelman, NH, Cherniack, NS, Lahiri, S, et al: The effects of abnormal sympathetic nervous function upon the ventilatory capacity of hypoxia. J Clin Invest 41:1153, 1970.

22. Fischbeck, KH, Ionasesch, V, Ritter, AW, et al: Localization of the gene for X-linked spinal muscular atrophy. Neurology (Cleveland) 36:1595, 1986.

23. Fischbeck, KH, Rushidi, N, Pericak-Vance, M, et al: X-linked neuropathy: Gene localization with DNA probes. Ann Neurol 20:527, 1986.

24. Fryns, JP and Van Der Berghe, H: Sex-linked recessive inheritance in Charcot-Marie-Tooth disease with partial clinical manifestations in female carriers. Hum Genet 55:413, 1980.

25. Gal, A, Mucke, J, Theile, H, et al: X-linked dominant Charcot-Marie-Tooth disease: Suggested linkage with a cloned DNA sequence from the proximal Xq. Hum Genet 7:38, 1985.

26. Guzzetta, F, Ferrière, G and Lyon, G: Congenital hypomyelination polyneuropathy. Brain 105:395, 1982.

27. Hahn, AF, Brown, WF, Koopman, WJ, et al: X-linked dominant hereditary motor and sensory neuropathy. Brain, 113:1511, 1990.

28. Hardie, R, Harding, AE, Hirsch, N, et al: Diaphragmatic weakness in hereditary motor and sensory neuropathy. J Neurol Neurosurg Psychiatry 53:348, 1990.

29. Harding, AE and Thomas, PK: Clinical features of hereditary motor and sensory neuropathy types I and II. Brain 103:259, 1980.

30. Harding, AE and Thomas, PK: Genetic aspects of hereditary motor and sensory neuropathy (Types I and II). J Med Genet 17:329, 1980.

31. Harding, AE, Thomas, PK, Baraitser, M, et al: X-linked recessive bulbospinal neuronopathy: A report of 10 cases. J Neurol Neurosurg Psychiatry 45:1012, 1982.

32. Harding, AE and Thomas, PK: Peroneal muscular atrophy with pyramidal features. J Neurol Neurosurg Psychiatry 47:168, 1984.

33. Harding, AE and Thomas, PK: Autosomal recessive forms of hereditary motor and sensory neuropathy. J Neurol Neurosurg Psychiatry 43: 669, 1980.

34. Harding, AE: Hereditary Ataxias and Related Disorders. Churchill Livingstone, Edinburgh, 1984.

35. Jacobs, JM, Scaravilli, F, Duchen, LW, et al: A new neurological rat mutant 'mutilated foot.' J Anat 132:525, 1981.

36. Joy, JL and Oh, S: Tomaculous neuropathy presenting as acute recurrent

polyneuropathy. Ann Neurol 26:98, 1989.

37. Kennedy, WR, Alter, M, and Sung, JH: Progressive proximal spinal and bulbar muscular atrophy: A sex-linked trait. Neurology (Minneapolis) 18:671, 1968.

38. Kennedy, WR, Sung, JH, and Berry, JF: A case of congenital hypomyelination neuropathy. Clinical, morphological and chemical studies. Arch Neurol (Chicago) 34:337, 1977.

39. Madrid, R and Bradley, WG: The pathology of neuropathies with focal thickening of the myelin sheath (tomaculous neuropathy): Studies on the formation of the abnormal myelin sheath. J Neurol Sci 25:415, 1975.

40. Meier, C and Moll C: Hereditary neuropathy with liability to pressure palsies. Report of two families and review of the literature. J Neurol 228:73, 1982.

41. Nukuda, H, Pollock, M, and Haas, LF: The clinical spectrum and morphology of type II hereditary sensory neuropathy. Brain 105:647, 1982.

42. Ouvrier, RA, McLeod, JG, and Conchin, TE: The hypertrophic forms of hereditary motor and sensory neuropathy. A study of hypertrophic Charcot-Marie-Tooth disease (HMSN type I) and Dejerine-Sottas disease (HMSN type III) in childhood. Brain 110:121, 1987.

43. Ouvrier, RA, McLeod, JG, Morgan, GJ, et al: Hereditary motor and sensory neuropathy of neuronal type with onset in early childhood. J Neurol Sci 51:181, 1981.

44. Pearson, J, Axelrod, F, and Daniels, J: Current concepts of dysautonomia: Neuropathologic defects. Ann NY Acad Sci 228:288, 1974.

45. Pearson, J: Familial dysautonomia (a review). J Auton Nerve Syst 1:119, 1979.

46. Pearson, J and Pytel, B: Quantitative studies of ciliary and sphenopalatine ganglia in familial dysautonomia. J Neurol Sci 39:123, 1978.

47. Prineas, JW, Ouvrier, RA, Wright, RG, et al: Giant axonal neuropathy—

generalized disorder of cytoplasmic microfilament formation. J Neuropathol Exp Neurol 35:458, 1976.

48. Rosenberg, RN and Chutorian, A: Familial opticoacoustic nerve degeneration and polyneuropathy. Neurology (Minneapolis) 17:827, 1967.

49. Rozear, MP, Pericak-Vance, MA, Stajich, J, et al: X-linked Charcot-Marie-Tooth disease: Clinical evaluation and genetic linkage analysis. Neurology (New York) (Suppl)35:192, 1985.

50. Salisachs, P and Lapresle, J: Argyll-Robertson-like pupils in the neural type of Charcot-Marie-Tooth disease. Eur Neurol 16:172, 1977.

51. Satya-Murti, S, Cacace, AT, and Hanson, PA: Abnormal auditory evoked potentials in hereditary motor-sensory neuropathy. Ann Neurol 5:445, 1979.

52. Schoene, WC, Asbury, AK, Åström, KE, et al: Hereditary sensory neuropathy: A clinical and ultrastructural study. J Neurol Sci 11:463, 1970.

53. Sobue, G, Hashizume, Y, Mukai, E, et al: X-linked recessive bulbospinal neuronopathy: A clinicopathological study. Brain 122:209, 1989.

54. Stebbins, NB and Conneally, PM: Linkage of dominantly inherited Charcot-Marie-Tooth neuropathy to the Duffy locus in an Indiana family. Am J Hum Genet 34:195A, 1982.

55. Swanson, AG: Congenital insensitivity to pain with anhidrosis. Arch Neurol (Chicago) 8:299, 1963.

56. Thomas, PK and Calne, DB: Motor nerve conduction velocity in peroneal muscular atrophy: Evidence for genetic heterogeneity. J Neurol Neurosurg Psychiatry 37:68, 1974.

57. Thomas, PK, Calne, DB, and Steward, G: Hereditary motor and sensory polyneuropathy (peroneal muscular atrophy). Ann Hum Genet 38:111, 1974.

58. Vance, JM, Nicholson, GA, Yamaoka, LH, et al: Linkage of Charcot-Marie-Tooth neuropathy type Ia to chromosome 17. Exp Neurol 104:186, 1989.

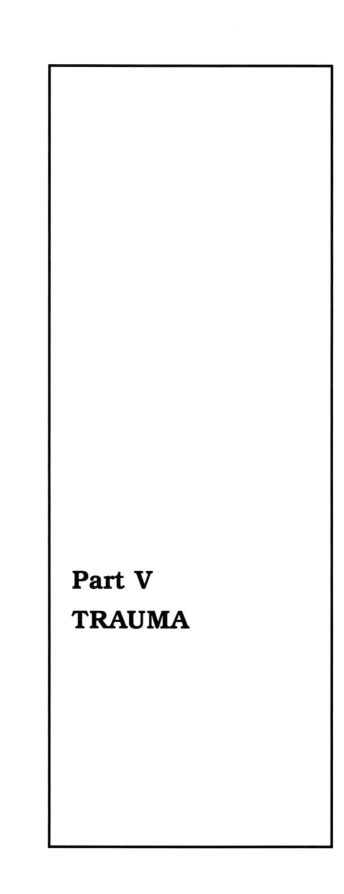

Part V
TRAUMA

Chapter 16

ACUTE AND CHRONIC FOCAL NERVE INJURY LESIONS

ACUTE NERVE INJURY
CHRONIC NERVE ENTRAPMENT
  AND COMPRESSION
CLINICAL FEATURES OF ACUTE
  AND CHRONIC INJURIES OF
  SPECIFIC NERVES

ACUTE NERVE INJURY

Definition and Classification

Acute nerve injury refers to damage resulting from sudden compression, transection, or stretching.

The nomenclature of acute nerve injury continues to be a source of controversy.[16,42,43,48,49,51] Experimental studies indicate five basic types: mild injury where axonal integrity is maintained but myelin is damaged;[35] more severe injury where axonal continuity is lost but the connective tissue framework of the nerve is maintained;[15] and the most severe injuries in which the connective tissue framework as well as the nerve fibers are damaged to varying degrees.[49]

Table 16–1 shows the anatomic classification of acute nerve injury.

Pathology and Pathogenesis

CLASS 1 (NEURAPRAXIA)

This lesion is a blockade of nerve conduction commonly associated with mild or moderate focal compression of nerve. There are two types: one is a mild, rapidly reversible type of block, most commonly resulting from transient ischemia of nerve (sitting with legs crossed); there is no resultant anatomic lesion. The other type results in more persistent conduction block and is assumed to result from paranodal or internodal demyelination (although direct pathologic confirmation in humans is not available). This type of conduction block has been reproduced experimentally by inflating a pneumatic tourniquet around limbs of cats[28] and baboons.[35] Myelin sheaths are displaced away from the site of compression,

Table 16–1 ANATOMIC CLASSIFICATION OF ACUTE NERVE INJURY

Suggested Nomenclature	Previous Nomenclature	Anatomic Lesion
Class 1	Neurapraxia	
	Transient	Conduction block due to ischemia
	Delayed reversible	Demyelination
Class 2	Axonotmesis	Axonal interruption
Class 3	Neurotmesis	Axonal and endoneurial interruption
Class 4	Partial	Perineurium and epineurium disrupted, nerve continuity retained
Class 5	Complete	Complete transection

causing overriding and damage of adjacent myelin segments, which results in localized demyelination and conduction block (see Fig. 2–8). Large-diameter fibers are affected more than smaller. Recovery is a result of remyelination and begins on average by 6 weeks post injury; usually it is complete.

CLASS 2 (AXONOTMESIS)

This lesion is associated with closed-crush and percussion injuries. Axonal interruption occurs, but the Schwann-cell basal lamina around each fiber remains intact, as does the endoneurial connective tissue[52] (see Fig. 2–9). Although wallerian degeneration occurs distal to the lesion site, regeneration is usually effective. Nerve regeneration proceeds at a rate of about 1 to 2 mm per day. Neurologic recovery occurs in serial order with more proximal muscles reinnervated before distal muscles. All functions eventually are restored properly since regenerating axons are guided back to their former terminations at the periphery by the continuous columns of Schwann cells that develop distally during the degenerative process.[17] Class 2 nerve injury is readily produced in experimental animals by crushing a surgically exposed peripheral nerve with smooth-tipped forceps.

CLASSES 3, 4, AND 5 (NEUROTMESIS)

Class 3 injuries disrupt axons and endoneurium but preserve fascicular orientation; class 4 injuries include disruption of perineurium; class 5 is total loss of continuity. Severe nerve injuries of this type result from stab wounds, the close passage of a high-velocity bullet, or nerve stretching sufficient to rupture the connective tissue components of the nerve (see Fig. 2–10). As in class 2 injuries, nerve fibers undergo degeneration distal to the injury, and regenerating sprouts grow from the proximal ends.[26,29–32] Some axons may traverse the lesion site and successfully grow along distal Schwann-cell columns, but

misrouting of axons is frequent and inappropriate end-organ connections commonly are established. Many regenerating fibers will grow in random fashion at the injury site because their basal laminae and connective tissue framework have been disrupted and collagen production is profuse. Continued growth and sprouting of these aberrantly regenerating axons results in the formation of a neuroma at the site of injury.[44] Neuromas may be intrafascicular if the perineurium has not been breached, lateral, or, in the case of total neurotmesis with wide separation of the nerve stumps, a bulbous terminal neuroma may develop. Such structures display a characteristic resilient texture and must be surgically removed if reunion of the cut ends is to be attempted. Surgical repair of individual fascicles is now preferred by some surgeons over simple anastomosis of the entire nerve, although the issue remains controversial.[22]

Clinical Features of Acute Compressive Lesions

Focal nerve compression commonly follows habitual adoption of certain postures (leaning on elbows), prolonged abnormal posture as may occur during stupor from drug overdose, moderate constrictive lesions (tourniquet paralysis), or sudden blunt trauma.[16] It is widely held that subclinically diseased nerves are more susceptible to the effects of pressure; compression syndromes are allegedly more common in individuals with uremia, diabetes, malnutrition, and alcoholism.[6,15,37] Genetic factors may predispose to pressure neuropathy; for example, there is a rare hereditary form of multifocal neuropathy that becomes symptomatic when individuals have minor trauma (see Chapter 15).[2,7]

Mild or moderate acute compression injuries, such as sleeping with an arm draped over a bed rail, usually produce class 1 lesions. Severe compression (closed-crush) injury, such as being struck by a moving object, frequently results in a class 2 lesion. Mixed lesions combining class 1 and 2 injuries are common. Both closed-crush injuries and severe prolonged compression may be associated with disruption of subcutaneous tissue, causing connective tissue proliferation that impedes nerve regeneration.[13]

CLASS 1 INJURY (MILD COMPRESSION)

Strength usually is impaired severely and tendon reflexes are absent below the injured site in a class 1 lesion. Sensory loss is less obvious and confined to large-fiber modalities. Sympathetic function is spared. This constellation of findings probably reflects the low vulnerability of unmyelinated sympathetic and small myelinated sensory fibers, and the dependence of motor and tendon-reflex function upon larger myelinated axons that undergo focal demyelination.[13]

Careful nerve conduction studies often demonstrate conduction block when stimuli are applied proximal to the lesion. Since axonal continuity is retained, motor and sensory potentials remain normal with stimulation distal to the lesion. Nerve conduction studies are not helpful in distinguishing class 1 (demyelinative) injuries from class 2 (axon interruption) unless performed after an interval of 5 days, since distal portions of transected axons may continue to conduct for 4 days.[24]

Recovery from class 1 injuries usually occurs spontaneously within 3 months.[48] Recovery is usually simultaneous throughout the distribution of the affected nerve, although differing degrees of compression among fascicles may allow recovery in some muscles before others. Denervation atrophy of muscle does not develop because axonal continuity is maintained.

There is no specific treatment of class 1 lesions other than appropriate orthotic devices (e.g., wrist splint for radial nerve palsy) and behavioral advice aimed at preventing recurrences.

Case History: Class 1 Compression Injury

An 18-year-old narcotics addict became unconscious following an accidental overdose. For 6 hours he lay on his right side with his elbow pressed against a table leg. When aroused, he discovered that his right hand was weak and numb, and he was admitted to the hospital. Examination revealed complete paralysis of all muscles of ulnar innervation, and diminished touch sensation over the hand in an ulnar distribution, with splitting of the ring finger. Pain and thermal sensation were normal. A simple hand-wrist splint was applied to prevent claw-hand deformity. Three days later, sensory recovery began. Motor nerve conduction studies on the 6th day revealed normal conduction in the ulnar nerve below the elbow and total conduction block across the elbow. On the 10th day, motor recovery commenced in the interosseous muscles and flexor carpi ulnaris, and, 2 days later, in the flexor digitorum profundus and adductor pollicis; 2 months following the injury motor and sensory function was almost normal.

Comment. Preservation of pain and thermal sensation supported the clinical impression that the ulnar nerve was in continuity and unmyelinated axons were spared. Normal motor nerve conduction below the lesion of total blockade after 6 days suggested that most axons were not affected and the prognosis for full recovery was excellent. Recovery began in muscles innervated at multiple levels of the ulnar nerve almost simultaneously—typical for this class of injury.

CLASS 2 INJURY (CLOSED-CRUSH INJURY)

Classically, class 2 injuries are associated with closed-crush trauma, but may also follow stretch/traction. (Stretch/traction injuries are discussed later in this section.) Variable loss of sensory, motor, and sympathetic function occurs from closed-crush interruption of unmyelinated and myelinated axons. Muscle atrophy is apparent after 1 month and may become extreme if all motor axons degenerate. Areflexia is the rule. It is unusual in closed-crush injuries for unmyelinated axons to be severed. Nevertheless, anesthesia and total sympathetic dysfunction may appear distal to the site of injury.[13]

Electrodiagnostic studies, performed at least 5 days after injury, can reliably differentiate between class 1 and class 2 lesions. Due to wallerian degeneration, electrical stimulation of motor and sensory nerves, distal to the lesion, will either fail to evoke a response or elicit one of low amplitude. The reduction in the size of the sensory potential, relative to the contralateral side, is directly proportional to the degree of sensory axon loss. Since motor conduction studies record from the muscle, not the motor nerve, reduction in the amplitude of the compound muscle action potential only parallels the loss of motor axons. Electromyography performed 10 to 14 days after injury may demonstrate spontaneous activity indicative of active denervation (fibrillation, positive sharp waves). The presence of spontaneous activity is an indication that axonal degeneration has occurred and should not be used to estimate the degree of axon loss.

Although recovery is slow, the prognosis in closed-crush injuries is usually excellent since regenerating axons grow along their original Schwann-cell tubes, ensuring that the pattern of motor and sensory restoration is appropriate. Recovery varies from a few months in a distal lesion, to over a year with proximal lesions, since regenerating axons elongate slowly (1 to 2 mm each day). Partial injuries may recover faster due to intact motor axons reinnervating the orphaned muscle fibers via collateral sprouts. With complete lesions, the pattern of recovery is proximal to distal, reflecting the course of axonal regeneration. Recovery occurs first in muscles closest to the site of injury. Electromyography (EMG) will detect recovery earlier than clinical observation. Small polyphasic motor units (nascent units), representing reinnervation of a few muscle fibers in the recovering motor unit, can be recruited under voluntary control. In partial lesions, collateral sprouting results in increased motor unit territory and muscle fiber density. Many recovering nerve terminals are immature and poorly myelinated. Chronic motor unit reinnervation is reflected by needle EMG recordings of large amplitude, long duration, polyphasic motor units recorded under voluntary contraction. During regeneration, light percussion over advancing tips of regenerating axons may elicit a tingling sensation (Tinel's sign). With time, this sign can be elicited in more distal sites, indicating continuing advance of regenerating axons. Reinnervation of denervated skin frequently is accompanied by distorted sensations; a slight stimulus may produce unpleasant sensations. This state usually abates when reinnervation is complete.[48]

Treatment of class 2 lesions is similar to that previously described for class 1. Simple orthotic devices, physical therapy, and reassurance are usually all that is required. Surgical exploration and neurolysis (clearing of excess connective tissue to free nerve fascicles) are not indicated, except in cases where there has been severe trauma to adjacent subcutaneous tissues. Therapeutic electrical stimulation of denervated muscles is of unproven benefit. Stretch/traction injuries are more often encountered with closed injury to the brachial plexus or roots.[3,4] They are seen following birth injuries, motorcyle accidents, heavy blows with downward displacement of the shoulder, and traction during falls, such as when an individual hangs with an outstretched arm (straphanger's palsy). Roots may be torn from the spinal cord with especially severe trauma (nerve root avulsion), such as falling from a great height or from a moving vehicle. It is less well appreciated that stretch/traction injuries of nerve may also be related to bony dislocations. The site of nerve injury is sometimes remote from the bony injury (for example, common peroneal nerve at neck of fibula with displaced fractures at ankle). In addition, nerve injury from gunshot wounds is usually in part related to nerve displacement and stretch/traction because of sudden expansion of tissues in the path of the bullet.[13,38]

Stretch/traction trauma may also result in class 3–5 injury. Damage to axons and connective tissue is often diffuse, and may be initially accompanied by hemorrhage, and eventually by intraneural fibrosis. Frequently there is a mixture of class 1 and class 2 injuries. Severe head or shoulder displacement may cause one or more cervical nerve roots to be avulsed from their attachment to the spinal cord.

Recovery from stretch/traction injuries is often limited, owing to the severe axonal and connective tissue disruption and the diffuse nature of the lesion. Occasionally there is initial mild improvement, representing recovery from the component of class 1 injury. Further recovery is usually delayed and incomplete. When the spinal roots are avulsed from the cord, efficacious regeneration is impossible. Spontaneous recovery is generally satisfactory in some birth injury cases. In the proximal (Erb's) form of brachial plexus damage from birth injury, weakness of abduction at the shoulder and flexion at the elbow often persists, although there may be little residual sensory loss. Full recovery takes place in about one third of these cases.

Case History: Traction Injury of Brachial Plexus

A 41-year-old policeman fell from his motorcycle at high speed, striking his head and the point of the right shoulder. He was unconscious for 2 hours and on awakening in the emergency room, was unable to move his right arm. Examination revealed paralysis of all muscles of the right arm and shoulder and anesthesia of the entire arm below the axilla. Horner's syndrome was not present. X-rays of the chest and shoulder revealed no fractures, and he was discharged after 1 week in the hospital. Evaluations 6 and 12 months later showed paralysis and atrophy of right upper extremity and shoulder muscles, descent of the head of the humerus, and total sensory loss below the axillary line. Electrodiagnostic studies demonstrated absence of all motor and sensory potentials of the right arm and complete denervation of limb and scapular muscles. The patient declined myelography.

Comment. This case represents a severe stretch/traction injury to the entire brachial plexus, with an unusual degree of proximal damage as evidenced by scapular weakness. There was no Horner's syndrome which, if present, would suggest T_1 root avulsion. A myelogram might have been useful to determine if root avulsion had occurred. There was little to be gained by surgical exploration of the plexus in the absence of fracture, bony dislocation, or massive subcutaneous tissue damage. Failure of proximal muscles to recover after an interval of 1 year implies considerable endoneurial disruption, in this case accompanying nerve root avulsion, and a poor prognosis.

CLASS 3 INJURY

Injuries that penetrate nerves are commonly associated with high-velocity missiles, stab wounds, broken glass, and motor vehicle and industrial accidents. Rarely, nerves may be severely lacerated by fractures.

Penetrating wounds generally produce partial or complete transection (classes 3, 4, and 5) and are characterized by considerable disruption of connective tissue.[48] Stretch/traction injury to nerve often occurs together with penetrating wounds, additionally disrupting neural architecture over considerable lengths (see previous section). High-velocity-missile wounds may cause all degrees of nerve damage.[38,43,48]

Injuries associated with partial or complete nerve transection produce clearly defined areas of total motor, sensory, and autonomic dysfunction. Muscle atrophy proceeds rapidly for about 3 months, at which time about 80 percent of bulk is lost. Muscle can still recover function if reinnervated up to 3 years after nerve transection. Thereafter, there is progressive loss of muscle fibers, and the prognosis for recovery becomes increasingly less satisfactory.

Shrinkage of the autonomous sensory zone of a nerve, its area of superficial cutaneous sensation, follows within days of complete transection. Presumably, this phenomenon reflects overlap in normal innervation. Clinically, it is important to recognize that this does not reflect regeneration. Pain of a superficial "pins-and-needles" type is commonly present in the autonomous zone of a transected nerve and abates within several months. Persistent deep pain may indicate neuroma formation. Both neuromas in continuity with a partial injury, or stump neuromas may be associated with spontaneous severe pain. Neuroma pain may also be generated by cold or pressure.

With partial or complete lesions of median, sciatic, and tibial nerves, or the lower part of the brachial plexus, causalgic pain may be a troublesome consequence.[43,48] Although well described after missile wounds, causalgia may appear after a variety of nerve injuries, including minimal trauma. Causalgic pain may begin immediately or after an interval of up to 60 days or longer. The pain of causalgia is severe, intractable, and has a burning or smarting quality. Severe paroxysms of pain provoked by touching or jarring the hand or foot may be superimposed, or pain may result from emotional causes. Vasomotor and sudomotor changes may be associated: the skin usually becomes dry and scaly, but excessive sweating may also be a feature. The patient adopts a protective attitude toward the limb, so that fixation of the joints of the fingers and wrist may develop, together with atrophic changes in skin and subcutaneous tissue. About 80% of cases are relieved by sympathectomy. Untreated, the pain gradually subsides over months or years. The pathogenesis of causalgia is unclear, but may be related to hyperactivity of autonomic sympathetic axons. It is widely held that the painful sensations result from abnormal activity generated both in the peripheral nerve and within the central nervous system.

Dramatic relief often follows sympathetic block, although it is not currently believed that there is a primary disorder of the sympathetic nervous system. Regional intravenous guanethidine infusion can be used to block sympathetic activity chemically. Sympathetic block may also be achieved surgically. Transcutaneous electrical stimulation has been found effective in some cases, especially if the stimulator can be placed in close proximity to the affected nerve. Stimulation should be employed for hours at a time and continued over many months before it is deemed ineffective. Inadequate results often occur because of insufficient stimulation time. Other neurostimulating techniques are occasionally helpful, including dorsal column stimulators, and deep brain stimulators. Many drugs have been tried, including tricyclic antidepressants, anticonvulsants, neuroleptics, steroids, prazosin, and phenoxybenzamine. None is universally effective. Narcotics may provide some relief but should be used with caution.

Transection of autonomic fibers causes diminished sweating and impaired vasomotor responses. Diminished sweating usually corresponds to the zone of autonomous cutaneous sensory loss; it may be helpful to perform repeated sweat tests in detecting signs of recovery from nerve injury. Vasodilatation develops and persists for several weeks following injury. Moderate vasoconstriction then appears, resulting in a cold, pale extremity. Numerous "trophic" disturbances commonly affect denervated extremities, such as thickened fingertips and shiny, atrophic skin.[48] Minor trauma, acting in conjunction with vasomotor malfunction in anesthetic fingers and toes, probably accounts for most of these phenomena.

The prognosis for recovery, in class 3–5 injuries, even with optimal surgical care, is usually only fair because axons often fail to reach their original end-organs. The treatment of penetrating nerve injuries is by reconstructive surgery and is best carried out by expert teams. Surgical repair is indicated in the following situations: to improve motor and sensory function in patients without spontaneous recovery, to ameliorate neuropathic pain, and to halt progressive axonal loss due to pseudoaneurysm, arteriovenous malformation, or hemorrhage in a closed space.[22]

The decision to operate on a functionless injured nerve is facilitated by knowledge that the nerve has been severed. The initial clinical examination cannot disclose whether the loss of function is due to nerve severance or whether the nerve remains in total or partial continuity. This information is crucial in deciding the course of management. Loss of function following injury from sharp objects usually indicates complete or partial nerve severance. However, it is remarkable how often a nerve remains in continuity following lacerating missile wounds, probably reflecting the tensile strength of its connective tissue.

DECISIONS IN MANAGING TRAUMATIC NERVE INJURIES

The following general rules are helpful in managing traumatic nerve injuries.[22,48,49]

1. In any instance where the nerve is known to be transected, surgical intervention (at the appropriate time) is mandatory. Nerve lacerations due to penetrating injuries, such as those from glass, knives, and metal, are good candidates for acute repair. This is especially true if the site of nerve injury is very proximal, such as in infraclavicular brachial plexus injuries and high sciatic nerve injuries. In these cases, the delay resulting from the time needed for the regenerating nerve to transverse the distance between site of injury and the most proximal muscles results in retraction of scar and stump, thereby making secondary repair more difficult. In contrast to clean transections just described, blunt transections, as from complex fractures or injuries from chain saws, are best repaired after several weeks. This delay

allows the surgeon to better observe the degree of resection needed.

Surgical repair of transected nerves includes end–end suture in cases where opposition of nerve stumps can be achieved without undue stretch. In cases where insufficient nerve is available for end–end suturing, a neural graft can be inserted, commonly from the sural and antebrachial cutaneous nerves. The shorter the length of graft needed, the better the results. If possible, groups of fascicles, proximal and distally, should be approximated by the intervening graft segment. There is considerable interest in developing non-neural conduits (nerve guides) to enhance nerve repair; presently, most of these studies are confined to experimental animals.[41]

2. Lesions in continuity from blunt trauma, as described above, usually recover spontaneously and most surgeons will wait about 2 to 3 months before exploration. Nerve lesions in which an extended segment of nerve is likely to be injured, such as with contusion/stretch or gunshot injuries, are often given longer observation periods (e.g., 4 to 5 months) to allow adequate time for regeneration, because of less satisfactory results with nerve grafts of long length. Once it is apparent from clinical and electrophysiologic evidence that adequate reinnervation is not occurring, surgical exploration should be attempted. The surgeon must determine whether to perform an external neurolysis, in which the nerve is dissected free of its surrounding tissues, including scar, or an internal neurolysis with resection and repair of nerve fascicles. The use of intraoperative electrophysiologic recordings has proven to be extremely useful to some peripheral nerve surgeons in deciding which fascicles are not conducting and therefore should be resected. Stimulation is applied proximal to the injured site to identify the most distal site (still proximal to the injured nerve segment) that is capable of conducting an impulse. Resection of the injured nerve segment is carried distally from the most distal

site where the action potential was last recorded to an area of good neural architecture. Repair is performed by either end–end anastomosis, if sufficient length is available, or via interposed nerve grafts.

As with all peripheral nerve injuries, good postoperative physical and occupational therapy is needed to ensure continued joint mobility and a return of functional ability.

CHRONIC NERVE ENTRAPMENT AND COMPRESSION

Definition and Etiology

Three types of injury are commonly associated with chronic trauma to peripheral nerve:

1. Compression in a fibro-osseous tunnel (carpal, cubital, tarsal tunnels)
2. Angulation and stretch (over arthritic joints, anomalous fibrous bands, under ligaments)
3. Recurrent external compression (occupational trauma to hands and feet)

Pathology, Pathogenesis, and Animal Models

There is considerable variation among the three types of injury listed above. The histopathology of the tunnel syndromes has been extensively studied in humans and in experimental animals, but little is known about types 2 and 3.

In the human carpal tunnel syndrome, the nerve appears enlarged immediately above the site and constricted in the compressed zone. Demyelination is prominent at the site of compression.[33,34] The morphologic features of demyelination in chronic entrapment differ sharply from the picture of invaginated nodes of Ranvier described in acute compression (see Fig. 2–8). Studies of the naturally occurring carpal tunnel syndrome in the guinea

pig reveal that myelin segments are deformed in a tadpole-like manner—bulbous at one end and tapered at the other.[36] The bulbous ends point away from the site of compression, indicating the direction of mechanical force. In time, myelin slips away and bare axons become remyelinated. Eventually, in older animals, there is axonal destruction and wallerian degeneration.[12] Unmyelinated fibers resist degeneration until late. Examination of human median nerves at the wrist, ulnar nerves at the elbow, and lateral femoral cutaneous nerves under the inguinal ligament has confirmed the observations in the guinea pig. It appears that mechanical factors are dominant in the demyelination of chronic compression. The role of ischemia is less certain. It may be responsible for the acute attacks of pain in the carpal tunnel syndrome, and short periods of ischemia can reversibly block conduction in damaged fibers.[11] It is possible that ischemia has a role in the pathogenesis of longstanding chronic entrapment.

Experimental chronic nerve constrictions have been less adequately studied than acute constrictions. They have been examined by applying snug ligatures around the nerves in immature animals, leading to constriction developing pari passu with growth,[16] and also by the application of constricting devices around nerve in adult animals.[45]

CLINICAL FEATURES OF ACUTE AND CHRONIC INJURIES OF SPECIFIC NERVES[5,47]

Brachial Plexus Injuries

SOURCE OF INJURY[25]

Brachial plexus injuries may be acute or gradual and affect the plexus diffusely or in a restricted manner. (Figure 16–1 illustrates the anatomy of the brachial plexus.) Rarely is the severity of

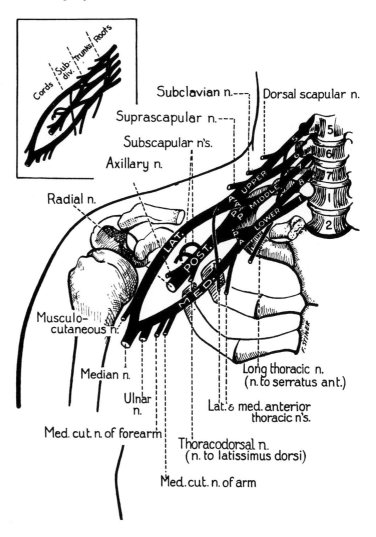

Figure 16–1. Diagram of the brachial plexus. The components of the plexus are depicted *(upper left detail).* (From Haymaker, W and Woodhall, B: Peripheral Nerve Injuries. WB Saunders, Philadelphia, 1953, with permission.)

damage similar in all parts of the plexus, and areas of segmental demyelination are often admixed with regions of significant axonal loss.

Motor vehicle accidents, especially involving motorcycles, cause most plexus injuries in North America. In various ways, the plexus is protected from stretch injury by the plexiform arrangement of its nerve bundles, the strength of the nerve funiculi, and the lack of fixed attachments. Unfortunately, the plexus's relative resistance to traction forces allows the stress to be directly transmitted to cervical roots, thereby increasing the likelihood of single- or

multiple-root avulsion.[48] Severe traction injuries commonly result in combined plexus and root damage, diminishing the likelihood of good functional recovery. The majority of avulsions involve the ventral roots, resulting in greater motor than sensory impairment.

Traction injuries also result from improper positioning during anesthesia, excessive sternal retraction during intracardiac operations, prolonged compression during narcotic-induced coma, and excessive supraclavicular pressure from the weight of heavy knapsacks (so-called rucksack paralysis).[25] Acute closed injuries and open injuries secondary to missiles and stab-bings usually affect the supraclavicular portion of the plexus, owing to its ex-posed position and thin protective fas-cial covering. Lung carcinoma often ini-tially affects the lower plexus, owing to the proximity of the pulmonary apex to the lower plexus. Irradiation more com-monly has been reported to affect the upper plexus, which is unprotected by the clavicle, but may also affect the plexus diffusely.

Infraclavicular injuries due to pene-trating wounds are less frequent than supraclavicular injuries, both because the plexus is less exposed and because fewer patients survive, owing to concur-rent injury to the lungs and great ves-sels at the root of the neck.[25] Fractures and dislocations of the shoulder may cause traction injuries to the lower plexus, which are usually milder than similar trauma to the supraclavicular structures and less likely to be compli-cated by root avulsions.[25,48] The axillary nerve is frequently severely affected by shoulder fractures and dislocations. In-fraclavicular injuries often have poor spontaneous recovery because of the long distances regenerating nerves must traverse before reinnervating dis-tal arm and hand muscles.

CLINICAL ASSESSMENT OF BRACHIAL PLEXUS LESIONS

A critical feature in managing bra-chial plexus injuries is determining whether coexistent nerve root avulsions are present. Widespread avulsions al-ways indicate poor motor recovery. Clinical evidence of root involvement in-cludes weakness of muscles whose in-nervation is directly from the root itself such as the serratus anterior and rhomboid muscles, Horner's syndrome indicating T_1 root involvement, corti-cospinal tract signs suggesting cord damage, and paroxysmal attacks of in-tense pain. Electrophysiologic investi-gations, myelography, and magnetic resonance imaging (MRI) are often use-ful for added verification.

The most frequent presentation of brachial plexus dysfunction is weak-ness of periscapular and proximal arm muscles, since most traumatic and in-flammatory lesions predominantly af-fect the upper plexus. Although the hand can function normally, it cannot be positioned for effective use. As most brachial plexus injuries result in axonal loss, there is usually significant wasting of affected muscles. The presence of dis-tal paresthesias, elicited by supraclavic-ular percussion, is an indication that the nerve roots have retained continu-ity with the spinal cord. Progressive distal advancement of Tinel's sign sug-gests that axonal regeneration is pro-ceeding past the site of injury. Unless prevented, severe limitations in shoul-der and elbow joint mobility often result from disuse, markedly interfering with functional recovery. As many brachial plexopathies are secondary to violent accidents, multiple distal peripheral nerve injuries may coexist, both confus-ing the initial assessment and compli-cating recovery.

THERAPY AND REHABILITATION OF BRACHIAL PLEXUS INJURIES

A prolonged period of time may elapse before the patient with a traumatic plexus lesion is capable of intensive re-habilitative efforts. Fractures, joint and vascular injuries, infections, and soft tissue damage often delay rehabilitative attention to the plexus injury. During this time, joint range-of-motion (ROM) exercises should be performed to pre-vent contractures that may significantly affect recovery.

Satisfactory spontaneous recovery oc-curs in some brachial plexus lesions. Infraclavicular plexus lesions, second-ary to fractures and dislocations around the shoulder, recover well, re-flecting the high percentage of class 1 and 2 lesions.[25,48] Onset of spontaneous recovery after mild injury ranges from 3 to 9 months; intrinsic hand muscles re-cover last. Even supraclavicular lesions may improve, as the distance from site of injury to proximal muscles is not ex-cessive.

Many peripheral nerve surgeons elect to explore traumatic brachial plexus injuries after about 3 to 6 months of observation. Recent advances in microsurgical techniques and use of intraoperative electrophysiologic studies, as described above, have improved the results expected with brachial plexus repair. Early exploration has been occasionally advocated for infraclavicular injuries, as the delay while regeneration reaches distal arm muscles has been deemed too long to allow favorable recovery. In addition, because it is rare that the hand muscles recover adequately after severe lower plexus injury, most surgical efforts are aimed at reinnervating the proximal arm and shoulder muscles, utilizing neuroma resection, end–end anastomosis if possible, and nerve grafts when needed.[22] Tendon and muscle transpositions are occasionally employed, with varying success, if nerve repair is unsuccessful.

Splints are useful both to protect the limb from a fixed deformity and to encourage function. A flail insensitive arm is susceptible to a variety of traumatic insults. Inappropriate positioning, resulting in pressure on ulnar and radial nerves, may amplify an already sizable functional deficit. It is especially important that the applied splint encourage limb function. Patients generally become adept at being one armed. In as little as 6 months, patients may become so accustomed to using only one limb that, even with reinnervation, the other arm never regains adequate function. The patient should be encouraged to use the affected arm. A simple sling, although adequate for protection, does little to encourage function and predisposes the shoulder and elbow to fixed contractures.

Upper plexus lesions predominantly impair arm abduction and elbow flexion, making it difficult to position the hand in a functional position. Splints for upper plexus lesions should include a forearm and upper arm support linked by an elbow ratchet lock; this allows the elbow to be placed in any one of five to six positions. The elbow locks automatically and can be released by active forearm extension or hand controls. Significant wrist extensor weakness may necessitate a hinged support that keeps the wrist extended but allows active wrist flexion.

Since a completely paralyzed and anesthetic limb may continue to suffer traumatic injury, many have chosen amputation rather than carry a useless appendage. Some will benefit from a flail arm splint. This device fits over the paralyzed limb, supporting the upper arm, forearm, and wrist. The wrist is stabilized in the neutral position and the elbow fixed via the previously described elbow lock. A pelvic band and rod with a posterior shoulder cap or a shoulder piece provides further support. Patients with intact shoulder adductors can learn to swing the arm across their body using pectoral muscles, allowing the spring load at the shoulder to return it to its resting place. The forearm piece and support platform may be used alone for injuries restricted to the posterior and medial cords or when the elbow flexors recover strength. The splint can be used until adequate functional recovery has returned.

Loss of shoulder abduction need not be permanent, despite axillary nerve injury. If the external rotators of the humerus are intact, a compensatory maneuver may be learned. The arm is initially externally rotated via the infraspinatus muscle. Abduction is begun by the supraspinatus with subsequent elevation by the long heads of the triceps and biceps. At about 70°, the clavicular fibers of the pectoralis major act as the principal abductor. Contraction of the serratus anterior continues elevation by bringing the arm forward and upward through its pull on the scapula. Most patients can learn the motion in 4 to 6 weeks and actually achieve good strength. The maneuver may be taught as soon as paralysis is evident and does not affect proper functioning of the prime movers once they recover. Should the infraspinatus be weakened, impairing external rotation of the humerus, this movement cannot be performed. An alternative movement learned by some is the *arm fling*, in which the limb is initially adducted across the middle of the body and then sharply swung back with a quarter twist of the trunk. The arm is elevated above the head in a pendular fashion. Although elevation cannot be maintained for long, the technique is useful for reaching high objects.

Brachial plexus lesions may be complicated by significant edema of the affected limb. Underlying causes include loss of sympathetic tone after root avulsion, pooling of venous blood owing to the loss of the pumping action usually supplied by the paralyzed muscles, and the pull of gravity on the frequently dependent arm. To counteract fluid accumulation, the arm should be elevated above heart level whenever possible; at night it should be supported on a stack of pillows. In refractory cases, a "Statue of Liberty splint" helps support the elevated arm. Elastic wrappings may help prevent fluid accumulation, but are ineffective once significant edema has occurred. Frequent massage of the affected limbs may encourage drainage.

Root avulsions frequently cause intractable pain. A background of burning or crushing discomfort may be interrupted by paroxysms of sharp incapacitating pain. With time, the intensity of the pain may diminish. Many therapeutic modalities have been advocated to relieve root avulsion pain. Some patients are afforded relief by prolonged transcutaneous electrical nerve stimulation (TENS) application. Relief with analgesics or sympathectomy is poor. Surgical lesions placed in the dorsal root-entry zone provide satisfactory relief in a high percentage of patients.

As many cannot return to their previous occupations, job counseling and retraining are critical. Traumatic plexus injuries are common in the young, and the emotional consequences may be devastating. Psychological assistance may be needed for coping with loss of body image. While awaiting recovery, motor re-education is imperative, since return of voluntary activity may be delayed for months.

Thoracic Outlet Syndrome

The branches of the eighth cervical and first thoracic roots to the brachial plexus may be damaged by angulation over an abnormal rib or, more usually, a fibrous band arising from the seventh cervical vertebra and attached to the first rib (neurogenic thoracic outlet syndrome). The subclavian artery may be affected by fibrous bands, giving rise to aneurysmal dilatation and vascular symptoms such as Raynaud's syndrome and embolic events (vascular thoracic outlet syndrome). The simultaneous occurrence of both neural and vascular phenomena is rare.

True neurogenic thoracic outlet syndrome is rare and has a characteristic constellation of signs and symptoms. Many patients with vague positional arm pain are referred to electromyography (EMG) laboratories to rule out thoracic outlet compression; few are found to have it.[53]

In neurogenic thoracic outlet, damage to the lower part of the brachial plexus causes weakness and wasting of the small hand muscles, and of the medial forearm wrist and finger flexors.[14] Thenar muscles tend to be more involved than hypothenar muscles, and occasionally there is selective wasting of thenar muscles mimicking carpal tunnel syndrome. Numbness, pain, and paresthesias occur along the inner border of the forearm and hand, extending into the medial two fingers. Carrying heavy articles in the hand on the affected side tends to provoke pain. Horner's syndrome may occur. Bedside vascular tests involving arm hyperabduction are invariably normal. Radiographs demonstrate either an elongated transverse process extending from the seventh cervical vertebra or a rudimentary cervical rib. Compression of the C_8, T_1 roots or the lower trunk of the brachial plexus in such cases is secondary to fibrous bands extending from these bony structures to the first rib. The anomalous fibrous bands are not evident on plain radiographs.

Some patients with cervical ribs and anomalous fibrous bands display vascular symptoms in the hand and arm. Symptoms include intermittent blanching of fingers, coldness, aching in the arm, and cyanosis with arm dependency. Arteriography demonstrates narrowing of the subclavian artery with a poststenotic dilatation. Symptoms are presumed to arise from local thrombus development and subsequent embolic events or vascular insufficiency. Adson's maneuver, in which the arm is hyperabducted and the head turned to the opposite direction, may demonstrate obliteration of the radial pulse. Its presence is not pathognomonic for thoracic outlet compression. A supraclavicular bruit is occasionally present.

Some thin women with long necks and low-set shoulders sometimes complain of posturally precipitated arm pain, sensory symptoms in the hand, or a feeling of weakness with objective findings (droopy shoulder syndrome).[50] Although vascular studies may occasionally demonstrate abnormalities of the axillary or subclavian arteries, in general, both radiographic and electrophysiologic studies are normal. The pathogenesis of these complaints is unclear.[50] The efficacy of surgical therapy is doubtful.

ELECTROPHYSIOLOGIC STUDIES

Electrophysiologic findings suggest dysfunction of the lower trunk of the brachial plexus. Carpal tunnel syndrome and ulnar nerve entrapment need to be excluded. Characteristic nerve conduction findings include absent or low-amplitude ulnar sensory potentials, normal median sensory potentials, reduced compound motor action potential amplitudes from thenar and hypothenar muscles, and prolonged ulnar F response latencies. Needle EMG demonstrates neurogenic abnormalities in muscles innervated by the lower trunk of the brachial plexus.

TREATMENT AND PROGNOSIS OF THORACIC OUTLET SYNDROME

Most patients with mild symptoms should be treated with conservative therapy involving exercises aimed at correcting slumping shoulder posture, encouraging full movement of the shoulder girdle and neck, strengthening the trapezius and rhomboid muscles, and straightening overall posture. Such exercises include shoulder shrugs, lateral arm lifts while holding light weights, and exercises that increase the neck's lateral and anterior-posterior range of motion. Positions that aggravate the symptoms, such as sleeping with the arms elevated, should be avoided. Nocturnal restrictive devices are useless. Serial electrophysiologic assessment of conduction velocity across the thoracic outlet is inaccurate and not worthwhile. Exercises are continued for at least a few months or as long as there is no objective muscle weakness, progressive objective sensory loss, or intractable pain.

Patients with true neurogenic thoracic outlet syndrome and muscle wasting and weakness usually require surgery. In most cases, surgical excision of the offending fibrous band relieves pain and halts the progression of weakness.[39] Established weakness and wasting may persist, probably because of the long distance regenerating nerves must travel before reinnervating the hand.

Radial Nerve (C$_5$–C$_8$)

ANATOMIC CONSIDERATIONS (Fig. 16–2)

The radial nerve is a continuation of the posterior cord of the brachial plexus; it innervates the triceps and anconeus muscles prior to or upon reaching the spiral groove, located in the midhumerus shaft. This proximal innervation explains sparing of forearm extension with radial nerve injuries due

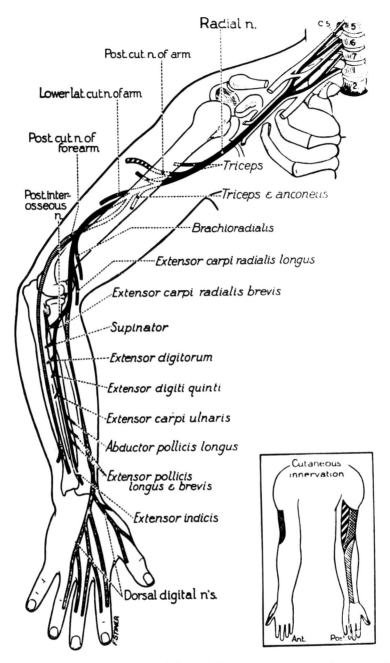

Figure 16–2. labels:

Radial n.

Post. cut. n. of arm

Lower lat. cut. n. of arm

Post. cut. n. of forearm

Post. interosseous n.

C 5
5
6
7
2

····Triceps

····Triceps & anconeus

····Brachioradialis

····Extensor carpi radialis longus

····Extensor carpi radialis brevis

····Supinator

····Extensor digitorum

····Extensor digiti quinti

····Extensor carpi ulnaris

····Abductor pollicis longus

····Extensor pollicis longus & brevis

····Extensor indicis

Dorsal digital n's.

Cutaneous innervation

Ant.   Pos.

Figure 16—2. The course and distribution of the radial nerve. (From Haymaker, W and Woodhall, B: Peripheral Nerve Injuries. WB Saunders, Philadelphia, 1953, with permission.)

to humerus fractures. Prior to traversing the spiral groove, the nerve passes posterior to the humerus, under the lateral head of the triceps. An anomalous fibrous band occasionally extends from the lateral head of the triceps to the humerus, creating a potential site of entrapment. The radial nerve is intimately associated with the humerus while traversing the spiral groove and, upon reaching the lateral aspect of the arm, gives off the posterior cutaneous nerve of the forearm. These sensory nerve fibers may branch from the main nerve proximal to the spiral groove or, in cases in which they accompany the parent nerve through the spiral groove, may occupy a more protected, posterior location within the nerve. These factors probably explain the frequent sparing of sensation with midhumerus radial nerve lesions.

After penetrating the intramuscular septum on the lateral aspect of the arm, the nerve innervates the brachioradialis, brachialis, and extensor carpi radialis muscles. In the cubital fossa it divides into the superficial branch that provides sensation to the skin over the dorsum of the hand and lateral three fingers, and a purely motor branch (posterior interosseous nerve), which, after innervating the supinator muscle, enters it and supplies the extensor muscles of the wrist and fingers.

SOURCE OF INJURY

The radial nerve may be injured in the axilla by lacerations, crutch use, missile wounds, or pressure from the head of a sleeping person ("lover's palsy").[48] The resulting triceps weakness impairs elbow extension. Nerve entrapment by a fibrous band derived from the lateral head of the triceps may occur in heavily muscled individuals recently involved in activities requiring prolonged elbow extension. Entrapment may occur upon entering the supinator muscle by fibrous bands of the flexor digitorum superficialis (arcade of Frohse) or within the muscle itself by tumors, ganglia, or inflammatory conditions. A painful elbow (resistant tennis elbow) may result from compression of the posterior interosseous nerve at this site. The superficial radial nerve's exposed position at the wrist makes it vulnerable to injury by either lacerations or compression from tight wristwatches, handcuffs, or plaster casts.

ASSESSMENT OF RADIAL NERVE FUNCTION

Most radial nerve injuries occur below the level of the axilla, sparing elbow extension. Lesions in the upper arm re-

sult in wrist drop owing to paralysis of the extensor carpi radialis. It is a feature of radial nerve lesions that all movements normally attributed to radial nerve innervation, except thumb abduction, may be simulated by other, nonradial-innervated muscles.

Total paralysis of supination is rare, for the biceps continues to act as a forearm supinator. Forearm extension, even in the face of triceps paralysis, may be achieved by gravity. Either gravity must be eliminated when testing the triceps or the examiner should test the patient's resistance to strong passive flexion of the forearm. The integrity of the brachioradialis must be determined by palpation, for the biceps will adequately flex the forearm even when the forearm is semipronated.

All extensor muscles of the wrist are affected by radial nerve lesions. Injury at the elbow spares the extensor carpi radialis, preserving adequate wrist extension. Radial deviation will occur with wrist extension, owing to extensor carpi ulnaris weakness. Even if all wrist extensors are paralyzed, extension may still be possible, as finger flexion can tighten the extensor digitorum communis, extending the wrist via tendon action.

Radial nerve lesions do not affect interphalangeal (IP) joint extension. Extension of the metacarpophalangeal (MCP) joint is often impaired, especially with the wrist in the neutral position. Finger extension may occur if the wrist is allowed to flex, from tightening of the extensor digitorum communis and resulting tendon action on the MCP joints.

Extension of the thumb's terminal phalanx, usually performed by the extensor pollicis longus, may be possible in radial nerve lesions via a tendon slip originating from the abductor pollicis brevis. In the presence of extensor pollicis longus weakness, thumb extension may still be performed by contractions of the abductor pollicis brevis, although simultaneous palmar abduction is usual. Concomitant thumb abduction is not invariable, however, so extension should be tested while the thumb lies adducted against the index finger. Terminal phalanx extension may also occur following strong contraction and sudden relaxation of the flexor pollicis longus or by moving the thumb in the palmar or ulnar direction, which causes a tightening of the paralyzed extensor muscle.

Paralysis of radial-nerve–innervated muscles exerts widespread effects on other hand muscles. Contraction of the long finger flexors becomes ineffective when deprived of the synergistic action of the wrist extensors. Grip is inefficient when thumb abductor paralysis exists. Appreciation of these relationships can avoid overestimating the extent of the injury.

The sensory deficit from radial nerve injuries rarely results in impaired hand function, as the maximum area of deficit lies within the first web space, an area not critical for stereognosis. The loss of wrist extension severely compromises power grip, although with mild recovery, grip power markedly improves.

POSTERIOR INTEROSSEOUS NERVE COMPRESSION

The posterior interosseous nerve may be compressed as it enters the supinator muscle under a fibrous arch (arcade of Frohse) or within the substance of the muscle. Compression may be caused by ganglia; tumors (especially lipomas); fibrous bands; fractures or dislocations of the radius; or spontaneously without an identifiable cause.[5] Posterior interosseous nerve lesions produce weakness of finger extension with variable degree of wrist drop, but sensation is spared over the dorsum of the hand. The degree of weakness is variable and may not affect all distal radial muscles to the same extent. In patients with rheumatoid arthritis, posterior interosseous nerve lesions must be distinguished from rupture of the extensor tendons to the thumb and fingers.

Electrophysiologic studies may demonstrate slowing of motor conduction across the elbow segment in severe cases. In most cases, however, either nerve conduction studies are unremarkable or distal motor potential amplitudes are slightly reduced. Needle EMG may demonstrate neurogenic change, either active denervation or chronic motor unit reinnervation, in forearm radial innervated muscles. The supinator muscle is usually spared.

TREATMENT AND PROGNOSIS OF SPECIFIC RADIAL NERVE INJURIES

Radial Nerve Injuries in the Upper Arm. The most frequent radial nerve lesion is compression of the nerve against the middle third of the humerus. This "Saturday night palsy" commonly occurs when intoxicated individuals fall asleep with their upper arms draped over a chair's edge or sleep with the head of another resting against the inner aspect of their upper arm. These are usually class 1 lesions with predominantly conduction block. Severe weakness with total paralysis of wrist and finger extensors should be treated with a wrist splint, as described below, that maintains wrist extension. Milder cases need no specific therapy. Complete recovery generally occurs in weeks, depending on the severity and duration of compression.

Patients with radial nerve entrapment within the triceps muscle usually recover spontaneously. Rarely, exploration and decompression may be needed in cases of progressive weakness or poor recovery.

Radial nerve injuries secondary to fractures of the humerus have an excellent prognosis for spontaneous recovery. Three quarters of radial nerve injuries resulting from humerus fractures recover spontaneously with the remainder requiring exploration.[48] Conservative therapy should be continued for at least 8 to 10 weeks before a decision is made that recovery is not proceeding.

Mild cases of posterior interosseous

nerve entrapment may recover spontaneously. A period of observation ranging from 3 to 6 months is generally prescribed to allow spontaneous recovery. More severe cases, in which weakness is either not recovering or is progressing, may require surgical exploration. Tumors, lipomas, ganglia, and constricting bands are removed. Entrapment under the arcade of Frohse, if present, is released. Nerve compression due to radial fractures should be observed initially for 2 to 3 months.

Regardless of the cause of the radial nerve palsy, most cases of wrist and finger drop require a proper splint to protect against hyperextension of paralyzed wrist extensor muscles and shortening of the flexor muscles. Mild wrist extensor weakness requires a static splint that maintains the wrist in approximately 30° of extension. If a palmar pad is used for support, it should not interfere with palmar sensation. The MCP joints may need to be supported in extension with the thumb extended and radially abducted. Finger joints must be regularly exercised, for they rapidly become stiff when immobilized for even a short time. This type of splint supports the hand in a position of function, thereby encouraging continued use. If the hand is able to be opened sufficiently by IP joint extension, MCP joint support may be unnecessary. Dynamic splints utilize a finger slip around the PIP joint to support the MCP joints in slight extension, and allow active finger flexion while offering protective positioning at rest.

During the stage of paralysis, daily passive exercises should be used to maintain an adequate range of joint movement, including large arm excursion. Once voluntary activity returns, specific exercises are employed to strengthen wrist and finger extensors. Intrinsic hand muscles become weak and inefficient, owing to dependence on the synergistic action of the wrist extensors, and may need strengthening to restore adequate hand function.

The initial sign of recovery in the wrist extensors is a slight flicker of extension upon attempted grasp.[48] The disappearance of thumb abduction on attempted extension, even without evidence of active extensor pollicis longus contraction, is often an early sign of reinnervation. Recovery in the extensor digitorum is heralded by disappearance of MCP flexion on attempted IP joint extension. Finally, disappearance of radial deviation of the wrist upon attempted extension indicates recovery of the extensor carpi ulnaris. Latent periods of reinnervation after lesions at the midhumerus level range from 2 to 10 weeks for class 1 injuries and from 8 to 40 weeks in class 2 injuries, depending on the severity;[48] class 3 lesions recover poorly.

Radial Nerve Injury at the Wrist. Wrist injuries due to lacerations or blunt trauma frequently involve the superficial radial sensory branch as a consequence of its exposed position. Compression by tight bands such as handcuffs and watch bands may also cause nerve dysfunction. Sensory loss varies according to site of injury and individual anatomy. It is often not loss of sensation that is troublesome but the development of painful paresthesias, dysesthesias, or neuromas.

Nonsurgical therapy involves the removal of precipitating or exacerbating causes. Radial nerve dysfunction from mild external compression often resolves spontaneously within weeks. Occasionally, immobilizing the thumb and wrist in a static splint may hasten recovery. Neither steroid injection nor release of the nerve from adhering scar tissue is usually indicated.

Axillary Nerve
(C₅–C₆)

This is a branch of the posterior cord of the brachial plexus. It supplies deltoid and teres minor and the skin over the deltoid. It is most commonly damaged in injuries to the shoulder; other causes include iatrogenic injection injuries, penetrating missile injuries, and as a common manifestation of brachial neuritis.[48] In the past, the axillary nerve was sometimes injured by pressure from a crutch ("crutch palsy"). The chief sign is an almost complete inability to raise the arm at the shoulder. Sensory loss is restricted to a small patch over the lateral shoulder but sensation is frequently normal, even in the presence of moderate weakness. Electrophysiologic studies should differentiate restricted axillary nerve damage from a posterior cord lesion. Needle EMG abnormalities are restricted to the deltoid and teres minor muscles. Most axillary nerve lesions recover spontaneously and completely, owing to the short distance between site of injury and the deltoid muscle. In traumatic cases, observation should be continued for about 3 to 4 months before surgical exploration and, if necessary, nerve repair.

Musculocutaneous Nerve (C$_5$–C$_6$)

This nerve is rarely damaged alone, but may be involved in injuries to the brachial plexus. It supplies the coracobrachialis, biceps, and brachialis, and the skin over the lateral aspect of the forearm through the lateral antebrachial cutaneous nerve. Flexion at the elbow is still possible by the brachioradialis, but is weak, and sensation may be impaired along the radial border of the forearm. The musculocutaneous nerve may be damaged, along with the axillary nerve, in shoulder dislocations and in brachial neuritis. The lateral antebrachial cutaneous nerve has been reported to be damaged by carrying objects of heavy weight supported at the elbow crease and spontaneously under the free margin of the biceps aponeurosis. Occasionally it suffers iatrogenic injury by misplaced antecubital injections.

Electrophysiologic studies should differentiate C$_6$ radiculopathy from isolated musculocutaneous nerve dysfunction by demonstrating neurogenic changes in the deltoid, brachioradialis, and occasionally, the supra- and infraspinatus muscles in the former. In addition, the lateral antebrachial cutaneous sensory potential is affected in musculocutaneous nerve injury but not C$_6$ radiculopathy. Therapy of musculocutaneous nerve lesions is usually conservative, with most achieving satisfactory recovery.

Median Nerve (C$_6$–C$_8$, T$_1$)

ANATOMIC CONSIDERATIONS
(Fig. 16–3)[48]

The median nerve arises from the medial and lateral cords of the brachial plexus. As it descends through the upper arm, it is intimately associated with the brachial artery, accounting for the high incidence of combined nerve-artery injuries with traumatic lesions. Prior to entering the forearm deep to the bicipital aponeurosis, the median nerve occupies a superficial position above the elbow, rendering it susceptible to external injury. At the elbow the median nerve passes, along with the brachial artery, beneath the supracondylar ligament, where it may become entrapped. In the cubital fossa, the nerve is separated from the shaft of the humerus by the brachialis muscle. It then passes between the two heads of the pronator teres muscle, dividing into a more superficial main branch and a deeper, purely motor, anterior interosseous branch.

The median nerve innervates the flexors of the wrist and fingers. While in the forearm, the main trunk supplies the pronator teres, flexor carpi radialis, palmaris longus, and flexor digitorum superficialis. The anterior interosseous branch separates from the parent nerve within the substance of the pronator teres and supplies the flexor pollicus longus, the lateral two flexor digitorum profundus muscles, and the pronator quadratus. At the wrist, the nerve and nine flexor tendons converge in the carpal tunnel, which is

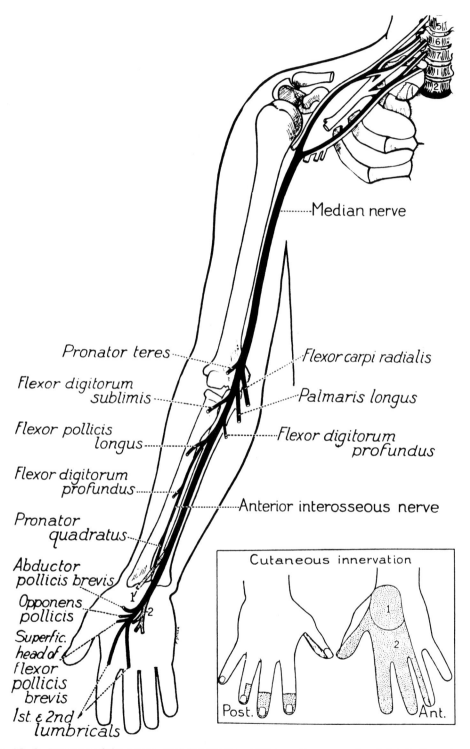

Figure 16–3. Diagram of the course and distribution of the median nerve. Inset details the palmar cutaneous branch *(1)* and the palmar digital nerves *(2)*. (From Haymaker, W and Woodhall, B: Peripheral Nerve Injuries. WB Saunders, Philadelphia, 1953, with permission.)

covered by the flexor retinaculum. The main nerve trunk distal to the flexor retinaculum usually supplies the abductor pollicis brevis, opponens pollicis, and the superficial head of the flexor pollicis brevis. It also innervates the lateral two lumbrical muscles and provides sensation to the palmar aspects of the lateral three and a half digits, and the dorsal surface of the distal phalanges of the index and middle fingers. Sensation to the lateral palm is supplied by the palmar cutaneous nerve, which branches from the main nerve trunk proximal to the carpal tunnel. Most of the sympathetic innervation to the hand is carried in the median nerve. The exposed location of the nerve at the wrist makes it liable to a variety of superficial injuries. A common anatomic variant is the Martin-Gruber anastomosis. In this condition, axons destined for the ulnar nerve travel with the median nerve until united with the main ulnar trunk via a forearm anastomosis.

SOURCE OF INJURY

The median nerve is vulnerable to missile wounds and lacerations at the upper arm and wrist where it is superficially positioned. In the forearm it is better protected, deep within the pronator teres muscle. Median nerve compression has been attributed to crutches pressing on the axilla, tourniquets, rifle sling palsy, and pressure from the head of a sleeping partner on the brachioaxillary angle or medial aspect of the arm.[48]

Median nerve damage secondary to bony fracture is less common than with other upper limb nerves because it is protected by overlying arm muscles. Although fractures of the elbow occasionally result in median nerve damage, only rarely is the nerve affected by fractures of the mid to upper one third of the humerus. The anterior interosseous nerve may be damaged by forearm fractures, since the nerve is closely related to the radius and ulna as it descends on the interosseous membrane.

Median nerve entrapments may arise from chronic compression in well-defined fibro-osseous tunnels or following repeated episodes of strenuous muscle activity. The most common nerve entrapment in the upper limb is compression of the median nerve under the transverse carpal ligament (carpal tunnel syndrome). Other median nerve entrapments include the pronator syndrome, the anterior interosseous syndrome, and entrapment by the supracondylar ligament.

ASSESSMENT OF MEDIAN NERVE FUNCTION

Severe dysfunction of the median nerve is devastating; it is far more disabling than any other single nerve lesion because the median nerve innervates muscles responsible for precision grip movements. Satisfactory precision grip depends on positioning the thumb as a post against which other fingers can press. Median nerve dysfunction prevents the thumb from assuming the post position by disabling abduction and opposition. Loss of the thumb's stabilizing function also affects power grip. Sensory loss in the palm also slows skilled finger movements.

The typical deformity resulting from median nerve injuries is thenar eminence wasting and loss of thumb abduction. The thumb is restricted to the plane of the palm, lying extended next to the index finger, unable to be moved in opposition (simian hand). Paralysis of median-innervated lumbrical muscles results in hyperextension of the MCP joints of the index and middle fingers owing to the unopposed action of the radially innervated extensor muscles. There is spindling of the index finger, owing to wasting of the terminal digit pad, and cracking of the nail bed. Upon attempting to close the fist, the thumb and index finger are unable to be folded into the palm. Lesions at or above the elbow result in additional weakness of forearm pronation, wrist flexion, and flexion of all proximal interphalangeal joints, as well as the distal interphalangeal joints of the index and middle fingers.

CARPAL TUNNEL SYNDROME

Symptoms and Signs. The median nerve may be compressed at the wrist as it passes deep to the flexor retinaculum. Carpal tunnel syndrome (CTS) is by far the most common entrapment neuropathy in the arm. It usually presents with acroparesthesias, numbness, tingling, and burning sensations in the lateral three digits.[5,19,47] Although paresthesias are usually restricted to the radial fingers, sensory symptoms may occasionally affect all the digits, as some fibers from the median nerve can be distributed to the fifth finger through a communication with the ulnar nerve in the palm.[21] Patients with CTS are frequently unable to describe accurately the exact distribution of their sensory disturbance; occasionally the best they can do is to report that the fifth finger is not involved. In our experience, patients with acroparesthesias who are unable to localize accurately the distribution of their sensory disturbance within their hand, most often have CTS. Nocturnal exacerbation of pain and paresthesias is characteristic and may either wake the patient from sleep or be prominent upon awakening in the morning. Although aching pain may involve the forearm and shoulder, it is rare to have numbness or paresthesias radiate proximal to the wrist. Shaking the hand frequently relieves pain. Symptoms may recur during the day following use, or when sitting with the hands immobile. Repetitive tasks such as knitting, sewing, and writing often precipitate symptoms. When specifically asked, patients often report dropping objects from the involved hand.

Acroparesthesias may persist for years without the appearance of other signs of median nerve damage. Objective sensory loss most consistently involves the second and third digits; often

there is splitting of the ring finger, with sensory loss on the lateral but not the medial aspect. Stroking the skin of the involved fingers and comparing the sensation to either the uninvolved fifth finger or the contralateral fingers often demonstrates sensory disturbance not evident on pin examination. Objective sensory loss rarely involves the palm.[10] Symptoms of Raynaud's phenomenon may occasionally develop. Most patients eventually develop weakness and wasting of median-innervated thenar muscles, particularly thumb abduction and opposition. Eventually, the thumb is unable to be maintained in the opposed and abducted position. As a result, the thumb is restricted to the plane of the palm, precluding it from acting as a post against which other fingers can push. Atrophy of the lateral aspect of the thenar eminence eventually occurs and occasionally may precede obvious weakness.

Percussion of the median nerve at the wrist may reproduce sensory symptoms in the median nerve distribution (Tinel's sign). Although more frequent in patients with CTS, Tinel's sign also occurs in normal subjects; its presence is not diagnostic. Forced hand flexion (Phalen's test) or extension (reverse Phalen's test) may reproduce the sensory symptoms and is highly correlated with the presence of CTS.

The majority of cases are in middle-aged women or people who use their hands in a repetitive and stereotypical fashion. As such, CTS most commonly involves the dominant hand, although bilateral involvement is extremely common. In many cases, only one side is symptomatic, the contralateral CTS being evident only by electrodiagnostic studies. CTS may develop acutely after a prolonged episode of unaccustomed hand use, such as weekend house-painting. CTS also results from any process that compromises the contents of the carpal tunnel. Most cases are probably due to nonspecific tenosynovitis of the flexor tendons. Other causes include tuberculous tenosynovitis, rheumatoid arthritis, osteoarthritis of the carpus, pregnancy, hemodialysis, myxedema, acromegaly, and infiltration of the transverse carpal ligament in primary amyloidosis.[5,46,47] It has been suggested that some patients may have a congenitally small carpal tunnel, thereby predisposing to later median nerve compression. Rarely, anomalous tendons and muscles, including the palmaris longus, the muscle belly of the flexor digitorum superficialis, and the lumbrical muscles may compromise the carpal tunnel.[5,47] Patients with mucopolysaccharidoses may develop median nerve compression. CTS is common with certain generalized peripheral neuropathies such as uremia and diabetes; it is believed that the generalized neuropathy predisposes the median nerve to compression. Acute CTS can be caused by hemorrhage or infection in the carpal tunnel and constitutes a medical emergency.

Differential Diagnosis. CTS should be considered for any unexplained pain or sensory disturbance in the hand, but it needs to be differentiated from the following clinical entities:

Cervical Radiculopathy. $C_{6,7}$ radiculopathies may cause sensory symptoms in the thumb and index and middle fingers, thereby simulating CTS. Sensory symptoms in $C_{6,7}$ radiculopathies often radiate along the lateral forearm and occasionally the radial portion of the dorsum of the hand. Unlike the CTS, pain in radiculopathy frequently involves the neck and may be precipitated by neck movement. Nocturnal exacerbation occurs with both CTS and radiculopathy but is more prominent with CTS. Although the distribution of sensory loss potentially may be identical, weakness, in radiculopathy, involves $C_{6,7}$ innervated muscles, not median innervated C_8 muscles. Patients with radicular pain tend to keep their arm and neck still, while patients with CTS shake their arms and rub their hands to relieve the pain. Electrodiagnostic studies are usually diagnostic although both $C_{6,7}$ root compression and

distal median nerve entrapment may coexist (double crush injury).

Thoracic Outlet Syndrome. Patients with nerve compression at the thoracic outlet may present with weakness of thenar muscles similar to CTS but, in contrast, have sensory symptoms involving the medial forearm and the medial two digits. Weakness and wasting also tend to involve the hypothenar muscles and arm discomfort is made worse with movement. Electrodiagnostic studies in true neurogenic thoracic outlet syndrome reveal evidence of lower trunk brachial plexus dysfunction.

Proximal Median Nerve Compression. Sensory symptoms not only involve the lateral fingers but also the palm. Weakness involves median forearm muscles that are innervated proximal to the wrist, such as the wrist and finger flexors and forearm pronators.

Electrodiagnostic Studies. In most cases, clinical signs and symptoms clearly indicate the diagnosis. Confirmation of atypical cases and an estimate of severity can be obtained by nerve conduction studies.[10] All patients undergoing surgery should have confirmatory electrodiagnostic studies. Sensory conduction studies are the most sensitive physiologic technique to diagnose CTS.

Digital median sensory potentials are often of low amplitude or absent, findings indicative of median nerve injury but not localizing to nerve compression at the wrist. Distal median sensory potential latencies and conduction velocities are frequently slowed. Sensitivity is enhanced by examining mixed nerve conduction in the shorter palm-to-wrist nerve segment. Occasionally, examination of sensory conduction from digit 3, rather than the usual digit 2, is often helpful. Other sensory conduction techniques that have proved useful include the comparison of median to ulnar mixed nerve conduction from palm to wrist, comparing median and ulnar sensory conduction to the ring finger, and comparing median to radial conduction to the thumb.

Median motor conduction is abnormal less often than sensory conduction but may have prolonged distal latencies and reduced compound motor action potential amplitudes, the latter usually reflecting axonal degeneration. When axonal degeneration has occurred, needle EMG reveals active denervation, chronic motor unit reinnervation, or both. Rarely, needle EMG may be abnormal when nerve conduction studies are normal. Nerve conduction abnormalities tend not to correlate with symptoms, although the size of the thenar evoked motor potential roughly parallels the degree of weakness.

Treatment and Prognosis. Therapy of CTS, in part, depends on etiology. If possible, underlying conditions should be treated. Space-occupying lesions (cysts, lipomas, ganglia, suppurative infection) may require surgery, while CTS due to underlying systemic conditions (myxedema, pregnancy, acromegaly, and so forth) may respond to specific systemic remedial therapy. Advanced lesions require directed therapy.

Conservative medical therapy of CTS is indicated for individuals with mild sensory symptoms without weakness or atrophy, intermittent symptoms, or an acute episode of CTS related to a specific injury. Initial treatment is nocturnal immobilization of the wrist in a position of rest by a volar wrist splint. The wrist is maintained in a neutral position; this is especially useful when symptoms are intermittent and nocturnally exacerbated. The splint should be worn for as long as it is effective; if symptoms recur, further splinting is usually unhelpful. Repetitive actions that precipitate symptoms should be avoided.

In general, splinting offers only temporary or minimal relief. Local injection of a mixture of lidocaine and methylprednisolone is indicated when splinting has failed in patients without marked sensory loss, thenar weakness, or muscle wasting. A 25-gauge needle, directed distally, is inserted proximal to the wrist crease and medial to the palmaris longus tendon, avoiding the median nerve; injection is deep to the transverse carpal ligament. A second injection should be administered only if improvement followed the first. Subsequent injections have decreasing efficacy: pain relief is of shorter duration than with the initial injection. No more than three to four injections should be administered, to avoid tendon rupture or median nerve injury. At times, the oral administration of low-dose steroids (i.e., prednisone 10 mg) or diuretics (especially in cases complicating pregnancy) is effective.

Surgery is indicated when conservative therapy fails to alleviate abnormal sensations or when thenar weakness or atrophy (evidence of axonal injury) develop. Surgery is usually effective in relieving pain and stopping the progression of weakness in almost all cases. If the lesion is not too advanced, recovery of strength usually occurs. Most cases of failed carpal tunnel release result from either incomplete transection of the transverse carpal ligament or faulty initial diagnosis. Preoperative electrodiagnosis is essential to avoid the latter mistake. Reported complications of surgery include neuroma formation, a tender dysesthetic scar, transection of the palmar cutaneous branch, infection, incomplete release, reflex sympathetic dystrophy, and damage to the superficial palmar arch. Electrodiagnostic abnormalities may persist for many months after surgery, although some improvement in motor conduction may be evident soon after surgery.

Case History: Carpal Tunnel Syndrome

A 60-year-old woman developed numbness and tingling over the palmar tips of the thumb, index, and middle fingers of the right hand. These sensations, initially most pronounced during vigorous house cleaning, in the course of the year became especially severe at night and were accompanied by an aching sensation of the right forearm.

The sensation of numbness spread over the entire medial palm, accompanied both by a "pins-and-needles sensation" and a "tight, bursting" feeling. The sensations were temporarily relieved by exercising the fingers or swinging the arm. She tolerated this condition until handwriting became difficult because of palmar dysesthesias and weakness of the thumb.

On physical examination there was diminished sensation to pin, touch, and thermal stimuli over the palmar surface of the thumb, index, and middle fingers, with splitting of the ring finger. Sensation was normal above the wrist and over the dorsal hand. The right thenar eminence was less prominent than the left, and strength in the right opponens pollicis and abductor pollicis brevis muscle was 4/5. Tapping the wrist at the palmar junction elicited a tingling sensation that radiated into the tips of the affected fingers (Tinel's sign). Flexing the hands at the wrist for one minute aggravated the paresthesias in the right hand (Phalen's sign) and had no effect on the left. The remainder of the physical examination was unremarkable. Radiologic and laboratory investigations disclosed no evidence of rheumatoid arthritis, dysproteinemia, or thyroid dysfunction. Electromyography revealed fibrillation potentials and positive sharp waves at rest in the right opponens pollicis and abductor pollicis brevis muscles. Motor nerve conduction velocities, measured from elbow to wrist in both right and left median nerves, were normal (48 m/sec). The wrist-muscle latency in the median nerve was 3.2 msec (normal) on the left and 6.3 msec on the right (prolonged). Sensory nerve action potentials, elicited by digital nerve stimulation from the index finger and recorded at the wrist, were of normal amplitude and latency (2.7 msec) on the left; none could be detected on the right. The patient refused surgery and was treated for 2 months with immobilization of the wrist with only moderate relief. An injection of steroid into the carpal tunnel was followed by rapid amelioration of paresthesias which persisted for 3 months. However, the abnormal sensations gradually returned, were not helped by a second local corticosteroid injection, and the patient consented to surgery. The day following operative transection of

the flexor retinaculum there was a dramatic lessening of the abnormal sensations. Nine months following surgery, sensation and strength were normal in the right hand, the right wrist-muscle median nerve motor latency was 5 msec, and the sensory potential was obtainable with prolonged minimal latency (4 msec).

Comment. This individual displayed many common features of CTS. The condition frequently occurs in women and is aggravated by repetitive movements of the wrist. Nocturnal worsening, proximal radiation of pain, and distressing acroparesthesias promptly relieved by surgery are typical. Muscle wasting was definitive clinical evidence of axonal degeneration and mandated surgery in this case.

MEDIAN NERVE ENTRAPMENT AT THE ELBOW (PRONATOR SYNDROME, ANTERIOR INTEROSSEOUS SYNDROME)

The pronator syndrome results from compression of the median nerve as it passes between the two heads of the pronator teres muscle and under the fibrous arch of the flexor digitorum superficialis muscle. Entrapment within the pronator teres is most frequently due to constriction by fibrous bands in the substance of the muscle, or passage through a tight flexor superficialis arch. Many affected patients have hypertrophied volar forearm muscles with symptoms precipitated by repeated pronation-supination activity.[5,47] Occasionally, the nerve is compressed under the lacertus fibrosus, a fascial band extending from the biceps tendon to the forearm fascia. The pronator syndrome is characterized by diffuse aching of the forearm and paresthesias in the median nerve distribution over the hand. Weakness varies, ranging from mild involvement of thenar and forearm musculature to none. Various tests have been advocated to localize the level of nerve entrapment within the pronator muscle mass. Pain in the proximal forearm upon forced wrist supination and wrist extension suggests compression at the level of the pronator teres.[5] Pain upon resisting forced forearm pronation of the fully supinated and flexed forearm suggests entrapment under the lacertus fibrosus. Pain upon forced flexion of the proximal interphalangeal joint of the middle finger suggests compression of the median nerve under the flexor superficialis arch.[5]

Compression of the anterior interosseous nerve causes weakness of the flexor pollicis longus, pronator quadratus, and the median-innervated profundus muscles. Impaired flexion of the terminal phalanx of the thumb and index finger is characteristic. There is no associated sensory loss. The anterior interosseous nerve may be damaged by forearm lacerations or fractures, fibrous bands within the pronator teres, the fibrous arch of the flexor superficialis, and as a manifestation of acute brachial neuritis.[48] Occasionally, anterior interosseous nerve dysfunction may follow vigorous muscular activity without an identifiable cause.

Entrapment of the median nerve at the elbow may occur due to an anomalous fibrous band extending from the medial epicondyle to a bony spur on the anteromedial surface of the humerus (ligament of Struthers). This produces weakness of median-innervated muscles, including the pronator teres, accompanied by the loss of the radial pulse when the arm is extended. X-rays may demonstrate the anomalous bony spur.

Electrophysiologic Studies. Nerve conduction studies in proximal median nerve compression syndromes are frequently normal. In more severe cases, the amplitude of distal motor and sensory potentials may be reduced. Prolonged distal motor latencies, from the elbow to the pronator quadratus, have been reported in anterior interosseous nerve entrapments but are inconsistent findings. Forearm median motor conduction velocities are rarely abnormal. The most consistent and sensitive physiologic abnormalities are neuro-

genic changes in median innervated forearm and hand median muscles on needle EMG. In mild cases, those manifesting pain but no weakness, all electrophysiologic studies may be normal.

Treatment and Prognosis. Conservative treatment of the pronator syndrome and spontaneous anterior interosseous nerve entrapment initially involves restricting arm motion by avoiding elbow flexion and pronation. Gentle splinting of the arm in supination occasionally relieves symptoms but also can aggravate the condition. Injection of steroids into the pronator teres muscle may offer transient relief. Nonsteroidal anti-inflammatory medication may be useful in mild cases. In both syndromes, persistent or progressive symptoms require exploratory surgery. Occasional cases are the result of schwannomas or other nerve tumors which can sometimes be visualized by MRI. Surgical excision of the spur and ligament in median nerve compression by the ligament of Struthers is usually successful.

Most compressive injuries to the median nerve do not require splinting. Traumatic injury, however, frequently results in extensive and severe weakness for extended time periods. In such cases, proper splinting of the thumb is essential to prevent deformity and preserve hand function. Regardless of the site of injury, movement of the thumb will be impaired. Loss of thumb abduction leaves it lying adjacent to the index finger, predisposing to thumb web adductor contractures, unless splinted in palmar abduction. Continued use of the hand facilitates functional recovery once reinnervation occurs. A C-bar or wooden dowel, inserted between the thumb and second metacarpal, maintains thumb abduction while an opponens bar on the proximal phalanx of the thumb stabilizes it in opposition to the index and middle fingers, thereby allowing full wrist movement and interphalangeal flexion of the thumb.

Proximal median nerve injuries, in addition to causing impaired thumb movements, weaken the long finger flexors. An effective dynamic splint will not only post the thumb but also permit full wrist movement so that strong wrist and finger extension can, via tenodesis effect, partially flex the terminal phalanges of the index and middle fingers. This allows the thumb to contact the index and middle fingers and perform pinch movements. As reinnervation of the abductor pollicis brevis occurs, the thumb, although still in the plane of the index finger, lies at an increasing distance. This early sign of median nerve recovery may be present before clinical evidence of voluntary contraction in the thenar muscles. With injuries above the elbow, reasonable strength returns to forearm muscles, although thenar musculature often fails to recover completely. In class 1 injuries, the latent period ranges from 1 to 3 months, depending on the severity of injury; class 2 injuries should show some evidence of improvement by 6 to 32 weeks, while in class 3 injuries, recovery is very prolonged.[48]

Ulnar Nerve (C_7, C_8, T_1)

ANATOMIC CONSIDERATIONS
(Fig. 16–4)

The ulnar nerve arises from the medial cord of the brachial plexus. As it descends into the upper arm, it passes around the elbow in the ulnar groove at the medial epicondyle and enters the forearm under an aponeurotic band between the humeral and ulnar heads of the flexor carpi ulnaris. The nerve is susceptible to acute and chronic compression as it passes above and behind the elbow within the ulnar groove and cubital tunnel. The nerve's vulnerability at these sites is attributed to: (1) crossing the extensor aspect of the joint; (2) being exposed in the postcondylar groove; (3) intimate contact with the bony bed of the medial epicondyle; and (4) its passage through a narrow opening in the cubital tunnel as it descends toward the forearm.[48]

At the wrist, the nerve runs atop the flexor retinaculum but is relatively fixed as it curves between the pisiform bone medially and the hook of the hamate laterally (Guyon's canal). After a superficial branch of the nerve supplies the hypothenar muscles, the deep branch crosses the palm and terminates in the flexor pollicis brevis and adductor pollicis. Because of their subcutaneous position within Guyon's canal, the superficial terminal branches are more susceptible to damage from lacerations and mild trauma than the deep branches, which, owing to their contact with the firm floor of the tunnel, are more vulnerable to chronic compression.

SOURCE OF INJURY

Ulnar nerve entrapment is the second most common nerve entrapment in the arm, ranking behind carpal tunnel syndrome. Ulnar compression may occur in the axilla/upper arm, elbow region (either ulnar groove or cubital tunnel), or distally at the wrist or hand. The exposed position of the ulnar nerve, as it passes around the elbow, and its superficial position at the wrist, make it vulnerable to external compression or penetrating injuries. In the forearm, the nerve is protected by the muscle mass of the flexor carpi ulnaris, which affords the nerve relative safety save for severe penetrating wounds. Most compression injuries of the ulnar nerve are class 1 and 2. Improper positioning during anesthesia may cause compression of the ulnar nerve.

The ulnar nerve is particularly susceptible to lepromatous invasion, especially in the cool zones at the elbow and wrist. Damage may occur after dislocations or fractures of the elbow or from chronic compression owing to habitual leaning against the elbows. Entrapment may occur in the cubital tunnel, where the nerve lies under the aponeurotic band between the two heads of the flexor carpi ulnaris. Occasionally, ulnar nerve dysfunction becomes evident many years after a supracondylar frac-

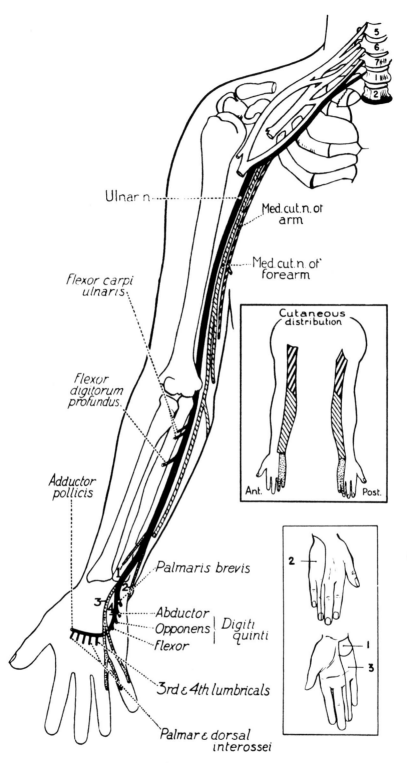

Figure labels within the illustration:

Ulnar n.
Med. cut. n. of arm
Med. cut. n. of forearm
Flexor carpi ulnaris
Flexor digitorum profundus.
Adductor pollicis
Palmaris brevis
Abductor
Opponens } Digiti quinti
Flexor
3rd & 4th lumbricals
Palmar & dorsal interossei

Cutaneous distribution
Ant.   Post.

Figure 16—4. The origin and distribution of the ulnar nerve, the medial cutaneous nerve of the forearm and medial cutaneous nerve of the arm. The numbered nerves are as follows: *(1)* palmar branch, *(2)* dorsal branch, *(3)* superficial terminal branch, *(4)* deep terminal branch. The fields of innervation of 1, 2, and 3 are detailed in the inset. (From Haymaker, W and Woodhall, B: Peripheral Nerve Injuries. WB Saunders, Philadelphia, 1953, with permission.)

ture of the humerus leading to an increased "carrying angle" at the elbows ("tardy ulnar palsy").[48]

ASSESSMENT OF ULNAR NERVE FUNCTION

Regardless of injury site or cause, the resultant deformity is hyperextension of the MCP joints of the ring and little fingers (about 30°) owing to the inability of weakened lumbrical muscles to oppose radial-innervated finger extensors. The resultant "claw hand" includes a variable degree of flexion at the IP joints of the ring and little finger, owing to the unopposed action of the flexor digitorum superficialis and, occasionally, the flexor profundus. Lesions at the elbow that weaken the flexor digitorum profundus result in less IP joint flexion deformity. An early sign of ulnar nerve regeneration after an elbow lesion is the gradual worsening of the ring and little finger flexor deformity from recovery of the profundus muscle. Paralysis of the flexor carpi ulnaris results in little deformity save for radial deviation of the wrist during flexion.

Ulnar nerve lesions result in marked weakness of power grip, but are not as disabling as median lesions because the thumb is spared. The ability to perform precise movements is also impaired as interossei weakness prevents finger spread. Patients are unable to hold a knife properly and tend to manipulate it between their index and middle fingers. Writing becomes awkward as sensory loss makes it difficult to appreciate the position of the hand. Weakness of ulnar-innervated intrinsic hand muscles impairs MCP joint flexion interfering with pinch grip.

Underestimation of the extent and severity of the injury stems from failure to recognize trick maneuvers. The tendons of the flexor digitorum profundus are aligned at the wrist so that only the one attached to the index finger is separate. Attempts at middle finger IP joint flexion results in contraction of the flexor digitorum profundus and tightening of both the tendon to the middle

finger and that of the terminal phalanx of the little finger. The accompanying flexion of the middle finger indicates this trick action.

Ulnar adduction of the thumb is a function of the adductor pollicis brevis, aided by the deep head of the flexor pollicus brevis. In the presence of adductor pollicus brevis paralysis, the most medial fibers of the abductor pollicis brevis may function as an ulnar adductor but, in doing so, must simultaneously bring the thumb forward away from the palm. This trick maneuver is avoided by testing thumb adduction with the thumb in contact with the palm.

Once the thumb is in the position of radial or palmar abduction, contraction of the extensor pollicis longus may simulate palmar adduction. In these instances, the flexor pollicis longus may aid thumb adduction. Adduction is accomplished by the action of the long thumb extensor working on a thumb fixed by the contraction of the long thumb flexor. Prominence of the tendons of the extensor pollicis and flexion of the terminal phalanx of the thumb proclaim this trick movement. This maneuver forms the basis of Froment's test, in which attempts at thumb adduction are accompanied by terminal phalanx flexion from compensatory action of the flexor pollicis longus. Strong adduction, however, may infrequently occur via the extensor pollicis longus, without concomitant action of the flexor pollicis. Recovery of thumb adduction is heralded by the disappearance of thumb interphalangeal flexion or MCP extension during the attempted movement.

Lumbrical paralysis causes inability to extend the IP joints. If the ring and little fingers' MCP joints are prevented from hyperextending, their terminal and middle phalanx may be extended by tendon action of the extensor digitorum communis. This action forms the basis of a dynamic splint which, by supporting the MCP joints, allows the long extensors to exert pull on the IP joints.

Dorsal interosseous weakness impairs finger abduction. Minimal abduction is still possible via the long extensors of the fingers, especially those of the index and little fingers, which have their own extensor muscles. Conversely, adduction is aided by relaxation of the extensors and contraction of the long flexors. In ulnar lesions, attempts at finger abduction result in prominence of the long extensor tendons on the dorsum of the hand and MCP joint extension. Proper testing of finger abduction requires the palm flat on the table and middle finger elevated. The long extensor tendon is now unable to act as an abductor and, in the presence of interosseous weakness, attempts at finger abduction result in obvious side-to-side hand movements.

ULNAR NERVE COMPRESSION AT THE AXILLA/UPPER ARM

This is a relatively uncommon site for ulnar nerve compression and usually results from blunt external trauma, such as pressure from improperly placed crutches, tourniquets around the upper arm, prolonged pressure from the head of a sleeping partner, or inadvertent posturing during a drunken stupor. Rarely, aneurysms of the brachioaxillary artery can compress the ulnar nerve.

The usual presentation involves weakness of all ulnar-innervated muscles, both proximal (flexor carpi ulnaris, flexor digitorum profundus) and distal (hypothenar muscles, interossei, adductor pollicis). Classically, there is co-injury to the median and radial nerves with consequent accompanying weakness. Diagnosis depends on demonstrating proximal involvement of median, radial, and ulnar nerves, along with electrodiagnostic studies showing neurogenic injury to appropriate muscles without focal slowing of ulnar conduction across the elbow region.

ULNAR ENTRAPMENT AT THE ELBOW

The association of the ulnar nerve with the medial epicondyle leaves the nerve in an exposed position within the bony canal and susceptible to minor trauma or chronic compression. The nerve enters the forearm via a narrow opening (cubital tunnel) formed by the epicondyle, the medial collateral ligament of the joint, and the firm aponeurotic band to which the flexor carpi ulnaris is attached.[9] It is estimated that up to one half of normal individuals have thickened and enlarged nerves within the ulnar groove. Elbow flexion reduces the size of the opening under the aponeurotic band, while extension widens it. Ulnar nerve dysfunction may occur secondary to cubital tunnel narrowing during elbow flexion, without additional external trauma or local pathology, or may be due to any condition that increases nerve size or reduces the space available for the nerve. Some patients with ulnar nerve lesions at the elbow give no history of recent injury but report a previous elbow fracture or traumatic injury. This "tardy ulnar palsy" is frequently insidious in onset and probably results from narrowing of the cubital tunnel secondary to osteoarthritis or an increased carrying angle at the elbow.

Ulnar nerve lesions at the elbow produce a variable deficit. Sensory symptoms usually precede weakness. Most common is numbness, paresthesias, or pain in the fourth and fifth fingers. Symptoms are occasionally positionally provoked, especially by prolonged elbow flexion, as during sleep or while talking on the phone. The distribution of sensory symptoms is better delineated by patients with ulnar entrapments than those with CTS. Although chronic elbow aching is common, sensory symptoms and signs do not extend proximal to the wrist. Sensory examination demonstrates impaired cutaneous sensation predominantly in the fifth finger with occasional splitting of the fourth finger. Objective sensory impairment may involve the dorsum of the hand but does not extend proximal to the wrist crease. In contrast to CTS, in which uncomfortable sensory symptoms bring the patient to medical attention, patients with ulnar neuropathies may present with marked muscle wasting and minimal sensory complaints.

Although weakness may affect any of the ulnar-innervated muscles, the first dorsal interosseus is usually the earliest and most severely affected; weakness and wasting of other ulnar hand and forearm muscles may follow. Early weakness of fifth-finger adduction may initially be manifested by patients catching their finger on their pant's pocket as they insert their hand. Severe weakness results in a claw-hand deformity with variable flexion of the distal digits, depending on the degree of profundus weakness. Ulnar neuropathy results in the loss of power grip and impaired precision movements.

The differential diagnosis of ulnar neuropathy at the elbow includes the following:

Cervical Radiculopathy. C_8/T_1 radiculopathy may cause sensory symptoms in the fourth and fifth fingers, but also along the medial forearm. Pain is more proximal, centering in the shoulder and neck, although the elbow is a common C_8 referral site. Electrophysiologic studies demonstrate intact ulnar sensory potentials in C_8 radiculopathies and no focal conduction abnormalities across the elbow segment. Needle EMG demonstrates denervation in C_8/T_1 median-innervated thenar muscles as well as ulnar-innervated muscles. Radiographic examination is often required: occasionally both cervical radiculopathy and ulnar neuropathy coexist (double crush injury).

Thoracic Outlet Syndrome/Lower Brachial Plexus. Sensory symptoms classically involve not only the fingers but also the medial forearm. Weakness affects hypothenar and thenar muscles with the latter actually being more severe. Electrodiagnostic studies fail to demonstrate focal slowing of conduction across the elbow and indicate a lower trunk brachial plexus lesion.

Syringomyelia. A syrinx affecting C_8 motor and sensory function may be confused with an ulnar neuropathy. Dissociated sensory loss is characteristic, with sparing of large-fiber sensation. Median-innervated C_8 motor function is impaired as well as ulnar, and often there are associated long tract findings in the legs. Despite extensive clinical sensory loss, ulnar sensory potentials are preserved due to the preganglionic nature of the lesion. MRI helps delineate the lesion.

Motor Neuron Disease. Weakness and wasting of intrinsic hand muscles are common in motor neuron disease. Sensory disturbance, however, is not found and accompanying upper motor neuron findings may be present. Fasciculations may be present diffusely, indicating the widespread nature of the disease. Thenar muscles are often affected in combination with hypothenar muscles.

Electrodiagnostic Studies. Slowed motor or sensory nerve conduction across the elbow, relative to the forearm segment, is a clear localizing sign; unfortunately, focal slowing of motor conduction is present in only 50% to 60% of cases. Absolute slowing of motor conduction across the elbow segment, regardless of forearm conduction, has been reported in 65% to 85% of patients with motor and sensory signs and about 50% of patients with only sensory impairment. Motor conduction studies to the first dorsal interossci muscle may be abnormal when that to the abductor digiti minimi is normal. Conduction velocity measurements across the elbow segment are very dependent upon elbow positioning; the arm should be bent at a 130° angle to minimize kinking of the nerve and allow more accurate distance measurements. Slowing of mixed nerve or sensory conduction across the elbow segment has been reported. Motor conduction block is occasionally demonstrated with stimulation proximal to the elbow, although conduction slowing is a more common finding. Sensory potentials are frequently reduced in amplitude, including the dorsal cutaneous branch. Motor potential amplitudes are diminished in lesions, resulting in axonal degeneration. Nerve conduction studies may be normal in mild cases without weakness; if axonal degeneration is present, needle EMG will demonstrate active and/or chronic denervation in ulnar innervated muscles. Neurogenic changes may or may not be present in the flexor carpi ulnaris muscle, depending upon fascicular arrangement and severity of lesion.

Treatment and Prognosis. Nonsurgical therapy is reserved for three groups of patients: those in whom symptoms are only posturally precipitated, those mild cases who have a recognizable occupational cause that can be remedied, and those who demonstrate only sensory symptoms without substantial progression. Conservative therapy includes avoiding aggravating movements such as repeated elbow flexion and extension or resting habitually on the elbows. Simply splinting the elbow in extension for prolonged periods, especially during sleep, may help.

Conservative therapy should be continued for at least two to three months, or as long as symptoms remain intermittent, mild, and only sensory. Careful follow-up is important to avoid missing a progressive lesion. Nonsteroidal anti-inflammatory agents occasionally help. Progression of sensory symptoms, appearance of weakness, or progressive electrophysiologic deterioration dictates surgical intervention. Most surgeons now recommend anterior transposition of the ulnar nerve deep to the flexor forearm muscle mass.[18] In some cases of entrapment within the cubital tunnel, simple resection and freeing of the nerve has been successful. Clinical improvement can be expected in about 75% of cases. In predicting recovery, the severity of the pre-operative lesion is important; earlier intervention results in better recovery. Complications include neuroma formation, recurrent scarring around the nerve, and persistent pain.

Traumatic injuries of the ulnar nerve

require specific splints designed to prevent MCP joint hyperextension of the ring and little fingers. Prolonged hyperextension results in fixation of collateral ligaments. A restraint on the dorsum of the proximal phalanx supports the MCP joint and keeps the collateral ligaments elongated. This enables the central slips of the common extensor tendons to exert their pull on the PIP joints rather than extending the MCP joints. Should IP joint flexion persist, a dynamic splint may be used to extend the DIP joint.

Combined median and ulnar nerve injuries are especially damaging to hand function owing to the marked weakness and sensory loss. Patients frequently compensate for their lack of pinching ability by using a "lateral pinch," in which the thumb contacts the side of the index finger. The movement should be discouraged, as it predisposes to thumb web contractures. A proper splint fixes the thumb in the posted position and prevents hyperextension of the MCP joints, thereby allowing the long extensors to aid in IP joint extension.

During the period of paralysis, ROM exercises are combined with splinting to maintain joint mobility. When active muscle function returns, progressive resistance exercises can be implemented to develop strength and endurance. Sensory re-education is less critical than with median nerve injuries.

ULNAR NERVE LESIONS AT THE WRIST

The ulnar nerve may be injured at the wrist level by lacerations or compressed within Guyon's canal or more distally within the palm. Although a large number of etiologic conditions may result in distal ulnar compression, by far the most common is chronic repeated trauma to the palmar area, such as occurs with heavy laborers or cyclists. Other etiologies include ganglia which compress the nerve either within Guyon's canal or distally along the deep terminal motor branch, fractures of the carpal bones, lipomas and other tumors, and rheumatoid arthritis. Acute trauma, such as a fall on the outstretched hand, may damage the nerve as it passes between the pisiform bone and the hook of the hamate.

Clinical expression is varied and depends on the nature and site of the injury. Compression within Guyon's canal may result in any combination of weakness in hypothenar and thenar ulnar-innervated muscles and sensory loss in the medial two fingers. The palmar and dorsal surfaces of the hand are spared due to sensory nerve branching proximal to the wrist level. Ulnar compression at the wrist should be suspected when weakness predominantly affects ulnar-innervated thenar muscles relative to the hypothenar muscles. Other clinical scenarios include selective weakness of the hypothenar muscles and isolated sensory loss to the medial two fingers. Rarely, a more proximal wrist lesion may additionally affect the palmar sensory branch.

The differential diagnosis is similar to that for ulnar neuropathy at the elbow. The most difficult diagnostic decision is distinguishing distal ulnar compression from entrapment at the elbow. The latter condition is marked by weakness of forearm muscles and clinical-electrophysiologic involvement of the dorsal cutaneous sensory branch.

Electrodiagnostic Studies. The most specific electrodiagnostic finding is a prolonged distal motor latency to the first dorsal interosseus compared to the abductor digiti minimi. The digital sensory potential to the fifth finger is often affected in contrast to sparing of the ulnar dorsal cutaneous sensory potential. Depending on the site of lesion, needle EMG may demonstrate active or chronic denervation in either hypothenar or thenar muscles with sparing of ulnar-innervated forearm muscles.

Treatment and Prognosis. Conservative therapy is indicated when there are sensory symptoms alone. Maneuvers responsible for or aggravating the injury should be avoided. Surgery is indicated when conservative therapy has failed to relieve discomfort, when motor or sensory dysfunction progresses, or when the deficit has no clear cause (i.e., possibly a mass).

Motor recovery after ulnar nerve injury at the wrist may be heralded by the return of involuntary abduction of the little finger on attempted thumb opposition. Early evidence of interosseous recovery is the disappearance of the side-to-side hand movements that occurred previously on attempted abduction of the elevated middle finger. Latent periods preceding recovery in the first muscle distal to the injury may range from 8 to 34 weeks, depending on the level and severity.[48]

Lumbosacral Plexus

Acute lesions of the lumbosacral plexus are uncommon. It may be compressed by a hematoma in patients receiving anticoagulant therapy or suffering from hemophilia, or be involved in fractures of the pelvis. Lumbosacral plexitis, believed to be analogous to acute brachial neuritis, is a rare occurrence. Proximal diabetic motor neuropathy may involve both the lumbosacral roots and plexus. The lumbosacral plexus may be compressed against the rim of the pelvis during parturition by the fetal head, with consequent weakness of tibial and peroneal muscles, and sensory impairment in the distribution of the fourth and fifth lumbar dermatomes. The superior gluteal nerve may also be affected. Recovery is initially good but may not be complete. Subacute injury to the plexus may stem from pelvic malignancy, such as from carcinoma of the cervix, bladder, prostate, or rectum, or it may be the site of a local neural tumor.

Femoral Nerve (L_2-L_4)

ANATOMIC AND ETIOLOGIC
CONSIDERATIONS (Fig. 16–5)

The femoral nerve originates in the lumbar plexus from branches of the

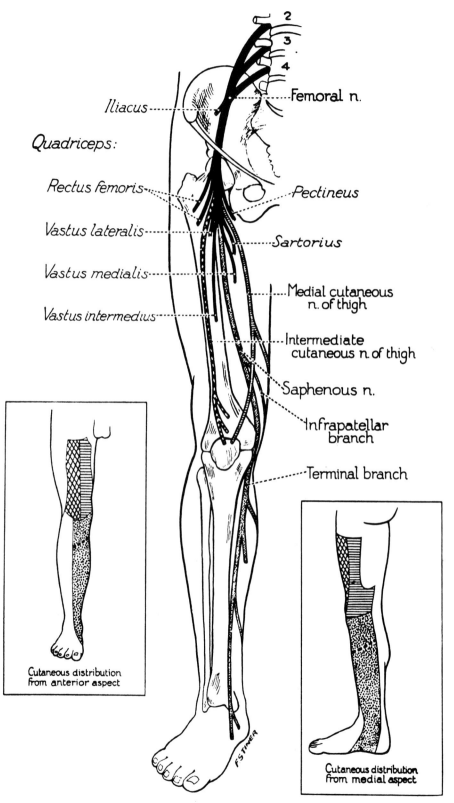

Figure 16–5. The course and distribution of the femoral nerve. The broken line in the cutaneous field of the saphenous nerve represents the boundary between the infrapatellar and terminal branches. (From Haymaker, W and Woodhall, B: Peripheral Nerve Injuries. WB Saunders, Philadelphia, 1953, with permission.)

posterior division of the $L_{2,3,4}$ roots. As it enters the iliac fossa, it passes between and innervates the psoas and iliacus muscles. The nerve may be damaged in this location by stabbings, missile wounds, or bleeding secondary to anticoagulation therapy or hemorrhagic disorders. As the nerve enters the femoral triangle in the thigh, it is positioned atop the iliacus muscle and lateral to the femoral artery. Muscles supplied include the pectineus, sartorius, and quadriceps. The nerve may be injured within the femoral triangle by penetrating lacerations or missiles and complications of femoral angiography. The nerve may also be damaged by retropcritoncal tumors or abscesses, irradiation, fractures of the pelvis or femur, stretch secondary to improper positioning on the operating table, and as a result of hip replacement. Femoral mononeuropathy may complicate diabetes or be part of a mononeuropathy multiplex. Rarely, no identifiable etiology can be found. The saphenous nerve originates just distal to the inguinal ligament and provides sensation to the patellar skin and medial aspect of the leg as far distally as the medial malleolus.

Femoral nerve injury usually produces weakness of knee extension due to quadriceps paresis. Proximal lesions may also impair hip flexion due to iliopsoas weakness. Mild weakness of knee extension is well compensated in walking by intact hip extensors that lock the knee in extension. Walking on uneven ground or up stairs is difficult, and there is significant impairment when accompanied by gluteal muscle weakness or when the quadriceps is totally paralyzed. Sensory loss may be present over the anterior and medial aspect of the thigh, at times extending to the medial malleolus and the great toe. Electrophysiologic studies demonstrate reduced motor potential amplitudes from the quadriceps muscles and low-amplitude or absent saphenous sensory potentials. Needle EMG demonstrates neurogenic changes in the quadriceps muscles with sparing of the adductor muscles and paraspinals. Neuroradio-graphic imaging with CT or MRI is frequently useful in excluding tumor or identifying hematoma.

TREATMENT AND PROGNOSIS

Traumatic femoral nerve lesions frequently recover spontaneously, even after lacerations and missile wounds. Conservative therapy should be continued for about seven months, unless the nerve has been observed to be transected.

Femoral nerve compression by hemorrhage is usually initially treated conservatively; spontaneous recovery is usual as the nerve remains in continuity, and the distances the regenerating nerve needs to traverse are short. Bleeding should be minimized by immediate reversal of anticoagulation and immobilization of the leg. Range-of-movement exercises to reduce fibrosis and heat to relieve muscle spasm are beneficial. Massive hematomas may require surgical removal.

Only in the minority of cases, when quadriceps paresis is very severe, does the knee require support. If the gluteus maximus is strong, walking often remains possible; however, with time, excessive hyperextension of the knee (genu recurvatum) may develop. This tends to occur infrequently in adults with recent lesions, but may complicate long-standing quadriceps weakness, (e.g., polio) and especially marked weakness in children. When quadriceps weakness is severe, or if it is accompanied by gluteal weakness, a brace may be beneficial. A typical knee brace involves the same type of shoes, stirrups, and uprights as with the Klensak ankle-foot splint, without a plantarflexion stop. Both uprights should be locked at the knee for stability. An upper thigh band connects the two uprights proximally. A second, lower thigh band is occasionally used. There are multiple techniques for knee restraint, including various combinations of lower thigh straps, suprapatellar and patellar tendon straps, and calf bands. A particularly efficient brace uses suprapatellar and patellar tendon straps to restrain the knee and lower thigh and calf bands to provide counterresistance. This arrangement allows the knee restraint straps to apply force on relatively pressure-resistant structures, while double strapping reduces the amount of force required. Variations include the use of knee restraints that lock on extension but are able to be opened for sitting.

Obturator Nerve (L₂–L₄)

The nerve emerges from the lateral border of the psoas muscle, crosses the lateral wall of the pelvis, and enters the thigh through the obturator foramen where it supplies gracilis, adductor longus and brevis, adductor magnus, obturator externus, and sometimes the pectineus, and the skin over the lower medial aspect of the thigh.

Damage to the obturator nerve results in weakness of adduction and internal rotation at the hip, pain in the groin, and sensory impairment on the medial part of the thigh. The nerve may be involved in neoplastic infiltration in the pelvis and can be acutely damaged by the fetal head or by forceps during parturition. Obturator nerve injury may complicate hip replacement or result from an obturator hernia. Electrophysiologic evaluation centers on differentiating obturator nerve injury from L₃,₄ radiculopathy and lumbar plexopathy. Needle EMG should demonstrate neurogenic changes restricted to the hip adductors. CT or MRI is often needed to exclude an intrapelvic tumor.

Lateral Cutaneous Nerve of the Thigh (L₂–L₃)

This nerve arises from the lumbar plexus, passes obliquely across the iliacus muscle, and enters the thigh under the lateral part of the inguinal ligament. It supplies the skin over the anterolateral aspect of the thigh.

Meralgia paraesthetica is an entrapment neuropathy resulting from compression of this nerve as it passes under the inguinal ligament.[20] It is most common in obese men or patients who have recently lost a substantial amount of weight, and may be unilateral or bilateral. The symptoms consist of numbness in the territory of the nerve combined with tingling or burning paresthesias provoked by prolonged standing, or following excessive walking. Weight reduction may help and in many instances the condition subsides spontaneously. Local injection of anesthetic agents is often of temporary relief; occasionally relief is longstanding. Decompression of the nerve is rarely necessary.[8] As with other focal neuropathies, it is important to exclude diabetes mellitus.

Sciatic, Tibial, and Peroneal Nerves

ANATOMIC CONSIDERATIONS
(Fig. 16–6)

The sciatic nerve is comprised of branches from the lumbosacral trunk (L₄,L₅) and the upper three sacral roots. The nerve enters the thigh through the sciatic notch, just below the piriformis muscle. Usually the nerve remains as one trunk until the lower thigh, where it divides into tibial and peroneal divisions. In 10% of cases, however, a separation into the two divisions is apparent at the sciatic notch. The sciatic nerve descends beneath the gluteus maximus, where it may be damaged by intramuscular injections. In the upper thigh, the tibial division innervates the hamstring muscles. The proximal innervation of the hamstrings, sometimes as high as the ischial tuberosity, explains the frequent sparing of these muscles in lesions of the thigh. The peroneal division is more susceptible to injury than the tibial division.

The sciatic nerve divides into its two component divisions immediately proximal to the popliteal fossa. The larger tibial nerve initially lies superficial,

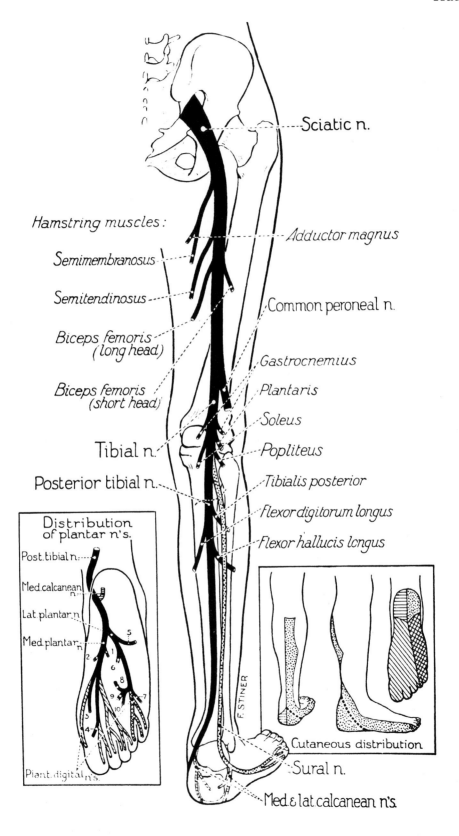

Sciatic n.

Hamstring muscles:

Adductor magnus

Semimembranosus

Semitendinosus

Common peroneal n.

Biceps femoris
(long head)

Gastrocnemius

Plantaris

Biceps femoris
(short head)

Soleus

Tibial n.

Popliteus

Posterior tibial n.

Tibialis posterior

Flexor digitorum longus

Flexor hallucis longus

Distribution
of plantar n's.

Post. tibial n.

Med. calcanean
n.

Lat. plantar n.

Med. plantar
n.

Plant. digital n's.

Cutaneous distribution

Sural n.

Med. & lat. calcanean n's.

F. STINER

making it vulnerable to minor lacerations, before it becomes embedded within the fossa's fatty contents. While in the popliteal fossa, the nerve is adjacent to the popliteal artery, rendering it susceptible to compression from aneurysmal dilatations. As the nerve exits the fossa, it runs between the gastrocnemius and soleus muscles. The sural nerve originates within the popliteal fossa and, after passing between the two heads of the gastrocnemius, is joined by a branch from the common peroneal nerve. The sural nerve descends down the lateral aspect of the calf, eventually winding posterior to the lateral malleolus before proceeding along the lateral aspect of the foot.

The tibial nerve passes posterior to the medial malleolus and dips beneath the flexor retinaculum at the ankle before dividing into lateral and medial plantar nerves. In the leg, the tibial nerve supplies the tibialis posterior, gastrocnemius, soleus, flexor hallucis longus, and flexor digitorum longus. Motor branches in the foot are distributed to the intrinsic foot muscles.

The common peroneal nerve exits the popliteal fossa by winding around the fibular head before dividing into superficial and deep peroneal branches (Figs. 16–7, 16–8). Its superficial position at the fibular head renders the nerve susceptible to compression. The nerve leaves the fibular head region through the fibular tunnel, the floor of which is the fibula itself and the roof the tendinous edge of the peroneus longus muscle. The superficial peroneal nerve passes between peroneus longus and extensor digitorum longus, innervates the former, and terminates as medial and lateral cutaneous sensory branches supplying the dorsum of the foot. There is a restricted area of sensory innervation to the web space between the first

and second toe that is supplied by the deep branch. The deep branch of the peroneal nerve descends the leg in the anterior compartment, supplying the tibialis anterior, extensor digitorum longus and brevis, and the extensor hallucis longus.

SOURCE OF INJURY

Within the pelvis, the sciatic nerve may be damaged by neoplasms, aneurysmal dilatations, or fractures. Improperly administered gluteal injections may damage the nerve. Rarely, the sciatic nerve is entrapped within the piriformis muscle. Thin individuals, or those with recent weight loss and depletion of gluteal fat, may experience acute or subacute compressions after prolonged sitting. Hip dislocations or arthroplastic surgery may cause sciatic nerve dysfunction.

The tibial nerve is occasionally compressed under the flexor retinaculum at the ankle (tarsal tunnel syndrome). This condition is associated with osteoarthritis, post-traumatic deformities at the ankle, tight shoes, and tenosynovitis.[23,27] Burning pain and tingling paresthesias occur in the sole, usually unilateral and following prolonged standing or walking. Severe cases have wasting of the intrinsic muscles of the foot, and sensory impairment over the sole. Nerve conduction studies help in diagnosis. Treatment is surgical section of the flexor retinaculum; occasionally local injection of steroids provides temporary relief.

Painful neuromas sometimes develop on the digital branches of the plantar nerves.[40] These give rise to the syndrome of "Morton's metatarsalgia" in which pain appears in the anterior plantar surface on standing. A localized area of tenderness is detectable on pal-

Figure 16–6. The course and distribution of the sciatic, tibial, and posterior tibial nerves. The cutaneous fields of the medial calcaneal and medial plantar nerves are indicated in the inset by lines, the sural nerve by dots, and the lateral plantar nerve by cross-hatching. The numbered branches of the plantar nerves supply intrinsic foot muscles. (From Haymaker, W and Woodhall, B: Peripheral Nerve Injuries. WB Saunders, Philadelphia, 1953, with permission.)

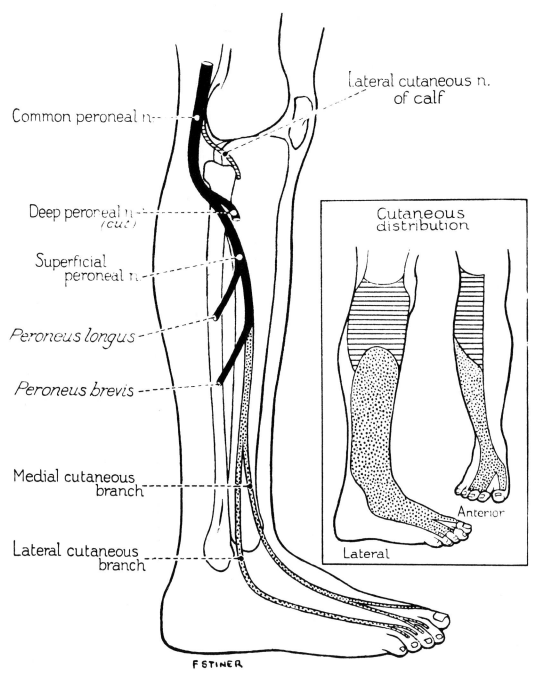

F STINER

Figure 16–7. The course and distribution of the superficial peroneal nerve. The inset details the cutaneous field of the superficial peroneal nerve *(dotted pattern)* and the lateral cutaneous nerve of the calf *(lined pattern)*. (From Haymaker, W and Woodhall, B: Peripheral Nerve Injuries. WB Saunders, Philadelphia, 1953, with permission.)

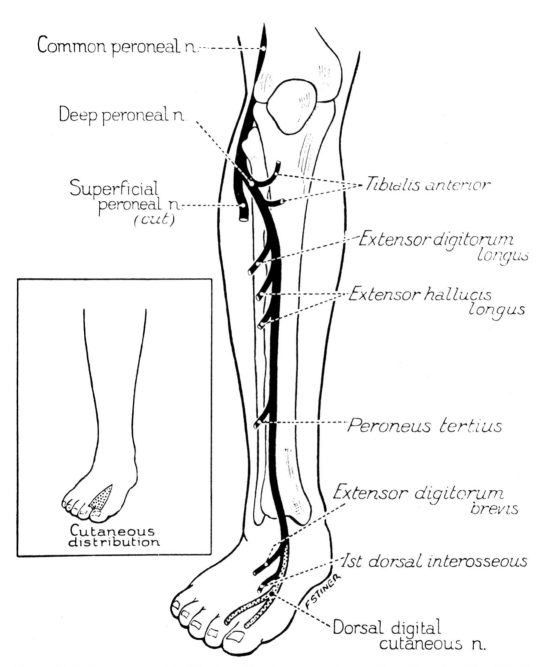

Figure 16−8. The course and distribution of the deep peroneal nerve. (From Haymaker, W and Woodhall, B: Peripheral Nerve Injuries. WB Saunders, Philadelphia, 1953, with permission.)

pation. Excision of the neuroma is effective treatment.

The tibial nerve may be lacerated in the popliteal fossa or entrapped under the flexor retinaculum at the ankle by trauma or nonspecific inflammation. The terminal divisions of the tibial nerve may be compressed between the heads of the metatarsal bones.

The peroneal nerve at the fibula head may be damaged by prolonged leg crossing, squatting, pressure during sleep, or improperly fitting casts. Occasionally it may be compressed within the fibular tunnel for no apparent reason. Forceful ankle inversion may damage the peroneal nerve at the fibular head, presumably as a result of stretch-induced nerve hemorrhage. In rare instances, the nerve may be compressed by bony tumors. Necrotizing vasculopathies, diabetic mononeuropathy, severe fibula fractures, and compression or ischemia from the anterior compartment syndrome may affect the peroneal nerves. The deep peroneal nerve may also be entrapped beneath the dense superficial fascia over the dorsum of the foot ("anterior tarsal tunnel syndrome").

ASSESSMENT OF SCIATIC, TIBIAL, AND PERONEAL NERVE FUNCTION

Sciatic nerve injury most commonly is manifested by paralysis of peroneal-innervated muscles. Weakness of ankle dorsiflexors and evertors cause foot drop with slight inversion; this has little localizing value since the same pattern of weakness may arise after injury in proximal or distal sites. Additional weakness of ankle inverters, toe flexors, or hamstring muscles helps localize the injury to a proximal site. Underestimation of the proximal extent of sciatic lesions is common, since there is a large section of unbranched nerve between the innervation of the hamstrings and the origin of the branch to the short head of the biceps femoris. Lesions of the sciatic or tibial nerve may result in paralysis of intrinsic foot muscles with hyperextension of the metatarsopha-

langeal joints and flexion of the IP joints (claw foot).

Trick maneuvers of leg musculature are not as prevalent as with arm muscles because of the limited number of possible movements. Despite hamstring paralysis, feeble knee flexion can be accomplished by contraction of the sartorius and gracilis muscles. Rebound after strong toe flexion may result in mild toe extension, simulating contraction of the extensor hallucis. Toe extension may also occur via tendon action after strong plantar flexion of the foot. Anomalous innervation of the extensor digitorum brevis by an accessory deep peroneal nerve may spare toe extension despite lesions of the deep branch. Slight plantar flexion of the foot is possible by contraction of the peroneus longus.

Sensory loss is extensive in complete sciatic lesions and the resultant hypesthesia of the sole predisposes it to traumatic injury. Lesions of the deep peroneal nerve result in sensory loss restricted to the web space between the first and second toe, sparing the rest of the foot. In superficial peroneal lesions, sensory loss extends to the midpoint of the lateral calf; the rostral portion of the calf is supplied by a separate branch originating from the common peroneal nerve in the popliteal fossa. Trophic changes in the leg and foot are common, since the majority of sympathetic fibers to the leg accompany the sciatic and tibial nerves.

ELECTROPHYSIOLOGIC STUDIES

The constellation of electrophysiologic findings in sciatic nerve injuries consists of varied combinations of reduced amplitudes in distal tibial and peroneal compound action motor potentials, as well as sural and superficial peroneal sensory potentials. The tibial H reflex is frequently absent or prolonged in latency. Needle EMG abnormalities may be restricted to muscles innervated by the peroneal division or be diffusely present. In such cases, demonstration of denervation in the short head of the biceps femoris indicates a lesion proximal to the fibular head. Occasionally, EMG abnormalities may be present in the adductor magnus due to co-innervation with the obturator nerve.

Peroneal nerve compression at the fibular head is often difficult to demonstrate electrophysiologically. Focal slowing of motor conduction across the fibular head, recording from the extensor digitorum brevis, is frequently absent. In many cases, motor potentials cannot be obtained from peroneal-innervated foot muscles; recording from the tibialis anterior muscle is useful in these patients. Focal conduction block or conduction slowing, recording from the tibialis anterior, may be demonstrated when stimulation at the fibular head is compared to that at the popliteal fossa. Approximately 14% to 22% of patients will have an accessory deep peroneal nerve, an anomalous motor branch from the deep peroneal through the superficial peroneal, which helps innervate the extensor digitorum communis. Common peroneal lesions often result in loss of the superficial peroneal sensory potential; selective involvement of the deep peroneal nerve leaves the sensory potential intact. Needle EMG demonstrates active and/or chronic denervation in peroneal-nerve-innervated muscles with sparing of the short head of the biceps femoris.

Distal tibial nerve entrapment under the flexor retinaculum at the medial malleolus is often difficult to demonstrate electrophysiologically. Tarsal tunnel entrapment is not the electrophysiologic counterpart of carpal tunnel syndrome. In carpal tunnel compression there is frequently slowing of motor and sensory conduction across the site of entrapment. Tarsal tunnel entrapment infrequently demonstrates focal slowing; instead motor and sensory axonal degeneration is the predominant physiologic abnormality. Digital sensory potentials are usually absent. Occasionally focal slowing is demonstrated either in motor nerves across the flexor retinaculum or in medial or lateral plantar nerves. In definite cases, EMG usually demonstrates active or chronic denervation in tibial-nerve-innervated intrinsic foot muscles.

TREATMENT AND PROGNOSIS

Mild stretch or compression injuries leave the nerve in continuity and generally recover spontaneously. In most cases, these lesions are a combination of class 1 and 2 injuries. Patients should be warned against habitual leg crossing, prolonged sitting on the toilet, or assuming unusual positions (i.e., those employed in yoga exercise).

Electrodiagnostic studies are useful in differentiating class 1 from class 2 lesions. Conservative therapy is continued until it is clear that regeneration is not occurring. The onset of recovery is expected with class 1 injuries by approximately one to five months.[48] Recovery from class 2 lesions depends on severity and location of injury, both of which affect the duration of the latent period. Most class 2 lesions recover spontaneously and should be initially treated conservatively. Two exceptions are: (1) chronic peroneal compression at the fibular head which often shows poor spontaneous recovery, possibly because of intraneural fibrosis, which retards regeneration. Such lesions, especially if progressive and/or painful, should be explored to free the nerve from adhesions. (2) Early surgical exploration is also warranted after gluteal or upper thigh trauma, which results in moderate to severe injury to the sciatic nerve. Because of the proximity of the injury to the cell body and the long distance the regenerated nerve must traverse, considerable time may elapse before it becomes obvious that recovery is not occurring. This may result in a poor response to surgery. We therefore recommend early exploration for proximal sciatic nerve lesions to ensure nerve continuity. Third-degree injuries rarely recover; they may be a consequence of sudden traction injury over a long length of nerve, resulting in rupture of nerve fibers and vessels.

Regardless of the etiology or site of injury, proper splinting of the foot protects weakened dorsiflexors from overstretch and avoids fixed plantar flexion deformities. The splint should ensure safe ambulation by providing adequate mediolateral stability of the ankle during the stance phase, thereby preventing ankle turning. It should also provide enough foot dorsiflexion during the swing phase to allow the toes to clear the ground.

In most cases, the only splint required is one that supports the ankle and foot. A standard metal double-upright Klenzak brace, attached to a firm shoe, compensates for weak dorsiflexors and gives adequate mediolateral ankle support. Excessive ankle inversion or eversion may be corrected by a T-strap that covers the malleolus and is attached to the metal upright (illustrated in Chapter 21). The plantarflexion stop at the ankle prevents toe dragging and stumbling during the swing phase. Patients requiring less mediolateral ankle support do well with a lightweight, plastic brace composed of a single posterior upright (i.e., Teufel brace). The splint fits into the shoe, thereby allowing a change of footwear as long as the relative heights of the sole and heel remain identical as when originally fitted. Greater mediolateral support is provided by a plastic brace that encases the sides of the ankle with reinforced plastic (i.e., Seattle brace).

The degree of dorsiflexion support provided must be a balance between the degree of lift required for toe clearance and the effect this lift will have on the knee joint. Splinting the ankle always affects the knee. The lack of plantarflexion during the stance phase of walking causes the patient to rock over the posterior portion of the heel to achieve the flatfooted position. The resulting knee flexion must be overcome by the knee extensors before the swing phase can be initiated. The greatest degree of knee bending and, therefore, the greatest stress on knee extensors, results when rigid plantarflexion stops are used. Use of a spring stop, instead of a rigid pin stop, allows gradual lengthening of dorsiflexors, which takes the strain off the knee extensors. Positioning the foot in 5° of plantarflexion significantly reduces knee bending. A cushion wedge or cutoff heel causes the foot to contact the ground proximal to the heel, and brings the line of force closer to the center of the knee with less knee bending.

The addition of a dorsiflexion stop often stabilizes the gait accompanying plantarflexion weakness which predisposes to forward instability. Patients with impaired pushoff may benefit from a rocker bar attached to the sole of the shoe. This facilitates shifting of their weight forward during initiation of the swing phase. An optimal brace provides the least amount of plantarflexion restriction needed for the toes to clear the floor during walking, while reducing effort by making walking more efficient.

Splints should be removed frequently for ROM exercises. The weight of blankets at night must not force the foot into plantarflexion and thus stretch the dorsiflexors. A metal brace is usually better for long-term support. Splinting should continue for as long as voluntary activity is impaired. Once recovery begins, strengthening exercises are initiated for inversion and eversion of the foot and extension and flexion of the ankle and toes.

Sural Nerve (L_5, $S_{1,2}$)

The sural nerve arises from the sciatic nerve and descends to the back of the calf, winds around posterior to the lateral malleolus, and reaches the lateral border of the foot (see Fig. 16–6). It supplies the skin in this distribution; it has no motor fibers. Sensory impairment occasionally results from pressure on the nerve as it lies in a superficial position in the back of the calf or from trauma around the ankle. Fracture of the base of the fifth metatarsal bone may damage the sural nerve within the foot. Hemorrhage or Baker's cyst in the popliteal fossa may damage the sural nerve, usually in combination with either peroneal or tibial nerves or both. The most common etiology of sural nerve injury is diagnostic nerve biopsy when the whole nerve is removed at the ankle.

REFERENCES

1. Aguayo, A, Nair, CP, Midgley, R, et al: Experimental progressive compression neuropathy in the rabbit. Arch Neurol 24:358, 1971.
2. Behse, F, Buchthal, F, Carlsen, F, et al: Hereditary neuropathy with liability to pressure palsies: Electrophysiological and histopathological aspects. Brain 95:777, 1972.
3. Bonney, G: Some lesions of the brachial plexus. Ann R Coll Surg 59:298, 1977.
4. Bonney, G: Prognosis in traction lesions of the brachial plexus. J Bone Joint Surg 41:4, 1959.
5. Dawson, D, Hallett, M, and Millender, L: Entrapment Neuropathies, ed 2. Little, Brown and Company, Boston, 1990.
6. Denny-Brown, D: Neurological conditions resulting from prolonged and severe dietary restriction. Medicine 26:41, 1947.
7. Earl, CJ, Fullerton, PM, Wakefield, GS, et al: Hereditary neuropathy with liability to pressure palsies. Q J Med 33:481, 1967.
8. Ecker, AD and Woltman, HW: Meralgia paresthetica: A report of one hundred and fifty cases. JAMA 110:1650, 1938.
9. Feindel, W and Stratford, J: The role of the cubital tunnel in tardy ulnar palsy. Can J Surg 1:287, 1958.
10. Fullerton, PM and Gilliatt, RW: The carpal tunnel syndrome. Lancet 2:241, 1965.
11. Fullerton, PM: The effect of ischaemia on nerve conduction in the carpal tunnel syndrome. J Neurol Neurosurg Psychiatry 26:385, 1963.
12. Fullerton, PM and Gilliatt, RW: Median and ulnar neuropathy in the guinea-pig. J Neurol Neurosurg Psychiatry 30:393, 1967.
13. Gilliatt, RW: Physical injury to peripheral nerves: Physiologic and electrodiagnostic aspects. Mayo Clin Proc 56:361, 1981.
14. Gilliatt, RW, LeQuesne, PM, Logue, V, et al: Wasting of the hand associated with a cervical rib or band. J Neurol Neurosurg Psychiatry 33:615, 1970.
15. Gilliatt, RW and Willison, RG: Peripheral nerve conduction in diabetic neuropathies. J Neurol Neurosurg Psychiatry 25:11, 1962.
16. Gilliatt, RW: Peripheral nerve compression and entrapment. In Lant, AF (ed): Eleventh Symposium on Advanced Medicine: Proceedings of a Conference Held at the Royal College of Physicians of London, 17–21 February, 1975. Pitman Medical, London, 1975, p 144.
17. Haftek, J and Thomas, PK: Electron-microscope observations on the effects of localized crush injuries on the connective tissues of peripheral nerve. J Anat 103:233, 1968.
18. Harrison, MJG and Nurick, S: Results of anterior transposition of the ulnar nerve for ulnar neuritis. Br Med J 1:27, 1970.
19. Heathfield, KWG: Acroparesthesiae and the carpal tunnel syndrome. Lancet 2:663, 1957.
20. Keegan, JJ and Holyoke, EA: Meralgia paresthetica. J Neurosurg 19:341, 1962.
21. Kendall, D: Aetiology, diagnosis and treatment of paraesthesiae in the hands. Br Med J 2:1633, 1960.
22. Klein, DG: Surgical repair of peripheral nerve injury. Muscle Nerve 13:843, 1990.
23. Lam, SJS: Tarsal tunnel syndrome. J Bone Joint Surg 49B:87, 1967.
24. Landau, WM: The duration of neuromuscular function after nerve section in man. J Neurosurg 10:64, 1953.
25. Leffert, RD: Brachial Plexus Injuries.

Churchill-Livingstone, New York, 1985.

26. Lewis, T, Pickering, GW, Rothschild, P, et al: Centripetal paralysis arising out of arrested bloodflow to the limb, including notes on a form of tingling. Heart 16:1, 1931.

27. Linscheid, RL, Burton, RG, Fredericks, EJ, et al: The tarsal tunnel syndrome. South Med J 63:1313, 1970.

28. Mayer, RF and Denny-Brown, D: Conduction velocity in peripheral nerve during experimental demyelination in the cat. Neurology (Minneapolis) 14:714, 1970.

29. Morris, JH, Hudson, AR, Weddell, G, et al: A study of degeneration and regeneration in the divided rat sciatic nerve based on electron microscopy. I. The traumatic degeneration of myelin in the proximal stump of the divided nerve. Z Zellforsch Mikrosk Anat 124:76, 1972.

30. Morris, JH, Hudson, AR, Weddell, G, et al: A study of degeneration and regeneration in the divided rat sciatic nerve based on electron microscopy. II. The development of the 'regenerating unit.' Z Zellforsch Mikrosk Anat 124:103, 1972.

31. Morris, JH, Hudson, AR, Weddell, G, et al: A study of degeneration and regeneration in the divided rat sciatic nerve based on electron microscopy. III. Changes in the axons in the proximal stump. Z Zellforsch Mikrosk Anat 124:131, 1972.

32. Morris, JH, Hudson, AR, Weddell, G, et al: A study of degeneration and regeneration in the divided rat sciatic nerve based on electron microscopy. IV. Changes in fascicular microtopography, perineurium and endoneurial fibroblasts. Z Zellforsch Mikrosk Anat 124:165, 1972.

33. Neary, D and Eames, RA: The pathology of ulnar nerve compression in man. Neuropathology and Applied Neurobiology 1:69, 1975.

34. Neary, D, Ochoa, J, Gilliatt, RW, et al: Sub-clinical entrapment neuropathy in man. J Neurol Sci 24:283, 1975.

35. Ochoa, J, Fowler, TJ, Gilliatt, RW, et al: Anatomical changes in peripheral nerves compressed by a pneumatic tourniquet. J Anat 113:433, 1972.

36. Ochoa, J and Marotte, L: The nature of the nerve lesion caused by chronic entrapment in the guinea-pig. J Neurol Sci 19:491, 1973.

37. Preswick, G and Jeremy, D: Subclinical polyneuropathy in renal insufficiency. Lancet 2:731, 1964.

38. Puckett, WO, Grundfest, H, McElroy, WD, et al: Damage to peripheral nerves by high velocity missiles without a direct hit. J Neurosurg 3:294, 1946.

39. Roos, DB: Experience with first rib resection for thoracic outlet syndrome. Ann Surg 173:429, 1971.

40. Scotti, TM: The lesion of Morton's metatarsalgia (Morton's toe). Arch Pathol 63:91, 1952.

41. Seckel, BR: Enhancement of peripheral nerve regeneration. Muscle Nerve 13:785, 1990.

42. Seddon, HJ: Surgical Disorders of Peripheral Nerves. Churchill-Livingstone, London and Edinburgh, 1972.

43. Seddon, HJ: Three types of nerve injury. Brain 66:327, 1943.

44. Spencer, PS: The traumatic neuroma and proximal stump. Bull Hosp Joint Dis 35:85, 1974.

45. Spencer, PS, et al: Reappraisal of the model for "bulk axoplasmic flow." Nature: New Biol (London) 240:283, 1972.

46. Staal, A: General discussion on pressure neuropathies. In: Vinken, PJ and Bruyn, GW (eds): Handbook of Clinical Neurology, Vol 7. North Holland-American Elsevier, Amsterdam and New York, 1970, p 276.

47. Steward, J: Focal Peripheral Neuropathies. Elsevier, New York, 1987.

48. Sunderland, S: Nerves and Nerve Injuries, ed 2. Churchill-Livingston, Edinburgh, 1978.

49. Sunderland, S: The anatomy and physiology of nerve injury. Muscle Nerve 13:771, 1990.

50. Swift, TR and Nichols, FT: The droopy shoulder syndrome. Neurology 34:212, 1984.

51. Thomas, PK: Nerve injury, In Bellairs, R and Gray, EG (eds): Essays on the Nervous System—A Festschrift for Professor J.Z. Young. Clarendon Press, Oxford, 1974, p 44.

52. Thomas, PK: Changes in the endoneurial sheaths of peripheral myelinated nerve fibers during Wallerian degeneration. J Anat 98:175, 1964.

53. Urschel, HG, Razzuk, MA, Wood, RE, et al: Objective diagnosis (ulnar) nerve conduction velocity and current therapy of the thoracic outlet syndrome. Ann Thorac Surg 12:608, 1971.

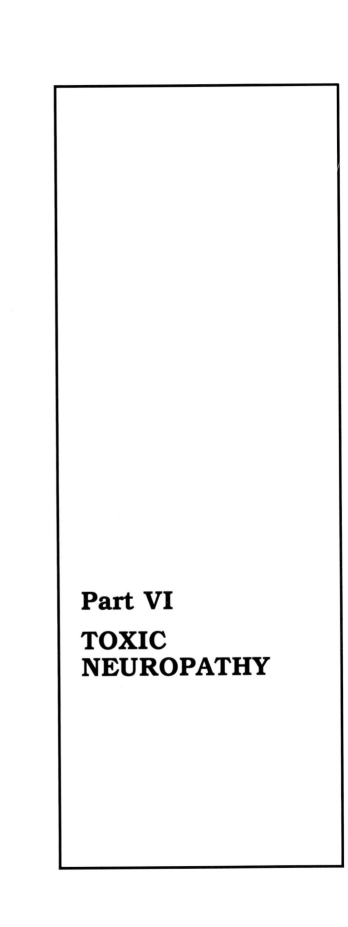

Part VI

TOXIC NEUROPATHY

Chapter 17

PHARMACEUTICAL AGENTS

AMIODARONE
CHLORAMPHENICOL
COLCHICINE
DAPSONE
NUCLEOSIDE NEUROPATHIES (ddC, ddI, d4T)
DISULFIRAM
DOXORUBICIN
ETHAMBUTOL
ETHIONAMIDE
GOLD
GLUTETHIMIDE
HYDRALAZINE
ISONIAZID
METRONIDAZOLE
MISONIDAZOLE
NITROFURANTOIN
NITROUS OXIDE
PERHEXILINE MALEATE
PHENYTOIN
PLATINUM
PYRIDOXINE
SODIUM CYANATE
TAXOL
THALIDOMIDE
VINCRISTINE

New pharmaceutical agents are constantly being identified or implicated as causes of human peripheral neuropathy; most such drugs appear to produce distal axonopathy, usually after prolonged use. There is more individual variation in nervous system vulnerability to some classes of drugs (antibiotics, anticancer agents) than to most occupational or environmental toxins. This likely reflects variation in absorption and integrity of the blood-brain barrier in systemic illness. There are few careful experimental studies of the neurotoxicity of these substances; clinical reports are often the sole basis for many of the alleged drug-induced peripheral neuropathies. Some instances doubtless reflect peripheral nervous system (PNS) dysfunction secondary to other, coincident conditions. This chapter discusses pharmaceutical agents consistently associated with neuropathy.

AMIODARONE

Amiodarone, a di-iodinated benzofuran derivative, is a potent anti-ventricular–arrhythmia agent; its neurologic adverse effects include tremor, optic neuropathy, and distal symmetric polyneuropathy.[44,75,97] Peripheral neuropathy usually occurs at high doses (550 mg per day) following prolonged administration (24 months); however, it has occasionally occurred in individuals taking a standard dose of 200 mg. Serum concentrations of 2.4 mg/L are associated with neuropathy and there is considerable variation in absorption.[44] Lower-limb weakness usually is the dominant symptom; paresthesias occur later. Gradually, a pattern of distal symmetric sensorimotor neuropathy evolves, often accompanied by tremor. Most recover if the drug is stopped soon after symptoms appear. Continued administration is associated with poor recovery, conduction velocity is markedly decreased, and distal denervation is present on electromyography (EMG), findings compatible with both demyelination and axonal degeneration. Nerve biopsy studies demonstrate predominant symmetric demyelination, loss of Schwann-cell organelles, and accumulations of lamellar inclusions in all cell types.[58] Experimental administration of amiodarone has not produced neuropathy, but the lysosomal lamellar cytoplasmic inclusions accumulate in tissues outside the blood-brain barrier.[26,27] Amiodarone likely causes the lamellar inclusions by its inhibition of phospholipases; possibly nerves in old age have increased vascular permeability to amiodarone, permitting access of the drug to Schwann cells. Amiodarone neuropathy shares clinical and histopathologic features with perhexiline maleate neuropathy. Perhexiline also causes a marked reduction in the activity of a lysosomal enzyme, sphingomyelinase.

CHLORAMPHENICOL

This antibiotic can produce a distal symmetric neuropathy, usually accompanied by optic neuropathy.[102] It is now rare, and most previous instances occurred in children with cystic fibrosis receiving prolonged, high-level therapy. It is generally held that the incidence of neuropathy parallels both dose and duration; the mean duration in reported cases is 229 days and the mean dose 255 g. Chloramphenicol is now given in short courses to avoid other, more serious hematologic complications.

The neuropathy is heralded by numbness of the feet, followed by moderate calf pain and tenderness.[60] Objective signs are few, save for diminution of pain and touch sensation, and loss of Achilles and patellar reflexes. The upper extremities are usually spared. Complete recovery of peripheral neuropathy and optic neuritis occurs if chloramphenicol therapy is stopped soon after initial sensory symptoms appear. Treatment with high doses of B vitamins is advocated, although its rationale is questionable.[22]

There is no animal model of this neuropathy and few morphologic studies of the human PNS disease. An instance of fiber loss in the gracile fasciculi is recorded in an individual without evidence of neuropathy.[23]

COLCHICINE

Chronic administration of colchicine at the usual dose of 0.6 mg twice daily can cause a mild sensory distal axonopathy.[107] It is usually overshadowed by a coincident debilitating vacuolar myopathy with elevated serum creatine kinase. The primary risk factor for colchicine myoneuropathy appears to be chronic renal dysfunction, common in gout.[67] Most cases have only minor distal acral sensory symptoms and a mild stocking and glove sensory loss. Myopathic proximal weakness usually dominates the clinical profile and an erroneous diagnosis of polymyositis is common. Sensory nerve conduction studies reveal diminished amplitude with near-normal velocity; sural nerve biopsy in one case displayed reduced numbers of myelinated fibers. Removal from medication is followed by a dramatic fall in creatine kinase and gradual clinical improvement. The pathogenesis of the axonopathy is suggested to be defective axonal transport resulting from impaired microtubule assembly. Vincristine, an agent associated with microtubular disaggregation, also causes distal axonopathy together with a vacuolar myopathy.

DAPSONE

Dapsone (Avlosulfon) is a sulfone derivative, 4,4-sulfanylbisbenzamine, used in treating leprosy, pneumocystis pneumonia, and dermatologic conditions. Instances of reversible neuropathy have been reported, most occurring after prolonged dermatologic treatment, usually 200 to 400 mg per day.[52,104] Most cases require months of treatment; one report describes onset following 16 years at 100 mg daily,[129] another after 5 days following massive, attempted suicidal ingestion of 6 g.[88] The low doses used for leprosy and brief courses used for pneumocystis infection are not associated with neuropathy. Weakness is the predominant feature in all cases, and it is widely held that dapsone produces a predominantly motor neuropathy, although two reports describe abnormalities in sural nerve biopsy specimens.[37,88] Weakness is symmetric and usually begins in distal extremities. Eventually, atrophy occurs. Electrophysiologic studies consistently reveal slowed motor conduction velocity and denervation of distal muscles, with no abnormalities of sensory conduction.

It is suggested that dapsone primarily affects the axons of motor neurons. Its sporadic occurrence may reflect variable acetylation, as with isoniazid (vide infra), although one case is described in a rapid acetylator.[48,115] There is no experimental animal model.[135]

NUCLEOSIDE NEUROPATHIES (ddC, ddI, d4T)

The nucleosides dideoxycytidine (ddC), dideoxyinosine (ddI), and d4T are used to treat the acquired immune deficiency syndrome. These promising agents display little hematopoietic depression, but may cause a dose-limiting dysesthetic, distal axonopathy; ddC is the best studied. High doses (greater than 0.03 mg/kg six times daily) for eight weeks cause a subacute onset of an intense aching and burning sensation over the soles.[82,117] Lancinating lower-limb pain soon follows and, if the drug is not stopped, the hands become involved. Few objective sensory findings accompany these symptoms, and nerve conduction studies are unremarkable at this stage. Subsequently a stocking-glove distribution of predominantly large-fiber–type sensory loss appears and the Achilles reflex is diminished. Mild weakness may accompany the extreme cases.[35] A coasting period of intensifying symptoms for six weeks may follow withdrawal of the drug. Lower doses cause a less-intense dysesthetic syndrome after variable intervals. All gradually improve following the coasting period; the degree of recovery depends upon the severity of neuropathy and the pre-existent debilitation of the patient. Antecedent subclinical HIV neuropathy appears to predispose to nucleoside neuropathy. Experimental studies in primates have produced slowed nerve conduction, but no histo-pathologic evidence of nerve fiber degeneration.[7,117]

DISULFIRAM

Disulfiram (Antabuse) is used as an adjunct in treatment of motivated chronic alcoholic patients. The principal neurotoxic syndrome associated with disulfiram is peripheral neuropathy of the distal axonopathy type.[15] Most cases occur at standard therapeutic doses (250 to 500 mg daily) and commence within several months of starting treatment; one report describes an onset after 30 years treatment with 250 mg daily.[13] Tingling paresthesias in the feet, followed shortly by unsteady gait, are initial complaints. Signs of diminished pain, temperature, and position sense in the feet, absent reflexes, and weakness of foot dorsiflexion are present in most cases. Eventually, distal upper extremities are involved. Cranial nerve palsies are not a feature of disulfiram neuropathy. Optic neuritis may occur independently of peripheral neuropathy. Drug withdrawal is followed by remission of signs within months in most cases.[13,15,85,110]

Mild slowing of motor nerve conduction, diminished amplitude of sensory action potentials, and electromyographic evidence of denervation in distal muscles are characteristic of disulfiram neuropathy. Sural nerve histologic changes include loss of myelinated fibers and axonal degeneration. Two biopsies have disclosed axonal swellings filled with intermediate filaments.[3,32] It is suggested that carbon disulfide, a metabolite of disulfiram, may have a role in the pathogenesis of the axonal swellings. The few experimental animal studies are compatible with distal axonopathy;[4] one describes local axonal degeneration related to local injection of the drug.[140]

Doxorubicin

Doxorubicin (Adriamycin) is an anthracycline antineoplastic agent that acts at the cell nucleus by intercalating between the base pairs of DNA, inhibiting messenger RNA synthesis. A single dose given to an experimental animal produces diffuse degeneration of limb sensory and gasserian ganglion cells; this causes limb ataxia and generalized hypesthesia; strength is preserved.[20] The distribution of this prototypic sensory neuropathy syndrome presumably is limited by the blood-brain barrier to the large, lipid-insoluble doxorubicin molecule. Experimental intraneural administration of minute amounts causes local Schwann-cell degeneration and demyelination;[39] larger amounts, following local intramuscular administration, are transported retrogradely to the anterior horn cells, which subsequently degenerate (suicide transport).[40]

There are no descriptions of human sensory neuropathy or neuronopathy from doxorubicin. Several factors may dictate the absence of reports of human neuropathy: species differences, the lower doses given in humans to avoid cardiac and hematopoietic effects, or its neurotoxic effects may be erroneously attributed to other commonly co-administered antineoplastic agents, e.g., vincristine, cisplatin, and podophyllin.

ETHAMBUTOL

Ethambutol, an antituberculous agent, may cause a severe optic neuropathy and mild sensory distal polyneuropathy.[87,122] These adverse effects are commonly associated with prolonged doses exceeding 20 mg/kg per day. The elderly are allegedly at greater risk for polyneuropathy. Numbness of the feet and fingers is customary, and is usually accompanied by mild, large-fiber—type sensory loss; weakness is rare. Recovery from the polyneuropathy generally occurs following drug withdrawal; recovery from optic neuropathy is more variable, especially in advanced cases.[100,122] An experimental murine study describes axonal degeneration in the sciatic nerve.[78]

ETHIONAMIDE

Ethionamide is used in the treatment of tuberculosis. Its role is limited because of serious gastrointestinal, dermatologic, and CNS side effects. Several clinical reports describe progressive, moderate distal symmetric sensory polyneuropathy; paresthesias of the feet, loss of Achilles reflex, and gradual recovery suggest that this is a distal axonopathy.[70,101,124] Neither electrophysiologic nor nerve biopsy studies are available, and there is no experimental animal model of ethionamide neuropathy.

GOLD

Organic gold compounds are employed in the treatment of rheumatoid arthritis, but potentially severe toxic reactions limit their use. Cutaneous and renal side effects are common; peripheral neuropathy is rare. None of the toxic side effects appear to be dose-related.

Peripheral neuropathy is often heralded by intermittent numbness in the legs, followed within months by progressive distal weakness, diminished pain and temperature sense, and areflexia in the distal extremities.[34,61,131,132]

Onset is usually subacute and may be accompanied by diffuse myokymia, muscle pain, and cutaneous hyperesthesia (grippe aurique). Following drug withdrawal, gradual recovery occurs, the degree of residual dysfunction usually being proportionate to the maximum disability. Motor and sensory nerve conduction velocities are moderately slowed, and cerebrospinal fluid (CSF) protein elevated. Nerve biopsies reveal a mixture of axonal degeneration and segmental demyelination.[61,131] Neither vasculitis nor endoneurial inflammatory cells are histologic features of this illness. Lymphocyte transformation studies are negative. The diagnosis of gold neuropathy may be especially difficult since the diffuse symptoms are difficult to interpret in patients with rheumatoid arthritis; in addition, rheumatoid arthritis is itself associated with peripheral neuropathy.

The nature and pathogenesis of this neuropathy are unclear. Neither the morphologic studies of nerve biopsies, nor an experimental animal study, clearly indicate whether the primary change is axonal or demyelinative.[61] It is also undecided if the PNS dysfunction results from a direct toxic effect of gold or has an allergic basis. Review of the experimental data and clinical phenomena does not strongly support either position, and the nature of gold neuropathy remains elusive. Gold therapy has also been associated with AIDP.

GLUTETHIMIDE

Glutethimide (Doriden) is a sedative-hypnotic drug that shares many of the pharmacologic properties of thalidomide, a structurally similar compound. Glutethimide neuropathy is rare, and usually follows long-term, high-dose therapy.[89] The clinical pattern suggests a distal axonopathy: slowly progressive distal sensory impairment, bilateral symmetric leg and foot paresthesias, and cramping and tenderness of the calf muscles. All sensory modalities are diminished in the feet, Achilles reflexes are lost, and mild ataxia occurs. Signs gradually remit within six months of drug withdrawal. There are no reports of nerve biopsies, and no experimental animal model.

HYDRALAZINE

Hydralazine (Apresoline) is widely used in the treatment of hypertension. It rarely produces peripheral neuropathy, presumably of the distal axonopathy type. Manifestations of neuropathy appear after widely varying intervals of treatment. In the few case reports, symptoms of distal-extremity numbness and paresthesias predominate; leg weakness is mild and less frequent.[62,105] Following withdrawal of the drug, neu-

rologic dysfunction gradually subsides and recovery is usually complete. There exist neither morphologic studies of human nerve nor an animal model of hydralazine neuropathy.

It is likely that hydralazine-induced PNS dysfunction is related to pyridoxine deficiency. Two factors support this notion: one is the similarity both in chemical structure and in clinical manifestations; the other is that both cause increased excretion of xanthurenic acid (vide infra) following a tryptophan load in humans.[105]

ISONIAZID

Isoniazid (INH) is one of the cheapest and most effective antituberculous drugs. Peripheral neuropathy of the distal axonopathy type is the most common toxic side effect. The primary route of INH metabolism is by acetylation. Individuals unable to acetylate normally (slow acetylators) maintain prolonged high blood levels of INH, and are more susceptible to neuropathy than are rapid acetylators. The basis for slow acetylation is genetic, inherited as an autosomal recessive trait,[11,41] and increases with age.[95]

INH exerts most of its toxic actions by interference in several ways with compounds of the vitamin B_6 group (pyridoxine, pyridoxal, and pyridoxamine). Most important for the pathogenesis of neuropathy is probably its inhibition of pyridoxal phosphokinase, the enzyme that phosphorylates pyridoxal to yield a coenzyme essential for several metabolic reactions.[11]

Pathology and Pathogenesis

Studies of human postmortem material, nerve biopsies, and experimental animal models strongly support the view that INH neuropathy is a distal axonopathy. Degeneration of myelinated and unmyelinated axons has been carefully documented in sural nerve biopsies,[90] while distal PNS axonal degeneration, denervation muscle atrophy, and fiber loss in dorsal columns are present at autopsy.[11] Peripheral neuropathy is readily produced in rats and its pattern closely mimics that seen in humans.[10,19,63] Both sensory and motor nerves undergo dose-dependent, distal axonal degeneration, and fiber loss is also present in the rostral gracile fasciculi.[63] One experimental study suggests that INH may also have a direct effect on the nerve fiber.[59] INH intoxication in other species (dog, chicken) produces widespread vacuolation of white matter, presumably reflecting a primary effect on the oligodendrocyte.[11]

The pathogenesis of peripheral neuropathy is clearly related to the overall interference with B$_6$ metabolism, and axonal degeneration can be prevented by the administration of vitamin B$_6$. The primary event that leads to axonal degeneration is uncertain, since the role of pyridoxal phosphate–dependent enzymes in axonal integrity is not known.[11] The overall pattern of axonal change in INH distal axonopathy is widely held to be similar to that of other metabolic axonopathies (see details in Chapter 2).

Clinical Features

Peripheral neuropathy is a dose-related effect, allegedly more likely to occur in the elderly, in malnourished individuals,[84] and slow acetylators. Common doses (3 to 5 mg/kg daily) are associated with a 2% incidence of neuropathy, 6 mg/kg daily with 17% incidence, and, with higher doses, the incidence increases still more. Symptoms of neuropathy may appear within three weeks in the latter group; conventional doses cause neuropathy after six months. Initial symptoms are tingling paresthesias in the feet, usually followed by complaints of weakness or unsteady gait. Rarely, paresthesias commence in the fingers. Loss of vibration, pain, and temperature sense is usually greater than position and deep pain.

Aching cutaneous pain in the calf muscles is an especially common complaint, and often accompanies distal leg weakness and reflex loss. The neuropathy evolves gradually with continued INH administration; such cases eventually develop distal muscle atrophy, ataxia, and profound sensory loss.[47]

Recovery usually commences within weeks of drug withdrawal and is gradual, taking months in mild cases and years with advanced involvement. Pyridoxine administration does not affect the rate of recovery.[11]

Peripheral neuropathy can be prevented in most instances by daily administration of 100 mg of pyridoxine. Fortunately, pyridoxine does not affect the antituberculous action of INH.

METRONIDAZOLE

Metronidazole (Flagyl), used in the treatment of anaerobic bacterial and protozoan infections and in Crohn's disease, causes a neuropathy of the distal axonopathy type.[14,28,123] Dysesthesias of toes and finger tips and a terminal intention tremor herald the onset of a predominantly large-fiber-type sensory neuropathy. Two reports describe axonal degeneration in sural nerve biopsies.[14,123] Recovery usually occurs if the drug is withdrawn soon after appearance of symptoms; delay may eventuate in prolonged sensory dysfunction.[56] Most cases have followed prolonged treatment with doses exceeding 30 g;[123] however, occasional instances have appeared at lower doses, and one report suggests the absence of a clear dose-effect relationship.[36]

MISONIDAZOLE

Misonidazole, used as a cell sensitizer for cancer radiotherapy, causes a neuropathy of the distal axonopathy type that is dose-limiting. A dose of 11 g/m^2 administered over a three-week interval causes a partially reversible neuropathy.[96] Cumulative doses of less than 6

g/m^2 weekly are recommended to avoid neuropathy. A painful, distal, lower-limb sensory neuropathy of subacute onset is described in most series.[81,96,116] Sural nerve biopsy and electrophysiology suggest a distal axonopathy.[81,96]

Experimental animal studies of misonidazole intoxication reveal distal axonal degeneration in intramuscular hind-foot nerves, accompanied by widespread necrotic changes in the CNS, somewhat resembling thiamine deficiency.[51] One study suggests that desmethylmisonidazole, a major metabolite, also causes peripheral neuropathy.[109]

NITROFURANTOIN

Nitrofurantoin (Furadantin) is a synthetic antimicrobial agent used to treat urinary tract infections. Peripheral neuropathy of the distal axonal type is among the most serious toxic effects of nitrofurantoin.[38,73] It is suggested that two factors predispose individuals with renal failure to nitrofurantoin neuropathy. One is the excessive tissue concentrations of nitrofurantoin, normally excreted by the kidney, that occur with renal failure; the other is the presence of subclinical neuropathy in uremia or diabetes.[133] Although neither suggestion is experimentally proved, it is considered prudent to avoid using this drug in uremic and diabetic patients. Nitrofurantoin neuropathy also occurs in individuals without pre-existing renal failure.[139] Normal adults treated with nitrofurantoin for two weeks may develop asymptomatic neuropathy with slowed nerve conduction.[127] Experimental animal studies have also demonstrated that intoxication can produce axonal degeneration in normal animals.[8,64]

Peripheral neuropathy usually commences within months, sometimes weeks, of beginning therapy. Numbness distally in the legs is usually the initial symptom, followed by the subacute onset of severe distal weakness in the limbs and profound sensory loss.

All reports stress the rapidity of development of this neuropathy, unusual among the distal axonopathies. Occasionally, a predominantly motor syndrome appears[86] and, in view of the subacute onset, it may superficially resemble AIDP. Electrophysiologic studies usually display mild slowing of motor nerve conduction, suggesting an axonopathy. The course is variable. If the drug is withdrawn immediately following initial symptoms, only mild impairment occurs and recovery is complete. Should severe changes occur before therapy is stopped, recovery is slow and incomplete.

Experimental animal studies and a human postmortem report[71] describe distal axonal degeneration. Postmortem changes in an especially advanced case include dorsal root degeneration and chromatolysis in dorsal root ganglion and anterior horn neurons. Spinal tract degeneration has not been convincingly demonstrated. The observations of axonal degeneration as far proximal as spinal roots and neuronal chromatolysis may correlate with poor recovery in advanced cases. The biochemical lesion underlying this unusual distal axonopathy is unknown, although it is suggested that nitrofurans may disrupt metabolism by competitive inhibition of pyruvate oxidation.[94]

NITROUS OXIDE

Nitrous oxide, widely used as a dental anesthetic and a food propellant, can cause megaloblastic anemia and toxic myeloneuropathy. Its abuse as a euphoriant occurs among young people and dentists in North America. Moderate abuse may result in signs of mild distal symmetric sensory polyneuropathy.[69,113] Prolonged high-level abuse also produces signs of myelopathy and optic nerve dysfunction.[55] The most common initial symptom is numbness in the distal arms and legs, often combined with poor finger dexterity, leg weakness, and gait imbalance. Signs of PNS dysfunction at this early stage include depressed tendon reflexes, and slight impairment of vibration and pain sensation in the hands and legs. Lhermitte's sign also occurs. Mild slowing of motor and sensory nerve conduction may accompany this early clinical stage. More severe cases develop lower extremity spasticity; severe loss of vibration, position, and pain sense in all extremities; hyperreflexia; and Babinski's responses. Sural nerve histology and CSF are normal.

The overall multifocal pattern, and evolution of the early stage of the clinical profile, suggest a toxic, predominantly distal axonopathy and a central myelinopathy; the fully developed syndrome mimics combined system disease (see Chapter 9). The well-known megaloblastic effect of this substance[2] reinforces the latter view, as does the demonstration of spongy degeneration in the dorsal and lateral columns in monkeys exposed to nitrous oxide. Administration of methionine allegedly prevents the myeloneuropathy in primates.[119] Animal and in vitro experiments indicate that nitrous oxide may give rise to this myelopathy by an interference with vitamin B_{12} metabolism. Experimental studies suggest that nitrous oxide produces megaloblastic anemia by inhibition of methionine synthetase.[43]

PERHEXILINE MALEATE

Perhexiline maleate (Pexid) is a calcium antagonist formerly used in the treatment of angina pectoris. Peripheral neuropathy and hepatic dysfunction are its principal toxic effects.

Initial complaints occur after treatment for months or years, and are not clearly dose-related. Numbness appears in the distal extremities, rapidly followed by diminution of all sensory modalities. Weakness affects proximal and distal muscles, the latter more severely. Facial diplegia and perioral numbness occasionally occur and dysphagia can be an early feature.[114] Autonomic neuropathy, including postural hypoten-

sion, may coexist and may be the presenting symptom. Reflexes are absent or diminished throughout. Once neuropathy is clearly established, recovery occurs slowly after discontinuing therapy.[134] Motor and sensory nerve conduction velocities are markedly slowed, and CSF protein values moderately elevated. Considerable weight loss may occur, and as the erythrocyte sedimentation rate is sometimes elevated, confusion with a paraneoplastic neuropathy can arise.

Muscle biopsy reveals denervation atrophy. Nerve biopsies demonstrate extensive segmental demyelination, remyelination, and axonal degeneration. Osmiophilic inclusions are abundant in Schwann cells, fibroblasts, and endothelial cells.[42] In sum, the clinical profile, electrophysiologic studies, and nerve biopsy findings indicate a combination of segmental demyelination and axonal degeneration. The nature of the primary event is unclear. Affected individuals have high plasma levels and a longer half-life of the drug. There is a genetically determined vulnerability to the neuropathy; at-risk individuals have an inability to hydroxylate debrisoquine.[120] Pretreatment phenotyping for this metabolic error may lower the incidence of neuropathy.[25] Administration of perhexiline maleate to the Dark Agouti rat, a poor hydroxylator of debrisoquine, caused neuronal lipid inclusion accumulation similar to those described in the human neuropathy.[80]

PHENYTOIN

Phenytoin (diphenylhydantoin, Dilantin), a drug widely used in the treatment of epilepsy and pain, causes peripheral neuropathy and cerebellar degeneration. Acute administration of phenytoin to experimental animals causes decreased nerve action potential amplitude and decreased nerve conduction velocity.[76] Neuropathy appears most often in individuals taking diphenylhydantoin for many years and is usually accompanied by elevated plasma levels of drug.[103] It is generally considered that while prolonged diphenylhydantoin therapy may result in diminished reflexes and minor electrophysiologic abnormalities, clinically significant peripheral neuropathy is rare.[74,121] The neuropathy is heralded by mild acral paresthesias and an unsteady gait. Sensory loss may be profound in the feet; weakness is less prominent. Gradual recovery follows drug withdrawal. A sural nerve biopsy in one case showed axonal atrophy and secondary demyelination.[103]

PLATINUM

Cis-diamine-dichlorplatinum 11 (cisplatin) is a widely used antineoplastic agent. Progressive, large-fiber, sensory, distal axonopathy is associated with its use. Polyneuropathy, frequently heralded by Lhermitte's sign, usually occurs after cumulative doses of 225 to 500 mg/m^2.[31,83,108] Rarely, polyneuropathy appears at levels as low as 100 mg/m^2.[50] There is diminished position and vibration sense, sometimes with pseudoathetoid finger movements, in concert with near-normal thermal pain perception. Weakness is mild, if present at all. Sensory nerve conduction velocity is not markedly altered; however, there are diminished sensory amplitudes and delayed sensory latencies.[106] Abnormal lower-limb somatosensory evoked potentials may occur early; motor conduction is near normal. Axonal loss is described in sural nerve biopsies and in the dorsal columns of the spinal cord at postmortem examination.[54,130] Most patients receiving doses lower than 500 mg/m^2 gradually improve following withdrawal; higher doses are associated with profound large-fiber acral sensory loss and poor recovery. Occasionally symptoms and signs do not commence until one or two months following drug withdrawal and worsen for another four months;[83] this scenario may be difficult to distinguish from the sensory neuronopathy associated with carcinoma. Local mononeuropathy and lumbar plexopathy can follow intraar-

terial administration of cisplatin.[18] Gastroparesis and vomiting frequently are associated with polyneuropathy, reflecting autonomic dysfunction.[24] Experimental administration to rats has produced subtle changes in dorsal root ganglion cell morphology, and widespread axonal degeneration in both peripheral and central nervous systems.[21,126] Cisplatin administration to rats also causes delay in the H reflex; co-administration of an ACTH analog, previously shown to enhance nerve repair, prevents these H-reflex changes. A preliminary clinical study of this ACTH analog suggests it may ameliorate human cisplatin neuropathy, and there is currently a multicenter study to investigate further its efficacy.[128] A recent study in mice suggests that daily administration of nerve growth factor may also prevent cisplatin neuropathy.[6] Cisplatin has been used to produce gastroparesis in the ferret[53] and seizures in amphibians.[12]

PYRIDOXINE

Pyridoxine, an essential, water-soluble vitamin (B$_6$) is a co-enzyme for many decarboxylation and transamination reactions. The recommended human daily requirement of pyridoxine is 2.5 mg. An acute, diffuse, irreversible sensory neuronopathy syndrome follows massive intravenous administration,[1] while a gradually progressive sensory distal axonopathy is associated with prolonged consumption of lower doses.[93,118]

Acute sensory neuronopathy is described in two individuals receiving 180 g of pyridoxine intravenously over a period of 3 days as treatment for mushroom poisoning.[1] One week following the injection, they experienced the onset of diffuse paresthesias and appendicular ataxia. Progressive whole-body sensory loss, incapacitating four limb ataxia, and autonomic dysfunction steadily developed. Strength was only slightly and transiently diminished. Nerve conduction studies disclosed absent sensory potentials and mildly diminished motor potentials of uncertain significance. Recovery from autonomic dysfunction has been good; however, they remain disabled from upper-limb sensory ataxia; neither can walk.

Gradually, progressive, reversible, sensory neuropathy is associated with oral consumption of high doses of pyridoxine (200 mg to 10 g daily), usually as part of a self-administration regimen for the premenstrual syndrome.[93,118] The onset of symptoms and course of the illness has been remarkably stereotyped; both are closely related to dose and duration of treatment. Levels of less than 1 g daily usually elicit symptoms after a year or longer, higher levels within months of commencement. An uncritical report claims onset of symptoms after 2.9 years following consumption of a daily mean dose of 117 ± 19 mg.[29] The illness is heralded by unsteady gait and numb feet; most initially report inability to wear high heels. Numbness of the hands and impaired finger dexterity follow within months. All eventuate with a stocking-glove distribution of sensory loss; large-fiber modalities appear especially affected, and strength is preserved. Distal limb tendon reflexes are absent. A study of deliberate, controlled administration to normal volunteers has demonstrated that subtle elevations in acral sensory thresholds precede symptoms.[9] Nerve conduction studies indicate profoundly diminished sensory amplitudes and normal motor conduction and amplitudes. Sural nerve biopsy reveals widespread, nonspecific axonal degeneration of myelinated fibers. Neurologic disability gradually improves following withdrawal; those examined after a prolonged follow-up period make a satisfactory recovery.[118]

Both the acute neuronopathy and the chronic axonopathy syndromes are readily reproduced in experimental animals. Dogs, rats, and guinea pigs develop an acute sensory neuronopathy syndrome characterized by sensory limb ataxia days following massive administration.[65,66,138] Necrosis of dorsal root ganglion cells is accompanied by centrifugal axonal atrophy and degeneration of peripheral and central sensory fibers. Lower doses, chronically administered, have little effect on ganglion cell morphology, but produce distal axonal atrophy and degeneration. The pathogenesis and biochemical basis of pyridoxine neurotoxicity are unknown. The purely sensory syndrome from megadoses may reflect the anatomic vulnerability of the dorsal root ganglion cells. A tissue culture study of pyridoxine analogs suggests that its effects are specifically related to its co-enzyme activity.[136]

SODIUM CYANATE

Sodium cyanate was formerly used to treat sickle-cell anemia. Peripheral neuropathy is a serious toxic effect and commonly occurs following prolonged, conventional levels of therapy. A study of 27 randomly chosen patients receiving sodium cyanate describes nerve conduction abnormalities in 16, sensory symptoms in 5, and signs consistent with PNS dysfunction in 10. All improve following drug withdrawal.[99] Severe neuropathy also may occur and is characterized by gradual development of distal lower extremity sensory loss and foot-drop. Morphologic studies of human sural nerve biopsies[91] and experimentally intoxicated primates and rodents are consistent with distal axonopathy, although proximal demyelination has been a prominent feature late in the experimental disease.[125]

TAXOL

Taxol is a plant alkaloid used as an antineoplastic agent for solid tumors. It is administered in a series of intravenous courses at three-week intervals. Doses above 200 mg/m^2 are associated with sensory neuropathy.[72] Symptoms begin suddenly within days of administration, rapidly increase, and gradually subside before the next dose. Subsequent treatments are accompanied by more intense symptoms. In some instances, treatment is discontinued because of neuropathy.[72] Initial symptoms are paresthesias and dysesthesias in hands and feet with proximal spread. Distal sensory loss for all modalities accompanied by absent tendon reflexes is present in all; weakness is unusual. Nerve conduction studies indicate both axonal degeneration and demyelination. Taxol has a unique mechanism of action; it binds to tubulin and promotes microtubule assembly.[92] This mechanism contrasts with that of vinca alkaloids, which inhibit microtubule assembly. In tissue culture, taxol causes the formation of bundles of microtubules throughout the cytoplasm, disrupting mitosis, neurite initiation, and branching.[77] Nerve growth factor can attenuate these in vitro cytotoxic effects of taxol,[98] and can delay the onset of neuropathy in experimental animals.[5] This finding has potential clinical significance for human taxol neuropathy, since adequate supplies of human nerve growth factor are available for efficacy trials. Direct injection of taxol into peripheral nerve also produces microtubular aggregates in both axons and Schwann cells.[111]

THALIDOMIDE

Thalidomide, formerly used as a sedative-hypnotic, is featured by serious side effects of peripheral neuropathy and embryopathy. Its current pharmaceutic use is confined to the treatment of erythema nodosum leprosum, discoid lupus, aphthosis, and prurigo hodolaris. Thalidomide neuropathy is characterized by unusual clinical features that may reflect widespread nervous system degeneration.[45,46,68,137]

Initial symptoms are always sensory, usually numbness and tingling in the feet, then in the hands. Sensory loss to pain and touch is more profound than to vibration or for joint position, and usually progresses proximally to yield a sizable stocking-glove pattern. Weak-

ness develops subsequently, and may be proximal or distal. Tendon reflexes may be depressed or increased, and Babinski's responses are occasionally present. Characteristic somatic findings of brittle nails and palmar erythema appear together with the neuropathy. Electrophysiologic studies reveal absent or reduced amplitude of sensory potentials and normal motor conduction. Recovery of sensory function may not commence for years following drug withdrawal, is generally poor, may take years, and some patients have residual dysesthesias. Motor recovery is usually satisfactory. Neuropathy may appear within six months at a daily dose of 100 mg; lower doses of 25 to 50 mg daily are associated with onset between 6 to 14 months.[137] It is suggested that frequent measures of sural nerve amplitude are effective in detecting the early onset of neuropathy in asymptomatic individuals.[68]

VINCRISTINE

Vincristine (Oncovin), an alkaloid derived from the periwinkle plant *Vinca rosea*, is widely employed in the treatment of leukemia, lymphoma, and some solid tumors. Peripheral neuropathy and gastrointestinal dysfunction constitute the principal toxic effects and may limit its use.[16,17]

Gastrointestinal dysfunction, constipation and adynamic ileus, and bladder atony, occasionally occur. All are widely held to result from dysfunction of the autonomic nervous system.[16]

Pathology

Human nerve biopsies consistently display axonal degeneration as the prominent feature; segmental demyelination is rare.[16]

Regenerating axons are present in recovering cases. Biopsies of distal muscles reveal denervation changes; proximal muscles contain mild, primarily myopathic, alterations. There is no satisfactory postmortem study of the human nervous system. Reports of animal studies contain descriptions of changes in widely varied areas of both the CNS and the neuromuscular system, and correlate poorly with the clinical findings. Species variation and differing experimental protocols may account for some of these discrepancies. Proximal axonal swelling, neuronal neurofilamentous proliferation, and primary diffuse myopathy are all described.[16,49] Local application to nerves produces diminished numbers and malorientation of neurotubules, maloriented neurofilaments, and fragmentation of the smooth endoplasmic reticulum.[112]

Pathogenesis

Vinca alkaloids function as mitotic spindle inhibitors by binding to tubulin, the microtubule subunit protein. It is proposed that tubulin binding impairs axonal transport and results in nerve fiber breakdown. Experimental animal studies that demonstrate abnormalities in fast axoplasmic transport and cytoskeletal abnormalities support this view.[79,112]

Clinical Features

Vincristine neuropathy occurs in a stereotyped manner, is manifest to some extent in most treated patients, and has clinical features that suggest an unusual distribution and evolution of axonopathy.[16,17]

Paresthesias, often starting in fingers before feet, are the most common initial symptom, and antecede sensory signs by several weeks. Sensory loss is usually mild and remains confined to the most distal parts of the extremities. Weakness, clumsiness, and muscle cramps subsequently appear, may evolve rapidly to severe motor impairment, and are occasionally most pronounced in distal, upper-limb muscles. Eventually, distal leg weakness occurs. Motor signs dominate vincristine polyneuropathy

and may seriously disable affected individuals. Tendon reflexes are absent or diminished in most cases; loss of Achilles reflexes may precede the first sensory symptoms, probably reflecting involvement of afferent fibers from muscle spindles. Numbness and sensory loss in the trigeminal distribution may occur, and jaw pain of obscure etiology is present in 5% to 10% of cases.

Electromyographic findings reflect denervation in distal muscles and sensory nerve conduction studies demonstrate reduced action potential amplitudes.

The prognosis of vincristine neuropathy varies. Withdrawal from therapy at the early stage leads to rapid disappearance of sensory symptoms. Weakness also wanes, but more slowly. Mild residual weakness, impairment of superficial sensation, and absent Achilles reflexes persist in severe cases. Neuropathy may also improve if the dose is reduced, enabling some patients to continue therapy without this disabling side effect. One report claims that co-administration of glutamic acid ameliorates vincristine neuropathy.[57]

REFERENCES

1. Albin, RL, et al: Acute sensory neuropathy-neuronopathy from pyridoxine overdose. Neurology 37:1729, 1987.
2. Amess, JAL, et al: Megaloblastic haemopoiesis in patients receiving nitrous oxide. Lancet 1:339, 1978.
3. Ansbacher, LE, Bosch, EP and Cancilla, P: Disulfiram neuropathy: A neurofilamentous distal axonopathy. Neurology 32:424, 1984.
4. Anzil, AP and Duzic, S: Disulfiram neuropathy: An experimental study in the rat. J Neuropathol Exp Neurol 37:585, 1978.
5. Apfel, SC, Lipton, RB, Arezzo, JC, and Kessler, JA: Nerve growth factor prevents toxic neuropathy in mice. Ann Neurol 29:87, 1991.
6. Apfel, SC, Arezzo, JC, and Kessler, JA: Nerve growth factor prevents

cisplatin neuropathy. Neurology 41 (Suppl):341, 1991.

7. Arezzo, J, et al: Dideoxycytidine (ddC) neuropathy: An animal model in the cynomologus monkey. Neurology 40(Suppl):428, 1990.

8. Behar, A, et al: Experimental nitrofurantoin polyneuropathy in rats. Arch Neurol 13:160, 1965.

9. Berger, A, et al: A prospective study of pyridoxine intoxication. Ann Neurol 26:167, 1989.

10. Beuche, W and Friede, RL: Remodeling of nerve structure in experimental isoniazid neuropathy in the rat. Brain 109:759, 1986.

11. Blakemore, WF: Isoniazid. In Spencer, PS and Schaumburg, HH (eds): Experimental and Clinical Neurotoxicology. Williams & Wilkins, Baltimore, 1980, p 476.

12. Blisard, KS and Harrington, DA: Cisplatin-induced neurotoxicity with seizures in frogs. Ann Neurol 26:336, 1989.

13. Borrett, D, et al: Reversible, late-onset disulfiram-induced neuropathy and encephalopathy. Ann Neurol 17:396, 1985.

14. Bradley, WG, Karlsson, IJ, and Rasso, ICG: Metronidazole neuropathy. Br Med J 2:610, 1977.

15. Bradley, WG and Hewer, RL: Peripheral neuropathy due to disulfiram. Br Med J 2:449, 1966.

16. Bradley, WG, et al: The neuromyopathy of vincristine in man: Clinical electrophysiological and pathological studies. J Neurol Sci 10:107, 1970.

17. Casey, EB, et al: Vincristine neuropathy: Clinical and electrophysiological observations. Brain 96:69, 1973.

18. Castellano, AM, Glass, JP, and Yung, WK: Regional nerve injury after intra-arterial chemotherapy. Neurology 37:834, 1987.

19. Cavanagh, JB: On the pattern of change in peripheral nerves produced by isoniazid intoxication in rats. J Neurol Neurosurg Psychiatry 30:219, 1964.

20. Cho, L, et al: A single intravenous injection of doxorubicin (Adriamycin) induces sensory neuropathy in rats. Neurotoxicol 1:583, 1980.

21. Clark, AW, et al: Neurotoxicity of cisplatinum: Pathology of the central and peripheral nervous systems. Neurology 30:429, 1980.

22. Cocke, JG: Chloramphenicol optic neuritis: Apparent protective effects of very high daily doses of pyridoxine and cyanocobalamin. Am J Dis Child 114:424, 1967.

23. Cogan, G, Truman, JT, and Smith, TR: Optic neuropathy, chloramphenicol and infantile agranulocytosis. Invest Ophthalmol 12:534, 1973.

24. Cohen, SC and Mollman, JE: Cisplatin-induced gastric paresis. J Neuroncol 5:237, 1987.

25. Cooper, RG, Evans, DA, and Whibley, EJ: Polymorphic hydroxylation of perhexiline maleate in man. J Med Genet 21:27, 1984.

26. Costa-Jussa, FR and Jacobs, JM: The pathology of amiodarone neurotoxicity. I. Experimental studies with reference to changes in other tissues. Brain 108:735, 1985.

27. Costa-Jussa, FR, et al: Changes in denervated skeletal muscles of amiodarone-fed mice. Muscle Nerve 11:627, 1988.

28. Coxon, A and Pallis, CA: Metronidazole neuropathy. J Neurol Neurosurg Psychiatry 39:403, 1977.

29. Dalton, K and Dalton, MJT: Characteristics of pyridoxine overdose neuropathy syndrome. Acta Neurol Scand 76:8, 1987.

30. DeKoning, P, et al: Evaluation of cis-diamine-dichloroplatinum (11) (cisplatin) neurotoxicity in rats. Toxicol Appl Pharmacol 89:81, 1987.

31. Dewar, J, et al: Cis-platin neuropathy with Lhermitte's sign. J Neurol Neurosurg Psychiatry 49:96, 1986.

32. Dibao, JM, Briggs, SJ, and Gray, TA: Filamentous axonopathy in disulfiram neuropathy. Ultrastruct Pathol 7:295, 1984.

33. Dick, DJ and Raman, D: The Guillain-Barré syndrome following gold therapy. Scand J Rheumatol 11:119, 1982.

34. Doyle, JB and Cannon, EF: Severe poly-

neuritis following gold therapy for rheumatoid arthritis. Ann Intern Med 33:1468, 1950.

35. Dubinsky, RM, et al: Reversible axonal neuropathy from the treatment of AIDS and related disorders with 2',3'-dideoxycytidine (ddC). Muscle Nerve 12:856, 1989.

36. Duffy, LF, et al: Peripheral neuropathy in Crohn's disease patients treated with metronidazole. Gastroenterol 88:681, 1985.

37. DuViviev, A and Fowler, T: Possible dapsone-induced peripheral neuropathy in dermatitis herpetiformis. Proc R Soc Med 67:439, 1974.

38. Ellis, FG: Acute polyneuritis after nitrofurantoin therapy. Lancet 2:1136, 1962.

39. England, JD, et al: Schwann cell degeneration induced by doxorubicin (Adriamycin). Brain 111:901, 1988.

40. England, JD, et al: Lethal retrograde axoplasmic transport of doxorubicin (Adriamycin) to motor neurons. A toxic motor neuronopathy. Brain 111:915, 1988.

41. Evans, DAP: Pharmacogenetics. Am J Med 34:639, 1963.

42. Fardeau, M, Tomé, FMS, and Simon, P: Muscle and nerve changes induced by perhexiline maleate in man and mice. Muscle Nerve 2:24, 1979.

43. Frasca, V, Riazza, BS, and Matthews, RG: In vitro inactivation of methionine synthase by nitrous oxide. J Biol Chem 261:15823, 1986.

44. Fraser, AG, et al: Peripheral neuropathy during long term high-dose amiodarone therapy. J Neurol Neurosurg Psychiatry 48:576, 1985.

45. Fullerton, PM and Kremer, M: Neuropathy after intake of thalidomide (Distaval). Br Med J 2:855, 1961.

46. Fullerton, PM and O'Sullivan, DJ: Thalidomide neuropathy: A clinical, electrophysiological and histological follow-up study. J Neurol Neurosurg Psychiatry 31:543, 1968.

47. Gammon, GD, Burge, FW, and King, G: Neural toxicity in tuberculous patients treated with isoniazid (isonicotinic acid hydrazide). Arch Neurol Psychiatry 70:64, 1953.

48. Gelber, R, et al: The polymorphic acetylation of dapsone in man. Clin Pharmacol Ther 12:225, 1971.

49. Green, LS, et al: Axonal transport disturbances in vincristine-induced peripheral neuropathy. Ann Neurol 1:255, 1977.

50. Greenspan, A and Treat, J: Peripheral neuropathy and low dose cis-platin. Am J Clin Oncol 11:660, 1988.

51. Griffin, JW, et al: Neurotoxicity of misonidazol in rats. I. Neuropathology. Neurotoxicology 1:299, 1979.

52. Gutmann, L, Martin, JD, and Walton, W: Dapsone motor neuropathy—an axonal disease. Neurology (Minneapolis) 26:514, 1976.

53. Hawthorne, J, Ostler, KJ, and Andrews, PLR: The role of the abdominal visceral innervation and 5-hydroxytryptamine M-receptors in vomiting induced by the cytotoxic drugs cyclophosphamide and cisplatin in the ferret. Q J Exp Pathol 73:7, 1988.

54. Hemphill, M, et al: Sensory neuropathy in cis-platinum chemotherapy. Neurology 30:429, 1980.

55. Heyer, EJ, et al: Nitrous oxide: Clinical and electrophysiologic investigation of neurologic complications. Neurology 36:1618, 1986.

56. Hishon, S and Pilling, J: Metronidazole neuropathy. Br Med J 2:832, 1977.

57. Jackson, DV, et al: Amelioration of vincristine neurotoxicity by glutamic acid. Am J Med 84:1016, 1988.

58. Jacobs, JM and Costa-Jussa, FR: The pathology of amiodarone neurotoxicity. II. Peripheral neuropathy in man. Brain 108:753, 1985.

59. Jacobs, JM, et al: Studies on the early changes in acute isoniazid neuropathy in the rat. Acta Neuropathol 47:85, 1979.

60. Joy, RJT, Scalettar, R, and Sodee, DB: Optic and peripheral neuritis. Probable effect of prolonged chloramphenicol therapy. JAMA 173:1731, 1960.

61. Katvak, SM, et al: Clinical and morphological features of gold neuropathy. Brain 103:671, 1980.

62. Kirkendall, WM and Page, EB: Polyneuritis occurring during hydralazine therapy. JAMA 167:427, 1962.

63. Klinghardt, GW: Arzneimittelschädigungen des peripheren Nervensystems unter besonderer Berücksichtigung der Polyneuropathie durch Isonicotinsäurehydrazid (experimentelle und human pathologische Untersuchungen). Proceedings Fifth International Congress of Neuropathy: Exerpta Medica Found Int Cong Series No 100, 1966, p 292.

64. Klinghardt, GW: Schädigungen des Nervensystems durch Nitrofurane bei der Ratte. Acta Neuropathol (Berlin) 9:18, 1967.

65. Krinke, G, Naylor, DD, and Skorpil, V: Pyridoxine megavitaminosis: An analysis of the early changes induced with massive doses of vitamin B6 in rat primary sensory neurons. J Neuropath Exp Neurol 44:117, 1985.

66. Krinke, G, Schaumburg, HH, and Spencer, PS: Pyridoxine megavitaminosis produces degeneration of peripheral sensory neurons (sensory neuronopathy) in the dog. Neurotoxicol 2:13, 1980.

67. Kuncl, RW, et al: Colchicine myopathy and neuropathy. N Engl J Med 316:1562, 1987.

68. Laguney, A, et al: Thalidomide neuropathy: An electrophysiology study. Muscle Nerve 9:837, 1986.

69. Layzer, RB, Fishman, RA, and Schafer, JA: Neuropathy following abuse of nitrous oxide. Neurology (Minneapolis) 28:504, 1978.

70. Leggat, PO: Ethionamide neuropathy. Tubercule 43:95, 1962.

71. Lhermitte, R, Fardeau, M, Chedru, F, et al: Polynévrites au cours de traitments par la nitrofurantoine. Presse Med 71:767, 1963.

72. Lipton, RB, et al: Taxol produces a predominantly sensory neuropathy. Neurology 39:368, 1989.

73. Loughridge, LW: Peripheral neuropathy due to nitrofurantoin. Lancet 2:1133, 1962.

74. Lovelace, RE and Horwitz, SJ: Peripheral neuropathy in long term diphenylhydantoin therapy. Arch Neurol 18:69, 1968.

75. Mansour, AM, Puklin, JE, and O'Grady, R: Optic nerve ultrastructure following amiodarone therapy. J Clin Neurol Ophthalmol 4:231, 1988.

76. Marcus, D, Swift, T, and McDonald, T: Acute effects of phenytoin on peripheral nerve function in the rat. Muscle Nerve 4:48, 1981.

77. Masurovsky, EB, Peterson, ER, Crain, SM, et al: Morphologic alterations in dorsal root ganglion neurons and supporting cells of organotypic mouse spinal cord-ganglion cultures exposed to taxol. Neuroscience 10:491, 1983.

78. Matsuoka, Y, Takayanagi, T, and Sobue, I: Experimental ethambutol neuropathy in rats. Morphometric and teased fiber study. J Neurol Sci 51:89, 1981.

79. McLeod, JG and Penny, R: Vincristine neuropathy: An electrophysiological and histological study. J Neurol Neurosurg Psychiatry 32:297, 1969.

80. Meier, C, et al: Perhexiline-induced lipidosis in the Dark Agouti (DA) rat. An animal model of genetically determined neurotoxicity. Brain 109:649, 1986.

81. Melgaard, B, et al: Misonidazol neuropathy, a clinical electrophysiological and histological study. Ann Neurol 12:10, 1982.

82. Merigan, T, et al: Circulating p24 antigen levels and responses to dideoxycytidine in human immunodeficiency virus (HIV) infections. Ann Intern Med 110:189, 1989.

83. Mollman, JE, et al: Unusual presentation of cis-platinum neuropathy. Neurology 38:488, 1988.

84. Money, GL: Isoniazid neuropathies in malnourished tuberculous patients. J Trop Med 62:198, 1959.

85. Morki, B, Ohnishi, A, and Dyck, PJ: Disulfiram neuropathy. Neurology

(Minneapolis) 31:730, 1981.

86. Morris, JS: Nitrofurantoin and peripheral neuropathy with megaloblastic anemia. J Neurol Neurosurg Psychiatry 29:22, 1966.

87. Nair, VS, LeBrun, M, and Kass, I: Peripheral neuropathy associated with ethambutol. Chest 77:98, 1980.

88. Navarro, JC, et al: Acute dapsone-induced peripheral neuropathy. Muscle Nerve 12:604, 1989.

89. Nover, R: Persistent neuropathy following chronic use of glutethimide. Clin Pharmacol Ther 8:283, 1967.

90. Ochoa, J: Isoniazid neuropathy in man: Quantitative electron microscope study. Brain 93:831, 1970.

91. Ohnishi, A, Petersen, CM, and Dyck, PJ: Axonal degeneration in sodium cyanate-induced neuropathy. Arch Neurol 32:530, 1975.

92. Parnass, JB and Horowitz, SB: Taxol binds to polymerized tubulin *in vitro*. J Cell Biol 91:479, 1981.

93. Parry, GJ and Bredesen, DE: Sensory neuropathy with low dose pyridoxine. Neurology 35:1466, 1985.

94. Paul, MG, et al: Inhibition by furacin of citrate formation in testis preparations. J Biol Chem 206:491, 1954.

95. Paulsen, O and Nilsson, LG: Distribution of acetylator phenotype in relation to age and sex in Swedish patients. A retrospective study. Eur J Clin Pharmacol 28:311, 1985.

96. Paulson, OB, et al: Misonidazol neuropathy. Acta Neurol Scand 100 (Suppl):133, 1984.

97. Pellissier, JF, et al: Peripheral neuropathy induced by amiodarone chlorohydrate. J Neurol Sci 63:251, 1984.

98. Petersen, E and Crain, S: Nerve growth factor attenuates neurotoxic effects of taxol and spinal cord-ganglion explants from fetal mice. Science 217:377, 1982.

99. Petersen, CM, et al: Sodium cyanate induced polyneuropathy in patients with sickle-cell disease. Ann Intern Med 81:152, 1974.

100. Petrera, JE, Fledelius, HC, and Trojaborg, W: Serial pattern evoked potential recording in a case of toxic

optic neuropathy due to ethambutol. Electroencephalogr Clin Neurophysiol 71:146, 1988.

101. Poole, GW and Schneeweiss, J: Peripheral neuropathy due to ethionamide. Annu Rev Respir Dis 84:890, 1961.

102. Ramilio, O, Kinane, BT, and McCracken, GH: Chloramphenicol neurotoxicity. Ped Inf Dis J 7:358, 1988.

103. Ramirez, JA, et al: Phenytoin neuropathy: Structural changes in the sural nerve. Ann Neurol 19:162, 1986.

104. Rapoport, AM and Guss, SB: Dapsone-induced peripheral neuropathy. Arch Neurol 27:184, 1972.

105. Raskin, NH and Fishman, RA: Pyridoxine-deficiency neuropathy due to hydralazine. N Engl J Med 273:1182, 1965.

106. Riggs, J, Ashraf, M, Snyder, RD, et al: Prospective nerve conduction studies in cisplatin therapy. Ann Neurol 23:92, 1988.

107. Riggs, JE, et al: Chronic human colchicine neuropathy and myopathy. Arch Neurol 43:521, 1986.

108. Roelofs, RI, et al: Peripheral sensory neuropathy and cisplatin chemotherapy. Neurology 34:934, 1984.

109. Rose, GP and Taylor, JM: The neurotoxicity of radiosensitizing drugs: A biochemical assessment of desmethymisonidazol (DMM) in the rat. Toxicology 34:43, 1985.

110. Rothrock, JF, et al: Fulminant polyneuritis after overdose of disulfiram and ethanol. Neurology 34:357, 1984.

111. Röyttä, M and Raine, CS: Taxol induced neuropathy: Chronic effects of local injection. J Neurocytol 15:483, 1986.

112. Sahenk, Z, Brady, ST, and Mendell, JR: Studies on the pathogenesis of vincristine-induced neuropathy. Muscle Nerve 10:80, 1987.

113. Sahenk, Z, Mendell, JR, Couri, D, and Nachtman, J: Polyneuropathy from inhalation of nitrous oxide cartridges through a whipped cream dispenser. Neurology (Min-

neapolis) 28:485, 1978.

114. Said, G: Perhexiline neuropathy: A clinicopathological study. Ann Neurol 3:259, 1978.

115. Sirsat, AM, Lalitha, VS, and Pandya, SS: Dapsone neuropathy—report of three cases and pathologic features of a motor nerve. Int J Lepr Mycobact Dis 55:23, 1987.

116. Saunders, MI, et al: The neurotoxicity of misonidazol and its relationship to dose half-life and concentrations in the serum. Br J Cancer 37(Suppl):268, 1978.

117. Schaumburg, HH, et al: ddC neuropathy, a study of 52 patients. Neurology 40(Suppl):428, 1990.

118. Schaumburg, HH, et al: Sensory neuropathy from pyridoxine abuse. N Engl J Med 309:445, 1983.

119. Scott, JM, et al: Pathogenesis of subacute combined degeneration: A result of methyl group deficiency. Lancet 2:334, 1981.

120. Shah, RR, et al: Impaired oxidation of debrisoquine in patients with perhexiline maleate. Br Med J 284:295, 1982.

121. Swift, TR, et al: Peripheral neuropathy in epileptic patients. Neurology (Minneapolis) 31:826, 1981.

122. Takeuchi, H, et al: Ethambutol neuropathy: Clinical and electroneuromyographic studies. Folia Psychiatr Neurol Jpn 34:45, 1980.

123. Takeuchi, H, et al: Metronidazol neuropathy: A case report. Jpn J Psychiatr Neurol 42:291, 1988.

124. Tala, E and Tevola, K: Side effects and toxicity of ethionamide and prothionamide. Ann Clin Res 1:32, 1969.

125. Tellez-Nagel, I, et al: An ultrastructural study of chronic sodium cyanate-induced neuropathy. J Neuropathol Exp Neurol 36:351, 1977.

126. Tomiwa, K, Nolan, C, and Cavanagh, JB: The effects of cis-platin on rat spinal ganglia: A study by light and electron microscopy and by morphometry. Acta Neuropathol 69:295, 1986.

127. Toole, JF, et al: Neural effects of nitro-

furantoin. Arch Neurol 18:860, 1968.

128. van dep Hoop, RG, Vecht, CJ, van der Burg, MEL, et al: Prevention of cisplatin neurotoxicity with an ACTH (4-9) analog in patients with ovarian cancer. N Engl J Med 322:89, 1990.

129. Waldinger, TP, et al: Dapsone-induced peripheral neuropathy. Case report and review. Arch Dermatol 120:356, 1984.

130. Walsh, TJ, et al: Neurotoxic effects of cis-platin therapy. Arch Neurol 39:719, 1982.

131. Walsh, JC: Gold neuropathy. Neurology (Minneapolis) 20:455, 1970.

132. Weiss, JJ, Thompson, GR, and Lazaro, R: Gold toxicity presenting as peripheral neuropathy. Clin Rheumatol 1:285, 1982.

133. White, WT, Harrison, L, and Dumas, J: Nitrofurantoin unmasking peripheral neuropathy in a type 2 diabetic patient. Arch Intern Med 144:821, 1984.

134. Wijesekera, JC, et al: Peripheral neuropathy due to perhexiline maleate. J Neurol Sci 46:303, 1980.

135. Williams, MH and Bradley, WG: An assessment of dapsone toxicity in guinea pig. Br J Dermatol 86:650, 1972.

136. Windebank, AJ, et al: Neurotoxicity of pyridoxine analogs is related to coenzyme structure. Neurochem Pathol 3:159, 1985.

137. Wulff, CH, et al: Development of polyneuropathy during thalidomide therapy. Br J Dermatol 112:475, 1985.

138. Xu, Y, Sladky, JT, and Brown, MJ: Dose-dependent expression of neuropathy after experimental pyridoxine intoxication. Neurol 39:1077, 1989.

139. Yiannikas, C, Pollard, JD, and McLeod, JG: Nitrofurantoin neuropathy. Aust N Z J Med 11:400, 1981.

140. Zuccarello, M and Anzil, AP: Localized model of experimental neuropathy of topical application of disulfiram. Exp Neurol 64:699, 1979.

Chapter 18

OCCUPATIONAL, BIOLOGIC, AND ENVIRONMENTAL AGENTS

ACRYLAMIDE
ALLYL CHLORIDE
ARSENIC (INORGANIC)
BUCKTHORN
CARBON DISULFIDE
CYANIDE
DIMETHYLAMINOPROPIONITRILE (DMAPN)
DIPHTHERITIC NEUROPATHY (POLYRADICULOPATHY)
ETHYLENE OXIDE
HEXACARBONS (n-HEXANE, METHYL n-BUTYL KETONE)
LEAD
MERCURY
METHYL BROMIDE
ORGANOPHOSPHORUS ESTERS
POLYCHLORINATED BIPHENYLS
SPANISH TOXIC OIL SYNDROME
THALLIUM
TRICHLORETHYLENE
VACOR (PNU)

An enormous number of potentially toxic chemicals are deployed in the workplace and general environment. Many have been identified or implicated as causes of peripheral neuropathy, usually of the distal axonopathy type. Except for acrylamide, ethylene oxide, thallium, vacor, the hexacarbons, and organophosphates, there are few careful experimental studies of these substances. Clinical reports, often isolated, are the sole basis for some of the alleged chemically induced peripheral neuropathies. Only agents consistently associated with neuropathy are included in this chapter. Specifically excluded are chemicals such as dioxin (TCDD), 2,4D,

carbon monoxide, methyl ethyl ketone, toluene, pyrethrins, DDT, and carbon tetrachloride, whose link with neuropathy seems tenuous. Diphtheritic neuropathy and buckthorn are included in this chapter as examples of biologic toxins consistently associated with neuropathy.

ACRYLAMIDE

Acrylamide monomer is important in several industries. Polymerization of the monomer produces substances that are useful flocculators and grouting agents. Acrylamide monomer is neurotoxic, while the polymer is not. The monomer is a white powder, soluble in water, and is absorbed readily by inhalation, ingestion, or through the skin. Most instances of neuropathy have occurred via dermal contact in industrial or occupational settings.[118] Episodic contact dermatitis of the hands usually antedates neuropathy, indicating that significant exposure has occurred. There is probably little risk of neurotoxic exposure for the population at large, except in rare instances where the monomer has contaminated well water.

Pathology

Acrylamide monomer clearly produces distal axonopathy in experimental animals,[107,120] and sural nerve biopsies from two humans recovering from acrylamide neuropathy revealed axonal degeneration predominantly affecting large fibers.[39] Detailed morphologic and electrophysiologic studies have been performed in various experimental species intoxicated with acrylamide, and it is widely held that acrylamide-induced changes are a valid model for the pattern of distal axonopathy induced by many neurotoxic agents.[119] The fundamental axonal change is accumulation of 10-nm neurofilaments. These accumulations initially appear at multiple sites in the distal regions of large-diameter myelinated axons of the PNS and CNS in the pattern outlined in Chapter 2. In cats and primates, the earliest morphologic changes have been detected in pacinian corpuscles of the hind and forefeet, primary annulospiral endings of muscle spindles, and preterminal axons of the gracile nucleus.[107,120] In dogs, megaesophagus and abnormal cough reflex[53] due to vagal neuropathy also develops.[101] In rats, with continued intoxication, proximal segments of peripheral nerves undergo change, and widespread tract-oriented degeneration occurs in the corticospinal, spinocerebellar, and optic tracts and in the mamillary bodies.[104] Other morphologic features of advanced neuropathy include abnormalities of unmyelinated fibers and loss of Purkinje cells.[19,93] Electrophysiologic studies of experimental animals confirm the pattern of PNS and CNS vulnerability.[7,103,123]

Pathogenesis

The metabolism of acrylamide is poorly understood; it is rapidly distributed throughout the body and some may persist in nervous tissue for 14 days. Small changes in the acrylamide molecule result in loss of neurotoxicity, and many analogues appear innocuous.[31] The biochemical mechanism of neurotoxic damage is unknown. Experimental studies have demonstrated profound early abnormalities of fast bidirectional axonal transport in the distal axon and modest abnormalities of anterograde slow transport in proximal regions.[42,43,76] The distal transport defect

may contribute to the subsequent multifocal distal degeneration.[45]

Clinical Features

SYMPTOMS, SIGNS, AND COURSE

Numbness and excessive sweating of feet and fingers and an unsteady gait are usual initial complaints.[40] Difficulty in walking is frequently disproportionate to the mild initial weakness. This characteristic of acrylamide neurotoxicity may reflect muscle-spindle afferent or cerebellar dysfunction. Excessive sweating is sometimes accompanied by skin peeling in the hands and other signs of exfoliative dermatitis.

Objective signs are present in all symptomatic individuals and stem from PNS motor, sensory, and, possibly, CNS cerebellar dysfunction. Early on, weakness of the hands and feet is usually combined with diffuse loss of tendon reflexes, unlike most toxic neuropathies that are featured by early selective loss of Achilles reflex. Sensory change inevitably includes depression of vibration sense. Other modalities are variously affected. Clumsiness and occasional intention tremor may be present in the upper extremities, and a broad-based swaying gait is a common early finding.[65,66] Cranial nerve palsy and autonomic dysfunction (aside from increased sweating) are not features of this condition.

Removal from exposure in the early stages results in gradual, complete functional recovery from both PNS and CNS involvement. Careful examination may reveal persistent depression of vibration sense as the sole abnormality. In more severely affected individuals, improvement usually continues for many months, but frequently there is residual distal weakness, gait ataxia, and vibration sense loss. In extreme cases of acrylamide neurotoxicity, other clinical findings have been prominent, including transient impaired vision (especially depth perception) and impairment of recent memory without loss of other cognitive functions. Upon recovery, in the extreme cases, we have observed truncal ataxia, lower-extremity spasticity, and residual recent memory loss. There is no specific treatment for acrylamide neurotoxicity.

Electrodiagnostic Studies

Electromyography of distal limb muscles usually reveals ample evidence of denervation.[66] Motor nerve conduction velocities characteristically are normal or only slightly reduced. Sensory nerve action potentials are almost always abnormal, and are suggested as a sensitive electrophysiologic screening test for early or subclinical neuropathy. Reduction in amplitude is the most common abnormality, and is present early even in distal portions of upper-extremity nerves.[65]

Differential Diagnosis

The diagnosis is not difficult in an individual with proven industrial or occupational exposure. The presence of gait ataxia, moist peeling hands, and peripheral neuropathy in such individuals leaves little room for doubt. A detailed occupational history is probably the most important diagnostic procedure, since acrylamide neurotoxicity almost never stems from unrecognized environmental sources.

Case History: Acrylamide Neuropathy

A 45-year-old male was hired as a grouter and had had no previous exposure to acrylamide. He received daily high-level skin exposure to acrylamide monomer, the principal component of the grouting compound. After 3 months, he complained of fatigue and a 7 pound weight loss, and also noticed that his hands were red and his skin peeling. By 4 months, he felt slightly unstable while walking and would occasionally lose balance when turning corners. Numbness of the feet and an aching pain in the calves de-

veloped by 5 months. He continued to work for another 2 weeks until his hands became numb and his gait was so unsteady that he was falsely accused of drinking on the job. He had no urinary bladder dysfunction or impotence.

His palms were red, moist, and peeling upon examination. Vital signs and general physical examination were normal. Neurologic examination revealed no abnormalities of cranial nerves or mental status. His speech was unremarkable and no nystagmus was present. Strength in the proximal upper extremities was normal but there was weakness, without wasting, of the intrinsic muscles of the hands and weakness of the wrist extensors (⅘ MRC scale). Neck, trunk, abdominal, and proximal lower-extremity muscles were of normal strength. There was weakness of dorsiflexion at the ankles (⅘ MRC scale). The tendon reflexes in the arms were sluggish, and there was loss of knee and ankle jerks, and absent plantar responses. A mild loss of pinprick sense was evident below the wrist and knees (glove and stocking distribution). Position sense was normal throughout. There was a striking loss of vibration sense (128 cps) over ankle, knee, and wrist. Coordinated movements in the upper extremities were unaffected. There was no past-pointing or loss of check of rapid movement, but truncal ataxia was evident. Heel-to-knee-to-shin maneuvers were unsteady and poorly performed. His gait was broad based and, on sudden turns, he reached for support. He was unable to stand with his feet together and eyes open without falling to one side. Motor nerve conduction in the median and ulnar nerves was 55 m/sec (normal), and 35 m/sec (mild slowing) in the common peroneal nerve. Lumbar puncture revealed clear, acellular fluid and a normal protein value.

Six months later, motor and sensory examinations were normal except for moderate impairment of vibration sensation in the lower extremities. The tendon reflexes were moderately increased throughout, except for absent Achilles reflexes. His gait was narrow-based without staggering, except when he was asked to make sudden turns. He could stand with feet together and Romberg's sign was not present.

Comment. The patient illustrates many of the cardinal features usually associated with a mild to moderate degree of toxic axonal neuropathy: insidious onset, gradual progression, symmetric distal (stocking-glove) motor and sensory loss, normal CSF protein, and mild impairment in motor nerve conduction. Unusual features of this illness included truncal and gait ataxia, and severe loss of vibration sensation. Early, severe vibration sense loss, disproportionate to impairment of other sensory modalities of weakness, and early gait ataxia have been consistent features in our experience with human acrylamide intoxication.

ALLYL CHLORIDE

Allyl chloride (3-chloropropene) is a reactive halogenated hydrocarbon used in the manufacture of epoxy resin and glycerin. Polyneuropathy of the distal axonopathy type is the principal manifestation of allyl chloride intoxication and several outbreaks are described in Chinese workers.[50] Prolonged high-level atmospheric exposure is associated with gradual onset of distal extremity numbness, stocking-glove distribution of sensorimotor deficit, and diminished ankle reflexes.[49] Electrophysiologic abnormalities include denervation potentials and prolonged distal motor latencies. Recovery commences following removal from exposure source. Experimental animal studies have demonstrated widespread multifocal accumulations of axonal neurofilaments, and fiber degeneration in the distal regions of peripheral nerves and dorsolateral pathways of the spinal cord.[48,49]

ARSENIC (INORGANIC)

Human neurotoxicity from arsenic is usually associated with ingestion of trivalent arsenic (arsenite). Pentavalent arsenic (arsenates) is less toxic. Arsenic compounds are not mined as such, but arsenates occur as by-products of smelting of copper and lead ores. Workers in smelting industries are especially at risk for chronic arsenic intoxication, as are miners, individuals whose wells are adjacent to mines containing arsenic-rich ores, and inhalant drug abusers whose agents are diluted with arsenic.[35,70] Trivalent arsenic remains an often-used vehicle for homicide and suicide, and survivors may develop a subacute neuropathy.

Pathology and Pathogenesis

Axonal degeneration is the predominant change in nerve biopsies from patients with arsenic neuropathy,[68] and it is probable that inorganic arsenic intoxication produces distal axonopathy. One autopsy report of a severe case of arsenic neuropathy, utilizing limited histologic techniques, describes changes in peripheral nerves and the dorsal columns of the spinal cord.[32] Unfortunately, there is no reasonable experimental animal model of inorganic arsenic neurotoxicity, and the nature and distribution of axonal changes remain to be elucidated.

The pathogenesis of arsenical distal axonopathy is unknown. It is suggested that the interaction of arsenates with the thiol group of lipoic acid is a link between arsenic neuropathy and the clinically similar axonopathy of thiamine deficiency.[91] It is assumed that arsenic acts on the lipoic acid component of the pyruvate dehydrogenase complex, inhibiting the conversion of pyruvate to acetyl CoA. The affinity of arsenic trioxide for keratin of hair and nail is attributed to similar thiol binding. BAL, the antidote, is dimercaptopropanolol (a dithiol), which forms a nontoxic stable ring with arsenic and is then excreted.

Clinical Features

Peripheral neuropathy is the predominant neurologic complication of exposure to inorganic arsenic.[24,35,55,57,68] Two varieties exist: a subacute type that appears within weeks of a massive over-dose (unsuccessful suicide or homicide),[68] and an insidiously developing type following prolonged, low-level environmental or industrial exposure.[35] Many reports of arsenic neuropathy do not differentiate between the two types of exposure, and this distinction is not widely appreciated.

CHRONIC EXPOSURE TYPE

Chronic low-level exposure to inorganic arsenic produces a consistent chronologic triad of conditions. The initial phase is characterized by weakness, malaise, anorexia, and vomiting. The second stage is featured by mucous membrane irritation, hyperkeratosis, darkened skin, white striae of the nails (Mees's lines), and pitting edema. It is likely that many individuals in this stage have subclinical neuropathy. The third stage is overt peripheral neuropathy, whose onset is heralded by numbness and burning sensations of the hands and feet. Diminished vibration and position sensation in the lower extremities is usually present and may result in a tabetic syndrome. Other modalities are consistently affected. Weakness is usually mild and confined to extensors of the feet and intrinsic hand muscles. Continued exposure may result in severe distal stocking-glove sensorimotor neuropathy. Recovery is usually excellent in mildly affected individuals, and less satisfactory in the more rare cases with severe involvement.

SINGLE-EXPOSURE TYPE

A single large dose of arsenic produces vomiting within minutes or hours. This may be followed by tachycardia, diarrhea, hypotension, vasomotor collapse, and death within a day. Survivors may develop neuropathy within 10 days to three weeks. Sensory symptoms appear first: usually these consist of numbness and intense paresthesias distally in the limbs. Weakness in the distal lower extremities soon follows and may involve the upper extremities as well. The illness has a subacute progression, usually evolves within two to five weeks, and the degree of impairment varies considerably. An AIDP-like syndrome occasionally occurs.[29] Some experience only mild, predominantly sensory, neuropathy. Others (probably the majority) develop severe distal sensorimotor polyneuropathy.[68] Early electrodiagnostic testing may yield findings suggesting a demyelinating polyradiculoneuropathy. Subsequent studies demonstrate an evolution into a distal axonopathy pattern. These studies suggest that in some cases, proximal segmental demyelination may precede distal axonal degeneration.[29] Impairment of position and vibration sense may be especially profound. Systemic signs of arsenic intoxication (skin, nails) are often not pronounced in this variety. Recovery is gradual and, in the mild cases, usually complete. Individuals with pronounced atrophy and severe sensory loss recover to some degree over a 2-year period, but often stabilize in a permanently disabled state.

Treatment of both varieties involves chelation therapy with either BAL or penicillamine. There is little evidence that treatment of fully developed neuropathy affects its course. It is suggested that chelation therapy be continued for months following exposure.[68]

Laboratory Studies

CLINICAL LABORATORY FINDINGS

Routine laboratory studies are usually unremarkable. Analysis of hair, fingernail clippings, and urine for arsenic reveal markedly elevated levels in chronic cases. Tissue arsenic levels are not consistently elevated in the subacute neuropathy following a single dose; however, urine levels may remain elevated for weeks. Levels in excess of 25 μg per 24 hours are abnormal unless the individual has recently ingested seafood, a source of pentavalent arsenic. The CSF is acellular and the protein level normal.

ELECTRODIAGNOSTIC STUDIES

Electromyography usually reveals evidence of denervation in distal lower-limb muscles. Motor nerve conduction is only mildly or moderately slowed, while the amplitude of sensory action potentials is profoundly depressed.[68]

Differential Diagnosis

Cases that follow massive single exposure are usually obvious. If the exposure history is unavailable, an erroneous diagnosis of AIDP may be entertained. Arsenical neuropathy following chronic low-level exposure may pose a formidable diagnostic problem. The neurologic syndrome is similar to those of uremia, nutritional disorders, or other nonspecific sensorimotor neuropathies. Furthermore, the systemic manifestations of arsenic intoxication are often mild and readily overlooked. Hair, urine, and fingernail arsenic levels are justified in evaluating polyneuropathies of obscure etiology.

BUCKTHORN

Ingestion of the fruit from the buckthorn shrub, *Karwinska humboldtiana*, is followed by a rapidly evolving, diffuse motor polyneuropathy that may be fatal.[15] Accidental ingestion of the fruit, which grows wild in the southwest, causes the death of children in Mexico and annually kills hundreds of cattle.

Symptoms commence 5 to 20 days after ingestion with weakness, initially in lower limbs, rapidly spreading and sometimes eventuating in quadriparesis and bulbar paresis. The illness is sometimes confused with poliomyelitis; its severity is proportional to the amount ingested. Neither pain nor sensory symptoms accompany this condition, but a subtle distal sensory loss

may appear. No electrodiagnostic studies are available; a sural nerve biopsy in one case showed segmental demyelination and axonal swelling. Experimental animal studies with both the isolated toxins and the fruit suggest widespread segmental demyelination in the peripheral nervous system to be the dominant pathology. An *in vitro* study indicates that axonal changes may also occur.[52] It is suggested that the clinical and pathological pattern resembles that observed in diphtheritic neuropathy.

CARBON DISULFIDE

Carbon disulfide (CS_2) is used in the production of viscose rayon fibers and cellophane films. CS_2 neurotoxicity is inevitably a result of airborne industrial exposure and has been a persistent problem in Scandinavia, Japan, and southern Europe. Subacute exposure of humans to high concentrations produces profound psychologic disturbance, while chronic low-level intoxication causes peripheral neuropathy of the distal axonopathy type in humans and experimental animals.[44,110]

Chronic exposure to low airborne levels (10 to 40 ppm) is reported to result in prolonged motor nerve conduction in the lower limbs, presumably evidence of subclinical neuropathy.[111] Exposure to levels of 170 ppm for 4 to 6 months produces symptoms of numbness distally in the lower extremities. Weakness, loss of Achilles and patellar reflexes, and diminished sensation for pin, touch, and vibration accompany these symptoms.[126] Upper-limb involvement follows continued exposure. Moderate slowing of motor and sensory nerve conduction and evidence of denervation in distal muscles are features of this neuropathy.[125] Recovery is slow and, in severe cases, incomplete. There are no detailed clinical studies of individuals with partial recovery, and it is possible that residual spinal cord damage may account for some of this disability. This notion is supported by experimental animal studies that demonstrate slowed conduction and tract-oriented distal axonal degeneration in the spinal cord as well as changes in distal fibers of the PNS.[97,110] There are no reliable reports of the morphologic changes in human CS_2 distal axonopathy. Experimental animal studies describe axonal swellings with accumulations of 10-nm neurofilaments, similar to the changes in acrylamide and hexacarbon neuropathies, as well as alterations in axonal transport of neurofilaments.[86]

CYANIDE

Acute cyanide intoxication is usually lethal, due to the rapid reaction of cyanide ion with trivalent iron of cytochrome oxidase. Chronic cyanide intoxication is not fatal, and may result in widespread CNS and PNS degeneration. Chronic cyanide poisoning from occupational sources is rare; however, environmental exposure may result from consumption of the seeds of certain stone fruits (apricot, peach, wild cherry). This condition is extremely rare in Europe and North America, but ingestion of large quantities of the cassava plant has been linked with a debilitating neurologic condition in Nigeria (Nigerian neuropathy), and cyanogenic plants are also implicated in other tropical neurologic disorders.[84] Affected Nigerians have significantly elevated plasma thiocyanate levels, are riboflavin-deficient, and are generally in poor health.[85]

Clinical Features

Initial symptoms of Nigerian neuropathy are painful paresthesias of the feet, followed by numbness of the hands. Subsequently, weakness develops in the distal lower limbs, along with a broad-based ataxic gait. Loss of sensation and weakness frequently display a stocking-glove distribution, with an especially severe impairment of proprioception in the legs. Severe atrophy of the lower legs, similar to HSMN type I,

occasionally is present. Tendon reflexes are usually exaggerated and hypertonicity may occur. Lower-extremity ataxia is attributed to sensory impairment rather than cerebellar involvement. Motor nerve conduction may be moderately slowed in the lower extremities.

Signs of CNS dysfunction are present in most cases. Optic atrophy and visual impairment are prominent features of this illness, and sensorineural hearing loss may occur. Many of the sensory symptoms are attributed to dorsal column dysfunction; hyperreflexia and hypertonicity probably reflect corticospinal tract involvement.[84]

The condition is slowly progressive and rarely fatal. There are no reports describing the effect of change of diet, nerve biopsies, postmortem examination, or a suitable animal model of this condition.

DIMETHYLAMINOPROPIO-NITRILE (DMAPN)

DMAPN, an effective catalyst in polymerization reactions, was introduced into the manufacture of polyurethane foams in 1967. A toxic distal axonopathy syndrome with atypical features occurred in many workers following its deployment.[34,88] DMAPN was soon recognized as the cause, withdrawn from use, and no new cases developed. Striking and unusual features accompanying DMAPN axonopathy are urinary hesitancy and sexual dysfunction. These antedate the usual symptoms of polyneuropathy, and affected individuals have been erroneously diagnosed as having prostatic hypertrophy.[87]

Pathology and Pathogenesis

Results of human nerve biopsy and experimental animal studies indicate that DMAPN produces distal axonal degeneration. Axonal swellings filled with neurofilaments and nonspecific organelles are described. Although the clinical profile suggests selective involvement of small nerve fibers, morphologic data do not strongly support this notion.[87] There are no reports describing DMAPN-induced changes in the autonomic nervous system of humans or animals. Reversible changes in autonomic function occur after short periods of exposure to DMAPN, suggesting that urinary and sexual dysfunction may, in part, be pharmacologic. The pathophysiology of the distal axonal degeneration is unknown.

Clinical Features

SYMPTOMS, SIGNS, AND PROGNOSIS

The latency between first exposure to DMAPN and appearance of symptoms varies with the concentrations or amounts used. Initial symptoms are usually urinary hesitancy and abdominal discomfort, followed by decreased stream, reduced frequency of urination, and occasional incontinence. Partial or complete impotence develops subsequent to the urinary dysfunction, and is often accompanied by numbness of the feet. Subsequently, proximal numbness and paresthesias appear in the legs and hands, accompanied by weakness in the legs. A distinctive feature of DMAPN neurotoxicity is diminished pain, temperature, and touch sensation in the lower sacral dermatomes.[87] Vibration sense is diminished in the feet. Weakness occurs in the distal extremities, and is most severe in the toe and foot extensors. Atrophy is uncommon. The tendon reflexes are usually preserved, an atypical feature for a distal axonopathy. Cranial nerve and autonomic functions, other than urinary and sexual, are spared. It is suggested that the unusual constellation of autonomic and sensory symptoms, together with relative sparing of tendon reflexes, indicate that DMAPN axonopathy selectively affects small nerve fibers.[87]

The prognosis is good for young individuals with mild or moderate involvement. Some older individuals have experienced persistent sexual and urinary dysfunction.

Electrodiagnostic Studies

Motor nerve conduction is usually normal; occasionally slight slowing is present in severely affected patients. Sensory nerve action potentials have diminished amplitudes. Sacral nerve latencies in three reported patients were prolonged.[87]

Urodynamic Studies

Cystometrograms may demonstrate a flaccid urinary bladder with increased residual volume, and intravenous pyelography usually reveals significant postvoiding residual urine.

Differential Diagnosis

The fully developed syndrome of DMAPN intoxication resembles no other neurotoxic condition. This constellation of findings, combined with epidemic occurrence in an industrial setting, readily establishes its presence. Isolated cases in the early stages, when urinary symptoms predominate, are difficult urologic diagnoses. Isolated cases in later stages may suggest diabetic autonomic and sensory neuropathy, amyloid neuropathy, or a lesion of the cauda equina.

DIPHTHERITIC NEUROPATHY (POLYRADICULOPATHY)

Definition and Etiology

Diphtheria is an acute infectious disease produced by *Corynebacterium diphtheriae*, and is usually a local inflammatory infection confined to the upper respiratory tract or sometimes the skin. A protein exotoxin is responsible for the cardiomyopathy and polyradiculopathy occurring in approximately 20% of cases. Diphtheritic neuropathy is now rare in North America and Europe, but remains a significant problem in countries without compulsory vaccination.

Pathology

Meticulous human postmortem and abundant experimental animal studies indicate that diphtheritic neuropathy results from widespread, noninflammatory demyelination of nerve roots and adjacent portions of somatic nerves.[38] The distribution of demyelinative lesions in individuals who die during an acute neuritic illness is consistently in the dorsal root ganglia and adjacent ventral and dorsal roots. Among cranial nerves, only the nodose ganglion of the vagus is frequently affected. Peripheral portions of spinal nerves appear to be largely spared in the acute illness. The distribution of lesions corresponds to zones of high permeability of the blood-nerve barrier. Dorsal root ganglia are vulnerable in several neurotoxic conditions (doxorubicin, methyl mercury), presumably because their capillaries are fenestrated. Experimental studies with isotope-labeled diphtheria toxin strongly support this localization, and may explain why the central nervous system and much of the PNS is spared.[128] Both human and experimental pathologic studies suggest that the demyelinative lesion is a direct effect of toxin on myelin or Schwann cells, and is not immunologically mediated.[75,129] Local injection of diphtheria toxin produces focal demyelination in many fibers, following a latent period, without accumulation of lymphocytes or plasma cells. Demyelination begins in the paranodal zones.[4] Once focal demyelination is complete, remyelination promptly begins. Axonal change is usually slight unless large doses of toxin are administered.

Pathogenesis

An *in vitro* biochemical study suggests that diphtheria toxin does not directly degrade myelin, but inhibits myelin protein synthesis by the Schwann cell.[89] Recent studies suggest that diphtheria toxin may activate additional cytotoxic mechanisms.[22] Since diphtheritic lesions undergo prompt remyelination, some Schwann cells remain viable. Once the toxin gains access to myelinated fibers, it is rapidly bound and its effect is not neutralized by antitoxin.

The close relationship of the loci of lesions to areas of permeable intraneural blood vessels has provided a key to understanding the pathogenesis of the diffuse polyneuropathy that appears five to eight weeks following the initial throat infection with *C. diphtheriae*. However, several of the most striking features of the clinical illness remain unexplained.[75] One is the prolonged latency following the initial faucial inflammation; exotoxin is released at this time but does not produce diffuse polyneuropathy. Another is the local pharyngeal and accommodation paralysis that appears early after faucial diphtheria. It seems likely that this reflects a local effect of exotoxin on peripheral nerve, but the mechanism is obscure and there are no adequate pathologic studies. Local paralysis may occur with cutaneous diphtherial infection.

Clinical Features

There are two distinct PNS syndromes associated with faucial diphtheria, a local pharyngeal-palatal neuropathy occurring soon after throat infection, and a diffuse sensorimotor neuropathy that develops after five to eight weeks. Presumably both conditions are caused by the exotoxin, one resulting from a local reaction on peripheral nerve, the other a blood-borne diffuse effect on somatic and autonomic ganglia and spinal roots. The overall incidence of local neuropathy is approximately 15%, generalized neuropathy about 10%. Individuals with local pharyngeal neuropathy are allegedly at greater risk of subsequently developing the generalized condition, and the frequency is proportionate to the intensity of the faucial infection. Prompt administration of antitoxin in faucial diphtheria sharply reduces the incidence of neuropathy.

LOCAL NEUROPATHY

Palatal paralysis, with nasal speech accompanied by impaired pharyngeal and laryngeal sensation, develops between 20 and 30 days of onset of the throat infection. Poor pharyngeal sensation and a weak cough may further compromise the respiratory status of individuals with inflammatory upper-respiratory lesions. Within another week, paralysis of ocular accommodation often appears, resulting in blurred vision. Pupillary reflexes to light are spared. Rarely, diaphragmatic, laryngeal, pharyngeal, facial, and oculomotor muscles then become weak; death from failure of respiration has occurred at this stage. Focal neuropathy in the limbs may appear with cutaneous diphtheria. A consistent feature of the early local neuropathy is the prevalence of weakness over sensory signs; this is predominantly a cranial motor neuropathy. Usually the condition gradually worsens for 2 or 3 weeks, then rapidly improves. The prognosis for complete recovery is excellent.

GENERALIZED NEUROPATHY

The generalized neuropathy develops within 8 to 12 weeks following the onset of the infective illness, although it may occur as early as 3 weeks. It is a mixed sensorimotor neuropathy of rapid onset with a predominantly distal distribution of weakness, paresthesias, and sensory loss. Tendon reflexes are depressed or lost. Occasionally, position sense is markedly impaired with relative preservation of muscle strength, giving rise to a sensory ataxia ("diph-

theritic pseudotabes"). Recovery is usually complete and begins within days or weeks of the onset of paralysis.[22]

Myocardial involvement may occur, resulting in cardiac failure or dysrhythmias, and may be fatal. It usually develops at about the third week of illness.

Laboratory Studies

BACTERIOLOGIC

Culture of *C. diphtheriae* from the fauces or from a cutaneous ulcer will establish the diagnosis.

CEREBROSPINAL FLUID

The protein content is often moderately elevated and a lymphocyte pleocytosis may be present.

ELECTRODIAGNOSTIC STUDIES

There is one report of clinical electrophysiology of human diphtheria using contemporary techniques. Serial electrophysiologic studies of a recent case indicate a demyelinating sensorimotor neuropathy.[115] Diphtheria toxin is widely used as an experimental model of PNS demyelination, and there is considerable knowledge about its electrophysiologic effects in experimental animals.[22]

ETHYLENE OXIDE

Ethylene oxide is a gas widely used in industry, especially in sterilizing heat-sensitive biomedical materials. Several reports describe distal symmetric polyneuropathy in individuals and experimental animals chronically exposed to low levels of ethylene oxide.[37,46,63,82,108,136] It is suggested that residual ethylene oxide in dialysis tubing may contribute to the peripheral neuropathy in patients on long-term hemodialysis.[131] Symptoms of distal-extremity numbness and weakness are accompanied by evidence of diminished sensation in the feet and hands. Tendon reflexes are diminished throughout and ankle jerks are absent. Motor and sensory nerve conduction velocities are mildly diminished. Encephalopathic symptoms and cognitive impairment may accompany the peripheral neuropathy.[25] Sural nerve biopsies reveal evidence of axonal degeneration. There are no reports of postmortem findings in humans with ethylene oxide intoxication. Gradual recovery from peripheral neuropathy commences upon withdrawal from exposure. Experimental animals chronically exposed to 250 ppm develop widespread sensory nerve fiber degeneration in the distribution of a central-peripheral distal axonopathy.[83]

HEXACARBONS (*n*-HEXANE, METHYL *n*-BUTYL KETONE)

n-Hexane and methyl *n*-butyl ketone are considered together in this section because each is metabolized to 2,5-hexanedione (2,5-HD),[28] the agent responsible for most, if not all, of the neurologic effects that accompany repetitive exposure to these compounds.[116,122]

n-Hexane is widely used as an inexpensive solvent and is a component of lacquers, glues, and glue thinners. World-wide human neurologic disease has been associated both with occupational exposure[54] and following deliberate inhalation of vapors containing *n*-hexane (glue sniffers).[5,61] Methyl *n*-butyl ketone (MnBK) has a greater neurotoxic potential than *n*-hexane, and was enjoying increasing use as a solvent until implicated in the 1973 outbreak of peripheral neuropathy in Ohio. Methyl ethyl ketone (MEK) is also present in some neurotoxic solvent mixtures and thinners containing either *n*-hexane or MnBK. Although some reports of human neuropathy have identified MEK as the causative agent, this solvent does not cause neuropathy in animals, but may have accelerated development of neurotoxicity in individuals exposed to *n*-hexane and MnBK.[116]

Pathology

The neurotoxic hexacarbons produce distal axonopathy in humans and experimental animals. Human nerve biopsies and postmortem nervous system tissue display massively swollen axons filled with 10-nm neurofilaments.[5] A distal-axonopathy-type distribution of giant axonal swellings is apparent in the PNS of experimental animals exposed to these compounds, in the distal ends of CNS tracts, and in cord-ganglion muscle combination tissue cultures exposed to 2,5-HD. Axonal swellings in the PNS appear to develop on the proximal side of nodes of Ranvier in distal, nonterminal regions of affected fibers. Organelles accumulate in these regions and may contribute to the breakdown of nerve fibers distal to the swelling.[121] Atrophy of proximal axonal segments accompanies this process. Axonal swellings develop more proximally with time.[120,122] They are associated with myelin retraction and focal demyelination to a degree that is unusual in an axonopathy. This demyelination may correlate with the marked distal slowing of nerve conduction characteristic of human hexacarbon neuropathy.[61]

Pathogenesis

There is strong evidence to suggest that the neurotoxic property of *n*-hexane and MnBK is largely attributable to the common gamma diketone metabolite 2,5-HD. Other gamma diketones such as 2,5-heptanedione and 3,6-octanedione cause neuropathy, whereas related compounds (2,4-hexanedione, 2,3-hexanedione, and 2,6-heptanedione), which lack the 1,4 spacing of the carbonyl groups, fail to produce experimental neuropathy.[28] Presumably, circulating 2,5-HD disrupts neuronal-axonal function and results in focal, massive accumulations of neurofilaments in distal regions of certain axons.[90] The mechanism of toxic action is not known; some studies suggest that pyrrole formation and direct cross-linking of neurofilaments by gamma diketone has a role;[41,100] others propose that an increase in neurofilament transport velocity contributes to proximal axonal atrophy and distal neurofilamentous swelling.[77]

Clinical Features

Sensory and motor dysfunction usually develops insidiously with occupational exposure.[54] Individuals sometimes consult an orthopedist on the erroneous assumption that instability of the knee reflects a joint disorder. Those deliberately inhaling high concentrations of solvent vapors containing n-hexane and MEK have experienced a subacute onset of motor neuropathy.[5] Weight loss, malaise, and abdominal pain frequently accompany the development of hexacarbon neuropathy.

SIGNS AND SYMPTOMS

The most common initial complaint, both in industrial cases and among glue sniffers, is numbness of the toes and fingers. This type of distal sensory neuropathy may be the only clinical illness in the least severe industrial cases.[54] The pattern of sensory abnormality is characteristically symmetric and involves only the hands and feet, rarely extending as high as the knee. A moderate loss of touch, pin, vibration, and thermal sensation is usually prominent and may be accompanied by loss of the Achilles reflex. In mild cases, other tendon reflexes are spared, and there is preservation of position sense and no sensory ataxia, cranial nerve abnormalities, autonomic dysfunction, or periosteal pain.

In the more severely involved industrial cases, weakness and weight loss occur, occasionally accompanied by anorexia, abdominal pain, and cramps in the lower extremities. Reflex loss is usu-ally less than that observed in other polyneuropathies, and even in the moderate to severe cases, may be confined to the Achilles reflex and finger-jerks.[122] Weakness most commonly involves intrinsic muscles of the hands and long extensors and flexors of the digits. A common complaint in these individuals is difficulty with pinching movements, grasping objects, and stepping over curbs.[3] Pure motor neuropathy is unusual in industrial cases. Vibration and position sense are only mildly impaired and sensory loss for pinprick and touch is usually confined to the hands and feet.

As the neuropathy becomes more severe, weakness and atrophy dominate the clinical picture and extend to involve proximal extremity muscles. Some glue-sniffing cases display a subacute distal to proximal progression of weakness early in the course of the disease. Glue sniffers with prolonged high exposure may display signs of "bulbar" or phrenic nerve paralysis. Blurred vision is an occasional complaint, but objective evidence of visual loss has not been documented.

No predisposing conditions have been proven to exist for hexacarbon neurotoxicity. Slowed motor nerve conduction is described in otherwise normal workers in a factory with documented cases of solvent neuropathy.[3] This strengthens the notion that subclinical and readily reversible nerve damage may be an unrecognized industrial problem.

Autonomic disturbances have been reported among the glue sniffers, but not in the industrial cases. Prominent among these disturbances is hyperhidrosis of the hands and feet, occasionally followed by anhidrosis.[5] Blue discoloration of the hands and feet, reduced extremity temperature, and Mees's lines are sometimes present in these patients. Impotence has occasionally occurred among glue sniffers with a moderate or severe neuropathy, but its relationship to nervous system dysfunction has not been established.

Laboratory Studies

CLINICAL LABORATORY FINDINGS AND CEREBRO-SPINAL FLUID

Routine clinical laboratory tests are normal. In the majority of cases of hexacarbon neuropathy, the CSF contains a normal level of protein, and cells are absent. The rare case with elevated CSF protein may reflect nerve fiber degeneration that has ascended to the spinal roots.

ELECTRODIAGNOSTIC STUDIES

The electromyographic abnormalities are usually symmetric and greater in distal than in proximal muscles. In patients with minimal involvement, sparse fibrillation and abnormal motor unit potentials are often the only findings. Nerve conduction times and clinical examination are usually normal at this stage.[3]

More severe cases generally display more frequent fibrillation potentials and positive sharp waves, and a reduced motor unit recruitment pattern on volition. With recovery or stabilization of the condition, the electromyographic changes disappear.

In cases with minimal involvement, motor and sensory nerve conduction is usually normal or in the low-normal range of velocity. As the clinical illness intensifies, a progressive slowing of conduction occurs,[61] and in the most extreme cases, distal peroneal nerve conduction cannot be elicited. Severe slowing of distal motor nerve conduction frequently appears disproportionate to the moderate weakness. Such profound slowing is characteristic of demyelinating neuropathies and may reflect paranodal demyelination associated with the giant axonal swellings.[61] A comprehensive study that analyzed visual, somatosensory, and brainstem auditory evoked potentials describes abnormalities in all modalities.[21] This finding may correlate with widespread distal axonal degeneration observed in the central nervous system tracts of animals experimentally intoxicated with 2,5-HD.[18,104]

NERVE BIOPSY

Nerve biopsies from mild cases may be normal, even when examined by electron microscopy. Occasionally, muscle biopsy from such cases reveals abnormalities of neuromuscular junctions or of intramuscular nerve twigs.[54] The most informative nerve biopsy results are usually obtained from moderately or severely involved individuals; in these cases, teased myelinated nerve fiber preparations have clearly illustrated paranodal giant axonal swellings accompanied by myelin retraction.[61]

Course and Prognosis

Insidious onset and slow progression are the hallmarks of the industrial cases. In most instances, this is a reflection of low-level intermittent exposure.[3,54] In some glue sniffers, especially those with excessive abuse, a subacute course develops, leading in severe cases to quadriplegia within 2 months of the first symptom. AIDP has been a serious diagnostic consideration in these patients.[5]

A universal feature of hexacarbon neurotoxicity is the continuous progression of disability after removal from toxic exposure. Progression ("coasting") usually lasts for 1 to 4 months and often occurs in the hospital. The degree of recovery in most cases correlates directly with the intensity of the neurologic deficit. Individuals with a mild or moderate sensorimotor neuropathy usually recover completely within 10 months of cessation of exposure. Severely affected industrial cases also improve, but sometimes retain mild to moderate residual neuropathy on follow-up examination as long as three years after exposure. Such individuals have, on occasion, developed hyperactive knee jerks and lower-extremity spasticity.[5,61] This reflex change probably reflects degeneration in the corticospinal tracts that accompanied the peripheral neuropathy.

Glue sniffers who sustain an extreme degree of distal atrophy may never recover full strength, and persistent hyperpathia and autonomic dysfunction are described.

Differential Diagnosis

The clinical features of hexacarbon neuropathy do not help to distinguish this condition from many other toxic or metabolic distal axonopathies. Profound slowing of motor nerve conduction disproportionate to moderate weakness, and nerve biopsy demonstration of paranodal giant axonal swellings strongly suggest hexacarbon neuropathy. The most useful diagnostic test is a detailed occupational and social history, specifically inquiring about solvent exposure and inhalant abuse.

Case History: Neuropathy from Exposure to n-Hexane

A 22-year-old female worked long hours in a small, poorly ventilated factory. She handled rags saturated with a solution containing n-hexane and inhaled large amounts of n-hexane vapor for two years before developing anorexia, weight loss, and a cramping sensation in the hands. Two weeks later, she noticed cramping in the calves and an unsteady gait that was improved by the use of high-heeled boots. Two months later, her legs felt much weaker and she noted numbness of the toes. These symptoms steadily increased in intensity over the next two months until she was unable to walk to work. She was admitted to the hospital. Neurologic examination revealed no abnormalities of cranial nerves or mental status. There was diffuse, symmetric, distal flaccid weakness of all extremities. The intrinsic muscles of the hands and dorsiflexors of the feet were 2/5 MRC scale. Proximal limb muscles were 3/5 MRC scale, and abdominal muscles were also severely weak. The vital capacity was normal. No tendon reflexes could be elicited. There was a moderate to severe loss of pin sensation in a stocking-glove distribution, with only a slight diminution in position and vibration sense. After four months in this state she gradually began to improve, was discharged from the hospital, and steadily regained strength over the next 18 months. Since that time, she noticed no further improvement of strength but began to walk again, although in an abnormal fashion. Neurologic examination two years after discharge from the hospital revealed a slow, stiff-legged, waddling gait. She could stand on a narrow base and Romberg's sign was not present. Rapid alternating movements of the upper extremities were performed well, as were finger-to-nose movements. Distal and proximal upper-extremity muscles were 4/5 MRC scale; distal lower-extremity muscles were 4/5 MRC scale. Many large, proximal lower-limb muscle groups (hamstrings, glutei, quadriceps) were 4/5 MRC scale. She was unable to rise from a chair without using her hands for support, unable to sit up from a supine position without rolling to one side and using her arms for assistance. There was a moderate diminution of pinprick and touch sensation below the ankle and wrist, but position and vibration sensation were normal. The tendon reflexes were brisk in the arms and very brisk in the legs, with bilateral Babinski's signs. There was increased resistance to rapid passive movements of the lower limbs, and she displayed a spastic catch.

Comment. This patient demonstrates two cardinal features associated with occupational axonal neuropathies: poor working conditions and gradual onset of symptoms. She developed a neuropathy of unusual severity with quadriparesis and near-total paralysis of the distal extremities because of the prolonged, high-level exposure to the hexacarbon toxicant. Motor impairment was far more severe than the sensory deficit, perhaps reflecting high levels of ambient n-hexane in the final months of exposure. Signs of corticospinal tract degeneration did not appear until the peripheral nerves regenerated.

LEAD

Symptomatic lead polyneuropathology is now rare in North America and Europe. There is disagreement as to whether subclinical nerve damage, defined by slight abnormalities in nerve conduction in asymptomatic individuals, occurs in lead-exposed workers.[13,51,112] Occupational exposure especially occurs in smelting industries, battery manufacture, demolition work, and in automobile radiator repair. Common environmental sources are paint ingestion and consumption of moonshine whiskey.[62] Lead is unusual among human neurotoxins, and its neuropathy so far resists the classification outlined in Chapter 2. Lead clearly can produce a demyelinating neuropathy in experimental animals and in tissue culture, yet the pathologic changes of the human neuropathy are poorly characterized. Furthermore, neither the experimental nor the available human neuropathologic material explain some of the bizarre clinical features of lead neuropathy.

Pathology and Pathogenesis

Wallerian-type axonal degeneration is the only well-described pathologic alteration in reports of human nerves from individuals with lead neuropathy.[62] Postmortem reports of cases in the older literature include descriptions of spinal-root degeneration, chromatolysis of anterior horn cells, and fiber-tract degeneration in the spinal cord. Inadequate descriptions and contradictory findings in these early reports, combined with the absence of contemporary postmortem data, have added to the confusion surrounding the neuropathology of lead neuropathy. Clinical features of some cases suggest primary dysfunction of anterior horn cells,[12] but neither human nor experimental animal studies firmly support this notion. There exists one contemporary report of a sural nerve biopsy from an individual with lead neuropathy, describing loss of large myelinated fibers.[113] The morphologic changes in the early stages of human neuropathy have not been reported; most descriptions apparently represent advanced stages of this condition.[62] Segmental demyelination is not a prominent feature in any reported case, but its existence has not been ruled out by these limited studies.

In contrast, segmental demyelination is clearly the predominant change in most models of experimental chronic lead neuropathy.[64] Several events occur prior to demyelination: there is an early rapid accumulation of lead in nerve, the blood-nerve barrier is abnormally permeable,[30] endoneurial pressure is increased,[73] nerves become grossly swollen, and intranuclear inclusions appear within Schwann cells. The role of these events in the pathogenesis of segmental demyelination is unclear. There are two leading hypotheses. One is that lead has an early direct toxic effect on Schwann cells[132,133] and accumulation of fluid has no primary role in demyelination. The second states that primary injury to the blood-nerve barrier leads to an accumulation of lead-containing edema fluid and elevated endoneurial pressure. This produces an abnormal, lead-rich endoneurial environment eventuating in Schwann-cell injury.

The cellular distribution of lead in the PNS is not known. In certain other organs it is selectively accumulated in nuclei and mitochondria, and lead-containing intranuclear inclusion bodies appear similar to the inclusions recently described in Schwann cells. This, in concert with tissue culture studies demonstrating demyelination and inhibition of Schwann-cell proliferates by lead, supports the hypothesis that lead directly affects Schwann cells.[114,130,133]

Clinical Features

Symptomatic lead neuropathy generally develops after prolonged exposure and is commonly preceded by weight loss, anorexia, fatigue, constipation, and episodic abdominal pains.

SYMPTOMATIC NEUROPATHY

Lead intoxication results in predominantly motor neuropathy of unusual variability. Sensory complaints and findings are minimal. Five distinct patterns of weakness are described in the older literature. In order of frequency they are:

1. Extensors of the fingers and wrists (wrist-drop type)
2. Proximal shoulder girdle muscles
3. Intrinsic hand muscles
4. Peroneal muscles
5. Paralysis of the larynx

Cases displaying such focal patterns of weakness are now rare.[12] The few recent studies of probable adult lead neuropathy describe progressive generalized weakness, mild distal atrophy, reflex loss, and occasional fasciculations.[13] Lower-limb weakness is prominent in childhood cases. Electromyography of weakened muscles usually reveals abundant fibrillation potentials.[12] Motor nerve conduction is normal. Sensory amplitudes in weakened extremities may be strikingly reduced. One contemporary report of a sural nerve biopsy describes loss of myelinated fibers.[13]

ASYMPTOMATIC NEUROPATHY

Electrophysiologic studies of neurologically normal, asymptomatic individuals with chronic occupational exposure to lead claim significant slowing of nerve conduction. Abnormal electromyographic findings indicative of denervation are also described. The degree of electrophysiologic abnormality sometimes correlates with elevated blood lead levels.[112] The interpretation of these studies has been seriously challenged in a careful study of Danish lead workers,[13] and the prevalence of subclinical lead neuropathy remains controversial.

Laboratory Studies

CLINICAL LABORATORY FINDINGS

Mild anemia, basophilic stippling of erythrocytes, and elevated urine coproporphyrin are characteristic of plumbism. Determination of urine lead is helpful, and levels in excess of 0.2 mg/L are usually regarded as significant.

Diagnosis can be confirmed after promoting lead excretion by dosing with a chelating agent and measuring the subsequent rise in urine lead levels. Excretion of over 600 μg in 24 hours indicates excessive lead stores in soft tissue. Blood levels are less helpful and, when low, may be misleading, because lead is rapidly cleared from the circulation. Elevated blood levels per se do not indicate intoxication, and controversy surrounds the safe level allowed in individuals with occupational exposure.[112] Mean blood levels in North American children are reportedly 15 to 28 μg/dl.[62]

CEREBROSPINAL FLUID

The CSF is acellular and the protein level normal.

ELECTRODIAGNOSTIC STUDIES

Described above under the heading "Clinical Features."

Treatment and Prognosis

The most important single feature of treatment is eliminating further exposure to lead. Once abnormal intake of lead is ended, much of the lead in soft tissue is gradually shifted into bone. As long as significant quantities of lead remain in bone, any illness associated with demineralization may mobilize it. This lead again circulates, becomes stored in soft tissues, and may exacerbate plumbism. The aim of chelation therapy is to bind the lead in soft tissue so that it can then be excreted. Either penicillamine (favored in mild cases) or calcium-disodium EDTA may be used as chelating agents. Chelating agents are administered in short courses, each associated with a rise in urine lead levels. Recovery may begin within two weeks following the initiation of therapy.[12] Strength gradually improves over a prolonged period, frequently more than a year. Complete recovery is usual except in extremely advanced cases.

Differential Diagnosis

Symptomatic lead neuropathy usually is readily identified·in exposed workers and in consumers of moonshine or illegally distilled whiskey who become diffusely weak or develop wrist-drop.

Clinical distinction of lead neuropathy from motor neuron disease is sometimes difficult, and may require urine lead levels with a chelation challenge. In general, lead intoxication should be suspected both in individuals with a generalized predominantly motor neuropathy, and in idiopathic plexus or isolated nerve lesions (especially the radial and peroneal).

MERCURY

Most of the neurologic effects of mercury are manifested by the central nervous system; symptomatic peripheral neuropathy is rare. Metallic mercury and mercury vapor may cause peripheral neuropathy; it is unclear if organic mercury does.

Mercury metal vapor exposure can cause a subacute, diffuse, predominantly motor neuropathy, mimicking AIDP.[135] Electrophysiologic studies and sural nerve biopsies indicate axonal degeneration with relative sparing of sensory functions. Gradual recovery follows removal from exposure. There are several reports of mild symptomatic or asymptomatic sensorimotor neuropathy in individuals with occupational exposure to elemental mercury.[1,2]

Alkyl mercury exposure can produce intense distal limb paresthesias, promi-nent features of both the Minamata and Iraqi outbreaks. Neither postmortem study nor careful electrophysiologic studies show evidence of peripheral nervous system involvement in these victims, and the sensory findings are held to stem from central nervous system dysfunction.[67,124] This is curious, since in experimental animals and tissue culture, the dorsal root ganglion cell is a major site of uptake and readily degenerates when exposed to organic mercury.[56,134]

METHYL BROMIDE

Methyl bromide has found use as a fumigant, fire extinguisher, refrigerant, and insecticide. Chronic exposure to high levels of methyl bromide may result in a syndrome characterized by signs of pyramidal tract, cerebellar, and peripheral nerve dysfunction. One report describes distal symmetric polyneuropathy in eight individuals following chronic low-level exposure.[60] The overall pattern of the clinical illness strongly suggests that methyl bromide polyneuropathy represents a distal axonopathy. This notion is unproven, as there are neither morphologic reports of human PNS tissue nor an adequate experimental animal study.

Symptoms of numbness and weakness distally in the legs appear after several months exposure. These symptoms gradually increase and are then accompanied by an unsteady gait and clumsy hands. Signs include variable flaccid weakness of lower-leg muscles, and moderately diminished sensation to pin and touch in a stocking-glove distribution. Tenderness of the calf muscles is usually prominent and may erroneously suggest the diagnosis of myositis. The Achilles reflex is usually diminished or absent. The CSF is normal. Electrophysiologic studies reveal denervation in distal muscles and delayed sensory conduction. Gradual improvement and complete recovery occurs within a year following withdrawal from exposure.[60]

ORGANOPHOSPHORUS ESTERS

Organophosphorus (OP) esters have major industrial applications: insecticides, petroleum additives, hydraulic fluids, flame retardants, and modifiers of plastics. Most OP esters inhibit acetylcholinesterase, a property exploited in their use as pesticides, helminthicides, and war gases.[26]

In addition, many OP esters can produce distal axonopathy characterized by widespread CNS and PNS degeneration (organophosphate-induced delayed polyneuropathy; OPIDP). This effect involves inhibition of a nervous system esterase, neuropathy target esterase (NTE), distinct from acetylcholinesterase.[59] OPIDP often has a curiously delayed onset following single exposure, and has been responsible for devastating epidemics of neurotoxic injury.[106,109]

The most infamous organophosphate is triorthocresyl phosphate (TOCP). Other organophosphates known to be responsible for human distal axonopathy include leptophos, mipafox, trichlorphonate, chlorpyrifos, metrifonate, methamidophos, parathion, and trichloronate. Many additional OP esters inhibit NTE and produce axonal damage in experimental animals. Distal axonopathy was formerly held to be a relatively uncommon neurotoxic result of exposure to organophosphorus compounds, which number more than 50,000, especially when compared with the well-known anticholinesterase effects. Experimental studies indicate that distal axonopathy may be a fairly pervasive effect of the organophosphate esters. NTE screening has revealed that many compounds previously regarded as innocuous may be capable of producing axonopathy under suitable conditions. The detailed biochemistry and pharmacology of the OP esters is complex and beyond the scope of this volume (see *Pathogenesis*, below) and was recently reviewed elsewhere.[58] Of all the organophosphorus compounds, TOCP has produced the most instances of neuropathy in humans, is extensively studied in animals, and will be discussed in this section as a paradigm for the other compounds of this class.

TOCP

TOCP is an oily substance, lipid-soluble, readily absorbed through skin and mucous membranes, which usually produces only mild cholinergic symptoms in humans. It is valued as a softener in the plastics industry and as a high-temperature lubricant. Many instances of neuropathy have resulted from adulteration or misuse of TOCP in food, drink, or cooking oil. Prominent outbreaks occurred in the United States from drinking adulterated Jamaica ginger extract (jake leg paralysis),[8] in Morocco due to cooking with contaminated cooking oil,[113] and in Sri Lanka from adulterated gingli oil.[109]

PATHOLOGY

TOCP is conclusively demonstrated to produce a distal axonopathy in humans and experimental animals. There are no contemporary morphologic studies of the human illness. Postmortem studies from the 1930 jake leg paralysis epidemic revealed wallerian degeneration in the PNS. CNS alteration in these cases included changes in the cervical levels of the gracile fasciculi and the lumbar levels of the corticospinal tracts, indicative of distal axonopathy.[8] Experimental animal studies in a variety of species have repeatedly confirmed this distribution of CNS findings, and clearly demonstrate a pattern of distal axonal degeneration in the PNS. Large-diameter, heavily myelinated nerve fibers appear most vulnerable, and lower-extremity nerves are most affected.[16,95] Ultrastructural studies of the axonal alterations demonstrate proliferation of smooth endoplasmic reticulum, but provide little insight into the pathogenesis of this condition.[95]

PATHOGENESIS

The biochemical mechanism underlying the most common form of OP neurotoxicity—inhibition of acetylcholinesterase—involves phosphorylation of the esterase. Since OP compounds are good phosphorylating agents, it is proposed that phosphorylation of additional nervous system esterases (other than acetylcholinesterase) may account for the delayed distal axonopathy produced by TOCP and related substances.[58,79] There are many esterases in the nervous system, and the physiologic function of most is unknown. Considerable biochemical evidence indicates that there is phosphorylation of a specific nervous system protein esterase, NTE, by OP.[59] The physiologic function of NTE has not been identified; in OPIDP it appears to behave like a receptor. NTE is present in brain, spinal cord, and peripheral nerve, and in nonneural tissues such as spleen and lymphocytes. Once phosphorylation of NTE is initiated, a second step, "aging" of the phosphoryl-enzyme complex, is required to produce neuropathy. Aging occurs rapidly and involves cleavage of a chain from the bound phosphorus atom, leaving a negatively charged monosubstituted phosphoryl residue at the active site. OP compounds that age the complex cause OPIDP; those OPs that do not age do not cause OPIDP.[79] The biochemical cascade that causes polyneuropathy weeks later is unknown. It appears that local inhibition of axonal NTE is critical,[79] and that increasing inhibition of retrograde axonal transport can be detected after only four days of intoxication with neuropathic OPs.[78]

It is suggested that monitoring of lymphocyte NTE in individuals routinely exposed to OP may prove useful for prevention of OPIDP, just as erythrocyte AChE activity is used to establish no-effect levels in occupational exposure.[72] In one case of severe chlorpyrifos intoxication, inhibition of lymphocytic NTE heralded the development of OPIDP.[71]

CLINICAL FEATURES

The inconsistent nervous system damage following exposure to TOCP probably reflects variabilities in dose and absorption. Most instances have involved oral ingestion; however, this substance is also readily absorbed through the skin and alveoli. Following a single large exposure, a transient, variable cholinergic response occurs, usually diarrhea, perspiration, and fasciculations. This response generally lasts a day and is rarely incapacitating (in contrast to the OP esters with potent anticholinesterase properties, whose muscarinic and nicotinic properties may be instantly fatal). An asymptomatic interval of 7 to 12 days then ensues before clinical evidence of neuropathy appears. Should repeated low-level exposure occur, typical in industry, then cholinergic symptoms may not be appreciated.

Symptoms and Signs. Initial symptoms of cramping pain in the calves are followed by tingling and burning sensations in the feet and occasionally in the hands. Weakness soon appears in the legs and rapidly involves the hands as well. In severe cases, weakness may spread proximally to involve muscles acting about the knees or hips. The course is subacute, in contrast to the chronicity of most distal axonopathies, and maximum involvement usually occurs within 2 weeks of the first symptoms.[8,109]

Weakness and flaccidity are prominent clinical signs, and despite the early paresthesias, TOCP neuropathy is predominantly motor. Sensory loss is present in every case, if carefully examined, but is usually trivial compared with the striking weakness. Foot-drop is common, and gait ataxia, disproportionate to weakness or sensory loss, occasionally occurs. The Achilles reflex is initially absent; patellar reflexes may be depressed or increased. Atrophy of the legs and intrinsic hand muscles is the rule. Cranial nerve involvement and autonomic dysfunction are not features of TOCP neuropathy. Signs of CNS dysfunction are present in most cases. These often are obscured by PNS degeneration in the early stages; however, as the weeks pass, progressive spasticity may develop in the lower extremities, yielding a characteristic mixture of upper and lower motor neuron involvement. A 30-year follow-up study of individuals with Jamaica ginger poisoning revealed spastic paraparesis and distal leg atrophy in some.[8]

The prognosis in mildly affected individuals, presumably with less exposure, is good. They generally make a near-complete recovery. Others with a more severe initial deficit are left with varying degrees of morbidity, including sequelae of both PNS (atrophy, claw hands, foot-drop) and CNS (spasticity, ataxia) damage. There is no specific treatment.

LABORATORY STUDIES

Clinical Laboratory Findings. Routine laboratory studies are usually normal. Erythrocyte cholinesterase levels are only mildly depressed by TOCP and several other OP esters that produce distal axonopathy.

Cerebrospinal Fluid. The CSF is usually acellular and the protein level normal or mildly elevated.

Electrodiagnostic Studies. Electromyography reveals changes of denervation in every case. These are confined to distal muscles in mild instances and generally involve proximal leg muscles with severe involvement. Motor nerve conduction is usually mildly slowed in the legs and may be normal in the arms, even in distal segments of mixed nerves in atrophic limbs. Sensory nerve potentials are usually diminished in the upper extremities, and it may not be possible to elicit responses in the legs. This curious phenomenon is an important factor in determining an appropriate test for screening at-risk individuals for subclinical neuropathy. It is suggested that the recording of lower-limb sensory nerve action potential amplitude is likely to be the most sensitive measurement.

DIFFERENTIAL DIAGNOSIS

The differential diagnosis of TOCP neuropathy is simple if there is clear evidence of its ingestion 2 weeks before the illness and it is accompanied by a cholinergic reaction. Should such evidence be lacking, the condition becomes almost impossible to establish with certainty. In the authors' experience, cases of organophosphorus neurotoxicity without obvious cholinergic reactions have been erroneously diagnosed as AIDP, encephalomyelitis, or multiple sclerosis. The subacute onset of a myeloneuropathy syndrome in an otherwise healthy individual should raise the suspicion of OP intoxication.

POLYCHLORINATED BIPHENYLS

Polychlorinated biphenyls (PCBs) have been extensively used as plasticizers and in electrical insulation. Although there has been considerable environmental pollution of waterways and marine life with PCB, neurotoxicity from this source is not described. Outbreaks of PCB neuropathy occurred in 1968 in Japan[80] and in 1978 in Taiwan,[23] when cooking oil became contaminated with tetrachlorobiphenyl.

Affected individuals develop acne, brown pigmented nails, and a discharge from the meibomian glands. Some have distal numbness, absent tendon reflexes, and diminished pin sensation in the feet. Motor nerve conduction and sural nerve conduction are significantly slowed. Rats intoxicated with tetrachlorobiphenyl develop slowed motor conduction.[81] The pathogenesis of this disorder is unknown.

Biphenyl itself has also been reported to induce sensorimotor neuropathy (thought to be demyelinating in type) and psychologic changes in individuals using this compound as an agricultural fungistatic agent.[47]

SPANISH TOXIC OIL SYNDROME

The Spanish toxic oil syndrome (STO) is a new entity that appeared as an epidemic in Spain in 1981 due to ingestion of adulterated rapeseed oil. STO is a multisystem disease that evolves in a distinct fashion. Initial manifestations are respiratory insufficiency, headache, eosinophilia, and malaise. Some die from acute respiratory insufficiency.[99] Survivors then experience severe myalgia for several months; its improvement heralds the onset of a diffuse neuromuscular syndrome of weakness, atrophy, and sensory disturbance. The neuromuscular syndrome involves 92% of individuals with STO, continues to progress for one year following ingestion, and is the main cause of disability. Patients develop patchy or diffuse sensory loss. In mild cases, a mononeuritic pattern is apparent. Myalgia and weakness are more diffuse, reflecting inflammatory and vasculitic changes in many muscles. Sural nerve biopsies and postmortem studies describe lymphocytic infiltration of epi- and perineurium in early cases, and marked perineurial fibrosis and thickening in advanced cases. Vasculitis, prominent in muscle biopsies, is not described in nerve biopsies.[74] Endoneurial inflammation is rare. STO appears to represent an immune-mediated, possibly vasculitic, toxic neuropathy; its pathophysiology is unclear, as is its etiology. The implicated oil contained anilides, peroxides, and precursors of arachidonic acid.

THALLIUM

Thallous salts have been widely employed in rodenticides and insecticides. This use is now discouraged in North America. Rare reports of intoxication usually stem from homicidal or accidental (especially in children) ingestion.[11]

Pathology and Pathogenesis

Thallium salts produce a distal axonopathy in humans. Postmortem exam-ination has clearly demonstrated axonal degeneration in the distal regions of long nerves with proximal fiber preservation.[17] Large-diameter fibers of sensory nerves appear especially vulnerable. Axonal swelling is described as an early change in one case. Sural nerve biopsy has yielded widely divergent findings; one report describes axonal vacuoles.[27] Experimental animal studies depict ultrastructural evidence of axonal damage, including swollen mitochondria, in the absence of clinical neuropathy.[119] Nervous system tissue cultures treated with thallium sulfate demonstrate similar mitochondrial swellings.[117]

The biochemical pathogenesis of thallium axonopathy is not known. It is suggested that thallium ions may combine with free sulfhydryl groups of proteins, or that thallium binds to mitochondrial membrane.[11] It is also proposed that some effects stem from thallium substitution for potassium,[10,27] particularly by replacing potassium in potassium-activated ATPase.

Clinical Features

There appear to be three distinct temporal varieties of peripheral neuropathy associated with thallium intoxication. These varieties, in order of frequency, are an acute type beginning within 1 to 2 days of a massive (sometimes lethal) single dose,[10,17,27] a subacute type with onset within weeks of a less massive single dose,[17] and a rare chronic type occurring after prolonged continued exposure to moderate levels of thallium.[94] All three presumably are distal axonopathies. The clinical variation may reflect differences in the spatiotemporal evolution of axonal degeneration in response to different levels of intoxication.

ACUTE TYPE

Gastrointestinal symptoms usually follow within hours of poisoning; occasionally they are delayed for a day.[10,17,27] Vomiting, diarrhea, and abdominal pain are common. Within 2 to 5 days, severe burning paresthesias begin in the legs and feet, often accompanied by intense joint pain. Sensory symptoms appear in the hands and sometimes the trunk. Weakness is not a prominent complaint in many instances, but is usually detectable on examination. All sensory modalities are affected. Tendon reflexes are usually present in the early stages, and may assist in differentiating thallium neuropathy from AIDP in the rare instances with progressive severe sensorimotor neuropathy affecting both cranial and peripheral nerves. Lethargy, coma, and cardiac and respiratory failure often develop, and death may occur within a week or two following ingestion. Thallium may interfere directly with myocardial function.[98] Recovery from neuropathy is gradual and may be incomplete. Sequelae of CNS anoxia (impaired mentation, seizures) may complicate rehabilitation. The diagnosis of acute thallium neuropathy is extraordinarily difficult unless there is clear evidence of homicidal, suicidal, or accidental ingestion. Alopecia, the classic indication of thallium poisoning, appears 15 to 39 days after ingestion. This characteristic sign, therefore, is not present in the early stages of acute neuropathy. Renal damage may occur.

SUBACUTE TYPE

This polyneuropathy usually develops 1 or more weeks after exposure, evolves more slowly, and frequently is accompanied by scalp alopecia.[17] Less constant clinical phenomena are hyperkeratosis, Mees' lines (white striae of the nails), ataxia, chorea, and various cranial nerve palsies.

Sensory symptoms and signs are predominant in this neuropathy. Painful, distressing paresthesias of the feet and hands are initial symptoms, and individuals are often unwilling to walk because of painful feet. Weakness is rarely a prominent symptom. Signs of sensory impairment include diminished pin, touch sensation, and proprioception over the distal extremities. Mild or moderate distal weakness always occurs. Tendon reflexes are usually either nor-

mal or slightly depressed. Tachycardia and hypertension, presumably reflecting autonomic dysfunction, frequently accompany this illness. The prognosis of the subacute neuropathy is excellent. Most individuals recover completely within 6 months of withdrawal from exposure. Hair regrowth usually begins within 10 weeks of withdrawal. The cerebrospinal fluid is usually normal. Motor and sensory nerve conduction may be slightly slowed.

CHRONIC TYPE

Most reports of the nervous system complications of chronic thallotoxicosis focus on extrapyramidal dysfunction; the sensorimotor neuropathy is not well characterized.[94]

Diagnosis and Treatment

Diagnosis is established by analysis of urine or organs for thallium. Microgram quantitation can be achieved. Alopecia is the distinctive clinical sign of this disorder, but is not inevitable. It appears likely that instances of thallium neuropathy, especially the acute variety, go undetected. The efficacy of treatment is controversial. Various regimens including administration of potassium chloride or Prussian blue are currently favored.[6]

TRICHLORETHYLENE

Trichlorethylene (TCE) is used in dry cleaning, rubber production, and as a degreasing agent. It was formerly widely employed as a general anesthetic. It is unlikely that chemically pure TCE is neurotoxic, and it is probable that cranial neuropathy stems from a decomposition product, dichloracetylene, resulting from interaction of TCE with alkaline materials. There is frequently an associated orofacial herpes simplex infection, suggesting that some of the findings may stem from reactivation of the virus.[20]

Acute industrial exposure or anesthetic exposure to TCE is associated with a unique neurotoxic syndrome, namely prompt dysfunction of the trigeminal, and, subsequently, to a lesser extent, the facial, oculomotor, and optic nerves. Peripheral limb nerves have never definitely been implicated.[14,33] The effect of impure TCE on the trigeminal nerve is so predictable, that, for a time, victims of tic douloureux were intentionally exposed to it. Accidental exposure to trichloracetylene has also caused facial sensory loss associated with orofacial herpes simplex, supporting the notion that dichloracetylene has an essential role.[102] Cranial nerves are affected soon after a period of exposure to high concentrations. Trigeminal neuropathy usually includes loss of sensory modalities in the entire distribution of the nerve, accompanied by weakness of muscles of mastication. Improvement occurs over a period of months, but mild facial sensory loss and pupillary reflex abnormalities may be permanent.[102] The pathophysiology of cranial neuropathy is obscure: it is suggested that the human clinical and electrophysiological data are compatible with demyelinating neuropathy.[36] One postmortem report describes extensive axon and myelin degeneration in the trigeminal nerve, sensory root, and brainstem nuclei and tracts.[14] Attempts to produce an experimental animal model of trichlorethylene or dichloracetylene trigeminal neuropathy have failed.[9,96]

VACOR (PNU)

Vacor, N-3-pyridylmethyl-N-p-nitrophenyl urea (PNU) is a structural analog of nicotinamide. It is used as a rodenticide, and its accidental ingestion causes an unusual, rapid onset, severe, distal axonopathy with prominent autonomic involvement.[69] The neuropathy is accompanied by acute diabetes mellitus secondary to necrosis of pancreatic beta cells.[92] Large suicidal ingestions are followed by limb weakness, and loss of postural reflexes within an hour.[69]

Rapid progression to severe weakness, cranial nerve dysfunction, and urinary retention follows, accompanied by diabetic ketoacidosis. Survivors experience a gradual return of strength over months, but remain diabetic with variable degrees of autonomic dysfunction. Limited postmortem studies reveal wallerian-like degeneration in peripheral nerves and dorsal roots and ganglia.

Experimental animals display a rapid and synchronous degeneration of distal peripheral terminal axons following a single dose of PNU, accompanied by a selective impairment of fast anterograde axonal transport in the distal portions of limb nerves.[127] It is suggested that the abnormality in fast transport in somatic and autonomic nerves accounts for the apoplectic onset of this unusual toxic neuropathy in humans. The biochemical mechanism of PNU neuropathy is unclear. Its effects can be prevented by high doses of nicotinamide, and it is possible that an NAD-dependent enzyme is inhibited by PNU with disruption of energy dependent processes of the axon.[127]

REFERENCES

1. Albers, JW, Kallenbach, LR, Fine, LI, et al: Neurological abnormalities associated with remote occupational elemental mercury exposure. Ann Neurol 24:651, 1988.
2. Albers, JW, Cariender, D, Levine, SP, et al: Asymptomatic sensorimotor polyneuropathy in workers exposed to elemental mercury. Neurology 32:1168, 1982.
3. Allen, N, Mendell, JR, Bellmaier, DJ, et al: Toxic polyneuropathy due to methyl n-butyl ketone. An industrial outbreak. Arch Neurol 32:209, 1975.
4. Allt, G and Cavanagh, JB: Ultrastructural changes in the region of the node of Ranvier in the rat caused by diphtheria toxin. Brain 92:459, 1969.

5. Altenkirch, HJ, Mager, J, Stoltenburg, G, et al: Toxic polyneuropathies after sniffing a glue thinner. J Neurol 214:152, 1977.

6. Andersen, O: Clinical evidence and therapeutic indications in neurotoxicology, exemplified by thallotoxicosis. Acta Neurol Scand 70:185, 1984.

7. Arezzo, JC, Schaumburg, HH, Vaughan, HG, Jr, et al: Hind limb somatosensory-evoked potentials in the monkey: The effects of distal axonopathy. Ann Neurol 12:24, 1982.

8. Aring, CD: The systemic nervous affinity of triorthocresyl phosphate (Jamaican ginger palsy). Brain 65:34, 1942.

9. Baker, AG: The nervous system in trichlorethylene intoxication, an experimental study. J Neuropathol Exp Neurol 17:649, 1958.

10. Bank, WJ, Pleasure, DE, Suzuki, K, et al: Thallium poisoning. Arch Neurol 26:456, 1972.

11. Bank, WJ: Thallium. In: Spencer, PS and Schaumburg, HH (eds): Experimental and Clinical Neurotoxicology. Williams & Wilkins, Baltimore, 1980, p 570.

12. Boothby, JA, DeJesus, PV, and Rowland, LP: Reversible forms of motor neuron disease. Arch Neurol 31:18, 1974.

13. Buchthal, F and Behse, F: Electrophysiological and nerve biopsy in men exposed to lead. Br J Ind Med 36:135, 1979.

14. Buxton, PH and Hayward, M: Polyneuritis cranialis associated with industrial trichlorethylene poisoning. J Neurol Neurosurg Psychiatry 30:511, 1967.

15. Calderon-Gonzalez, R and Rizzi-Hernandez, H: Buckthorn polyneuropathy. N Engl J Med 277:69, 1967.

16. Cavanagh, JB and Patangia, GN: Changes in the central nervous system of the cat as a result of tri-o-cresyl phosphate poisoning. Brain 88:165, 1965.

17. Cavanagh, JB, Fuller, NH, Johnson, HRM, et al: The effects of thallium salts, with particular reference to the nervous system changes. Q J Med 43:293, 1974.

18. Cavanagh, JB and Bennetts, RJ: On the pattern of change in the rat nervous system produced by 2, 5-hexanediol: A topographical study by light microscopy. Brain 104:297, 1981.

19. Cavanagh, JB and Nolan, CC: Selective loss of Purkinje cells from the rat cerebellum caused by acrylamide and the response to beta-glucuronidase and beta-galactosidase. Acta Neuropathol 58:210, 1982.

20. Cavanagh, JB and Buxton, PH: Trichlorethylene cranial neuropathy: Is it really a toxic neuropathy or does it activate latent herpes virus. J Neurol Neurosurg Psychiatry 52:297, 1989.

21. Chang, Y-C: Neurotoxic effects of n-hexane on the human central nervous system: Evoked potential abnormalities in n-hexane polyneuropathy. J Neurol Neurosurg Psychiatry 50:269, 1987.

22. Chang, MP, Baldwin, RL, Bruce, C, et al: Second cytotoxic pathway of diphtheria toxin suggested by nuclease activity. Science 246:1165, 1989.

23. Chia, L and Chu, F: Neurological studies on polychlorinated biphenyl (PCB)-poisoned patients. Am J Ind Med 5:117, 1984.

24. Chhuttani, PN, Chawla, LS, and Sharma, TD: Arsenical neuropathy. Neurology (Minneapolis) 17:26,9, 1967.

25. Crystal, H, Schaumburg, HH, Grober, E, et al: Cognitive impairment and sensory loss associated with chronic low-level ethylene oxide exposure. Neurology 38:567, 1988.

26. Davis, CS and Richardson, RJ: Organophosphorus compounds. In Spencer, PS and Schaumburg, HH (eds): Experimental and Clinical Neurotoxicology. Williams & Wilkins, Baltimore, 1980, p 527.

27. Davis, LE, Standefer, JC, Kornfeld, M, et al: Acute thallium poisoning: Toxicological and morphological

studies of the nervous system. Ann Neurol 10:38, 1981.

28. DiVincenzo, GD, Hamilton, ML, Kaplan, C, and Dedinas, J: Characterization of the metabolites of methyl n-butyl ketone. In Spencer, PS and Schaumburg, HH (eds): Experimental and Clinical Neurotoxicology. Williams & Wilkins, Baltimore, 1980, p 846.

29. Donofrio, PD, Wilbourn, AJ, Albers, JW, et al: Acute arsenic intoxication presenting as Guillian-Barré-like syndrome. Muscle Nerve 10:114, 1987.

30. Dyck, PJ, Windebank, AJ, Low, PA, et al: Blood nerve barrier in rat and cellular mechanisms of lead-induced segmental demyelination. J Neuropathol Exp Neurol 39:700, 1980.

31. Edwards, PM: The distribution and metabolism of acrylamide analogues in rats. Biochem Pharmacol 24:1277, 1975.

32. Erlicki, A and Rybalkin, A: Ueber Arseniklähmung. Arch Psychiatr Nervenkr 23:861, 1982.

33. Feldman, RG, Mayer, RM, and Taub, A: Evidence for peripheral neurotoxic effect of trichlorethylene. Neurology (Minneapolis) 20:599, 1970.

34. Feldman, RG, Siroky, M, Niles, CA, et al: Neurotoxic dysuria due to dimethylaminopropionitrile. Neurology (Minneapolis) 29:560, 1979.

35. Feldman, RG, Niles, CA, Kelly-Hayes, M, et al: Peripheral neuropathy in arsenic smelter workers. Neurology (Minneapolis) 29:939, 1979.

36. Feldman, RG, White, RF, Currie, JN, et al: Long-term follow-up after single toxic exposure to trichlorethylene. Am J Ind Med 8:119, 1985.

37. Finelli, PF, et al: Ethylene oxide induced polyneuropathy. A clinical and electrophysiologic study. Arch Neurol 40:419, 1983.

38. Fisher, CM and Adams, RD: Diphtheritic polyneuritis: A pathological study. J Neuropathol Exp Neurol 15:243, 1956.

39. Fullerton, PM: Electrophysiological and histological observations on peripheral nerves in acrylamide poisoning in man. J Neurol Neurosurg Psychiatry 32:1186, 1969.

40. Garland, TO and Patterson, MWH: Six cases of acrylamide poisoning. Br Med J 4:134, 1967.

41. Genter, MB, Szakal-Quinn, G, Anderson, CW, et al: Evidence that pyrrole formation is a pathogenetic step in gammadiketone neuropathy. Toxicol Appl Pharmacol 87:351, 1987.

42. Gold, BG, Griffin, JW, Price, DL: Slow axonal transport in acrylamide neuropathy: Different abnormalities produced by single-dose and continuous administration. J Neurosci 5:1755, 1985.

43. Gold, BG, Price, DL, Griffin, JW, et al: Neurofilament antigens in acrylamide neuropathy. J Neuropathol Exp Neurol 47:145, 1988.

44. Gottfried, MR, Graham, DG, Morgan, M, et al: The morphology of carbon disulfide neurotoxicity. Neurotoxicology 6:89, 1985.

45. Griffin, JW and Watson, DF: Axonal transport in neurological disease. Ann Neurol 23:3, 1988.

46. Gross, JA, Haas, ML, and Swift, TR: Ethylene oxide neurotoxicity: Report of four cases and review of the literature. Neurology (Minneapolis) 29:978, 1979.

47. Hakkinen, I, Hernberg, S, Karli, P, et al: Diphenyl poisoning in fruit paper production: A new health hazard. Arch Environ Health 26:70, 1973.

48. He, F, Jacobs, JM, and Scaravelli, F: The pathology of allyl chloride neurotoxicity in mice. Acta Neuropathol 55:125, 1981.

49. He, F, Lu, B, Zhang, S, et al: Chronic allyl chloride poisoning. An epidemiological clinical, toxicological and neuropathological study. G Ital Med Lav 7:5, 1985.

50. He, F and Zahng, S: Effects of allyl chloride on occupationally exposed subjects. Scand J Work Environ Health 11:43, 1985.

51. He, F, Zhang, S, Li, G, et al: An electroneurographic assessment of subclinical lead neurotoxicity. Int Arch

Occup Env Health 61:141, 1988.

52. Heath, JW, Ueda, S, Bornstein, MB, et al: Buckthorn neuropathy *in vitro*: Evidence for a primary neuronal effect. J Neuropathol Exp Neurol 41:204, 1982.

53. Hersch, MI, McLeod, JG, Sullivan, CE, et al: Abnormal cough reflex in canine acrylamide neuropathy. Ann Neurol 26:738, 1989.

54. Herskowitz, A, Ishii, N, and Schaumburg, HH: n-Hexane neuropathy. A syndrome occurring as a result of industrial exposure. N Engl J Med 285:82, 1971.

55. Heyman, A, Pfeiffer, JB, Willett, RW, et al: Peripheral neuropathy caused by arsenical intoxication. N Engl J Med 254:401, 1956.

56. Jacobs, JM, Carmichael, N, and Cavanagh, JB: Ultrastructural changes in the dorsal root and trigeminal ganglion of rats poisoned with methyl mercury. Neuropathol Appl Neurobiol 1:1, 1975.

57. Jenkins, RB: Inorganic arsenic and the nervous system. Brain 89:479, 1966.

58. Johnson, MK: Receptor or enzyme: The puzzle of NTE and organophosphate-induced delayed polyneuropathy. Trends Pharma Sci 8:174, 1987.

59. Johnson, MK: Organophosphates and delayed neuropathy—is NTE alive and well? Toxicol Appl Pharmacol 102:385, 1990.

60. Kantarjian, AD and Shaheen, AS: Methyl bromide poisoning with nervous system manifestations resembling polyneuropathy. Neurology (Minneapolis) 13:1054, 1963.

61. Korobkin, R, Asbury, AK, Sumner, AJ, et al: Glue sniffing neuropathy. Arch Neurol 32:158, 1975.

62. Krigman, M, Bouldin, T, and Mushak, P: Lead. In Spencer, PS and Schaumburg, HH (eds): Experimental and Clinical Neurotoxicology. Williams & Wilkins, Baltimore, 1980, p 490.

63. Kuzuhara, S, Kanazawa, I, Nakanishi, T, et al: Ethylene oxide polyneuropathy. Neurology 33:377, 1983.

64. Lampert, PW and Schochet, SS: Demyelination and remyelination in lead neuropathy—electron microscopic studies. J Neuropathol Exp Neurol 27:527, 1968.

65. LeQuesne, PM: Neurophysiological investigation of subclinical and minimal toxic neuropathies. Muscle Nerve 1:392, 1978.

66. LeQuesne, PM: Acrylamide. In Spencer, PS and Schaumburg, HH (eds): Experimental and Clinical Neurotoxicology. Williams & Wilkins, Baltimore, 1980, p 309.

67. LeQuesne, PM, Damlujl, SF, and Rustam, H: Electrophysiological studies of peripheral nerves in patients with organic mercury poisoning. J Neurol Neurosurg Psychiatry 37:333, 1974.

68. LeQuesne, PM and McLeod, JG: Peripheral neuropathy following a single exposure to arsenic. J Neurol Sci 32:437, 1977.

69. LeWitt, P: The neurotoxicity of the rat poison vacor. N Engl J Med 302:73, 1980.

70. Lombard, J, Levin, H, and Weiner, WJ: Arsenic intoxication in a cocaine abuser. N Engl J Med 320:869, 1989.

71. Lotti, M, Moretto, A, Zoppellari, R, et al: Inhibition of lymphocytic neuropathy target esterase predicts the development of organophosphate-induced delayed polyneuropathy. Arch Toxicol 59:176, 1986.

72. Lotti, M: Organophosphate-induced delayed polyneuropathy in humans: Perspectives for biomonitoring. Trends Pharm Sci 8:175, 1987.

73. Low, PA and Dyck, PJ: Increased endoneurial pressure in experimental lead neuropathy. Nature 269:427, 1977.

74. Martinez, AC, et al: Neuromuscular disorders in a new toxic syndrome: Electrophysiological study—a preliminary report. Muscle Nerve 7:12, 1984.

75. McDonald, WI and Kocen, RS: Diphtheritic neuropathy. In Dyck, PJ, Thomas, PK, and Lambert, EH (eds): Peripheral Neuropathy, Vol 2.

WB Saunders, Philadelphia, 1984, p 2010.

76. Miller, MS and Spencer, PS: Single doses of acrylamide reduce retrograde transport velocity. J Neurochem 43:1401, 1984.

77. Monaco, S, Autilio-Gambetti, L, Lasek, RJ, et al: Experimental increase of neurofilament transport rate: Decreases in neurofilament number and in axon diameter. J Neuropathol Exp Neurol 48:23, 1989.

78. Moretto, A, Lotti, M, Sabri, MI, et al: Progressive deficit of retrograde axonal transport is associated with the pathogenesis of Di-n-butyl dichlorvos axonopathy. J Neurochem 49:1515, 1987.

79. Moretto, A, Lotti, M, and Spencer, PS: In vivo and in vitro regional differential sensitivity of neuropathy target esterase to Di-n-butyl-2, 2-dichlorvinyl phosphate. Arch Toxicol 63:469, 1989.

80. Murai, Y and Kuroiwa, Y: Peripheral neuropathy in chlorobiphenyl poisoning. Neurology 21:1173, 1971.

81. Ogawa, M: Electrophysiological and histological studies of experimental chlorobiphenyl poisoning. Fukuoka Acta Med 62:74, 1971.

82. Ohnishi, A, Inoue, N, Yamamoto, T, et al: Ethylene oxide neuropathy in rats. Exposure to 250 ppm. J Neurol Sci 74:215, 1986.

83. Ohnishi, A, Inoue, N, Yamamoto, T, et al: Ethylene oxide induces central-peripheral distal axonal degeneration of the lumbar primary neurons in rats. Br J Int Med 42:373, 1985.

84. Osuntokun, BO: An ataxic neuropathy in Nigeria. Brain 91:215, 1968.

85. Osuntokun, BO, Aladetoyinbo, A, and Adeuja, AOG: Free cyanide levels in tropical ataxic neuropathy. Lancet 2:372, 1970.

86. Pappolla, M, Penton, R, Weiss, HS, et al: Carbon disulfide axonopathy. Another experimental model characterized by acceleration of neurofilament transport and distinct changes of axonal size. Brain Res 442:272, 1987.

87. Pestronk, A, Keogh, J, and Griffin, JG: Dimethylaminopropionitrile. In

Spencer, PS and Schaumburg, HH (eds): Experimental and Clinical Neurotoxicology. Williams & Wilkins, Baltimore, 1980, p 422.

88. Prestronk, A, Keogh, J, and Griffin, JG: Dimethylaminopropionitrile intoxication: A new industrial neuropathy. Neurology (Minneapolis) 29:540, 1979.

89. Pleasure, DB, Feldman, B, and Prockop, DJ: Diphtheria toxin inhibits the synthesis of myelin proteolipid and basic proteins by peripheral nerve in vitro. J Neurochem 20: 81, 1973.

90. Politis, M, Pellegrino, RG, and Spencer, PS: Ultrastructural studies of the dying-back process. V. Axonal neurofilaments accumulate at sites of 2, 5-hexanedione application: Evidence for nerve fibre dysfunction in experimental hexacarbon neuropathy. J Neurocytol 9:505, 1980.

91. Politis, M, Schaumburg, HH, and Spencer, PS: Neurotoxicity of selected chemicals. In Spencer, PS and Schaumburg, HH (eds): Experimental and Clinical Neurotoxicity. Williams & Wilkins, Baltimore, 1980, p 613.

92. Pont, A, Rubino, JM, Bishop, D, et al: Diabetes mellitus and neuropathy following vacor ingestion in man. Arch Intern Med 139:185, 1979.

93. Post, EJ: Unmyelinated nerve fibers in feline acrylamide neuropathy. Acta Neuropathol 42:19, 1978.

94. Prick, JJG: Thallium poisoning. In Vinken, PJ and Bruyn, GE (eds): Handbook of Clinical Neuroloy, Vol 36. North Holland Publishing, Amsterdam, 1979, p 239.

95. Prineas, J: The pathogenesis of dying-back polyneuropathies. Part I. An ultrastructural study of experimental tri-orthocresyl phosphate intoxication in the cat. J Neuropathol Exp Neurol 28:571, 1969.

96. Reichert, D, Leibolt, G, and Henschler, D: Neurotoxic effects dichloracetylene. Arch Toxicol 37:23, 1976.

97. Robert, CS and Becker, E: Effects of inhaled carbon disulfide on sen-

sory evoked potentials of Long-Evans rats. Neurobehav Toxicol Teratol 8:533, 1986.

98. Roby, DS, Fein, AM, Bennett, RH, et al: Cardiopulmonary effects of acute thallium poisoning. Chest 85:236, 1984.

99. Ricoy, JR, Cabello, A, Rodriguez, J, et al: Neuropathological studies on the toxic syndrome related to adulterated rapeseed oil in Spain. Brain 106:817, 1983.

100. Sager, PR: Cytoskeletal effects of acrylamide and 2, 5-hexanedione: Selective aggregation of vimentin filaments. Toxicol Appl Pharmacol 97:141, 1989.

101. Satchell, PM, McLeod, JG, Harper, B, et al: Abnormalities in the vagus nerve in canine acrylamide neuropathy. J Neurol Neurosurg Psychiatry 45:609, 1982.

102. Saunders, RA: A new hazard in enclosed environmental atmospheres. Arch Environ Health 14:380, 1967.

103. Schaumburg, HH, Arezzo, JC, and Spencer, PS: Delayed onset of distal axonal neuropathy in primates after prolonged low-level administration of a neurotoxin. Ann Neurol 26:576, 1989.

104. Schaumburg, HH and Spencer, PS: Environmental hydrocarbons produce degeneration in cat hypothalamus and optic tract. Science 199:199, 1978.

105. Schaumburg, HH and Spencer, PS: Clinical and experimental studies of distal axonopathy—a frequent form of brain and nerve damage produced by environmental chemical hazards. Ann NY Acad Sci 329:14, 1979.

106. Schaumburg, HH and Spencer, PS: Selected outbreaks of neurotoxic disease. In Spencer, PS and Schaumburg, HH (eds): Experimental and Clinical Neurotoxicology. Williams & Wilkins, Baltimore, 1980, p 883.

107. Schaumburg, HH, Wisniewski, HM, and Spencer, PS: Ultrastructural studies of the dying-back process. I. Peripheral nerve terminal and axon degeneration in systemic acrylamide intoxication. J Neuropathol Exp Neurol 33:260, 1974.

108. Schroder, JM, Hoheneck, M, Weiss, J, et al: Ethylene oxide polyneuropathy: Clinical follow up study with morphometric and electron microscopic findings in a sural nerve biopsy. J Neurol 232:82, 1985.

109. Senanayake, N and Jeyaratnam, J: Toxic polyneuropathy due to gingli oil contaminated with tricresyl phosphate affecting adolescent girls in Sri Lanka. Lancet 1:88, 1981.

110. Seppäläinen, AM and Haltia, M: Carbon disulfide. In Spencer, PS and Schaumburg, HH (eds): Experimental and Clinical Neurotoxicology. Williams & Wilkins, Baltimore, 1980, p 356.

111. Seppäläinen, AM and Tolonen, M: Neurotoxicity of long term exposure to carbon disulfide in the viscose rayon industry. A neurophyiological study. Work-Environmental Health 11:145, 1974.

112. Seppäläinen, AM and Hernberg, S: Subclinical lead neuropathy. Am J Ind Med 1:413, 1980.

113. Smith, HV and Spalding, JMK: Outbreak of paralysis in Morocco due to orthocresylphosphate poisoning. Lancet 2:1019, 1959.

114. Sobue, G and Pleasure, D: Experimental lead neuropathy: Inorganic lead inhibits proliferation but not differentiation of Schwann cells. Ann Neurol 17:462, 1985.

115. Solders, G, Nennesmo, I, and Persson, A: Diphtheritic neuropathy, an analysis based on muscle and nerve biopsy and repeated neurophysiological and autonomic function tests. J Neurol Neurosurg Psychiatry 52:876, 1989.

116. Spencer, PS, Couri, D, and Schaumburg, HH: n-Hexane and methyl n-butyl ketone. In Spencer, PS and Schaumburg, HH (eds): Experimental and Clinical Neurotoxicology. Williams & Wilkins, Baltimore, 1980, p 456.

117. Spencer, PS, Peterson, ER, Madrid, R, et al: Effects of thallium salts on

neuronal mitochondria in organo-typic cord-ganglia-muscle combination cultures. J Cell Biol 58:79, 1973.

118. Spencer, PS and Schaumburg, HH: A review of acrylamide neurotoxicity. I. Properties, uses, and human exposure. Can J Neurol Sci 1:143, 1974.

119. Spencer, PS and Schaumburg, HH: Central-peripheral distal axonopathy—the pathology of dying-back polyneuropathies. In Zimmerman, HM (ed): Progress in Neuropathology, Vol 3. Grune & Stratton, New York, 1976, p 253.

120. Spencer, PS and Schaumburg, HH: Ultrastructural studies of the dying-back process. IV. Differential vulnerability of PNS and CNS fibers in experimental central-peripheral distal axonopathy. J Neuropathol Exp Neurol 36:300, 1977.

121. Spencer, PS and Schaumburg, HH: Ultrastructural studies of the dying-back process. III. The evolution of experimental giant axonal degeneration. J Neuropathol Exp Neurol 36:276, 1977.

122. Spencer, PS, Schaumburg, HH, Sabri, MI, et al: The enlarging view of hexacarbon neurotoxicity. CRC Crit Rev Toxicol 7:273, 1980.

123. Sumner, AJ and Asbury, AK: Physiological studies of the dying-back phenomenon. Muscle stretch afferents in acrylamide neuropathy. Brain 98:91, 1975.

124. Takeuchi, T: Neuropathology of Minamata disease in Kumanamoto: Especially at the chronic state. In Roizin, L, Shiraki, H, and Grcevic, N (eds): Neurotoxicology. Raven Press, New York, 1977, p 235.

125. Vasilescu, C: Motor nerve conduction velocity and electromyogram in carbon disulphide poisoning. Revue Roumaine Neurologie et Psychiatri 9:63, 1972.

126. Vigliani, EB: Carbon disulphide poisoning in viscose rayon factories. Br Med J 2:235, 1954.

127. Watson, DF and Griffin, JW: Vacor

neuropathy: Ultrastructural and axonal transport studies. J Neuropathol Exp Neurol 46:96, 1987.

128. Waksman, BH: Experimental study of diphtheritic polyneuritis in the rabbit and guinea pig. III. The blood-nerve barrier in the rabbit. J Neuropathol Exp Neurol 20:35, 1961.

129. Webster, DeF, H, Spiro, D, Waksman, B, et al: Phase and electron microscopic studies of experimental demyelination. II. Schwann cell changes in guinea pig sciatic nerves during experimental diphtheritic neuritis. J Neuropathol Exp Neurol 20:5, 1961.

130. Windebank, AJ: Specific inhibition of myelination by lead: In vitro comparison with arsenic, thallium and mercury. Exp Neurol 94:203, 1986.

131. Windebank, AJ and Blexrud, MD: Residual ethylene oxide in hollow fiber hemodialysis units is neurotoxic in vitro. Ann Neurol 26:63 1989.

132. Windebank, AJ, McCall, JT, Hunder, HG, et al: The endoneurial content of lead related to the onset and severity of segmental demyelination. J Neuropathol Exp Neurol 38:692, 1980.

133. Windebank, AJ and Dyck, PJ: Localization of lead in rat peripheral nerve by electron microscopy. Ann Neurol 18:197, 1985.

134. Windebank, AJ and Dyck, PJ: Differential toxicity of lead, arsenic, mercury, and thallium on neuronal growth and myelination in vitro. Neurology 36:135, 1986.

135. Windebank, AJ, McCall, JT, and Dyck, PJ: Metal neuropathy. In Dyck, PJ, Thomas, PK, Lambert, EH, and Bunge, RP (eds): Peripheral Neuropathy, Vol 2, ed 2. WB Saunders, Philadelphia, 1984, p 2148.

136. Zampollo, A, Zacchetti, O, and Pisati, G: On ethylene oxide neurotoxicity: Report of two cases of peripheral neuropathy. Ital J Neurol Sci 5:59, 1984.

Part VII

**NEUROPATHIES
OF UNCERTAIN
CAUSE OR RARE
OCCURRENCE**

Chapter 19

NEUROPATHIES OF UNCLEAR ETIOLOGY

**IDIOPATHIC FACIAL PARALYSIS
(BELL'S PALSY)
TRIGEMINAL SENSORY NEUROPATHY
IDIOPATHIC CRANIAL
POLYNEUROPATHY (CRANIAL
POLYNEURITIS)
PLEXUS NEUROPATHIES
CRITICAL ILLNESS
POLYNEUROPATHY
IDIOPATHIC SENSORY
NEURONOPATHY SYNDROME
ACUTE PANDYSAUTONOMIC
NEUROPATHY**

These neuropathies comprise a group of common and uncommon disorders whose cause is either unknown or uncertain. Their importance lies mainly in their frequency.

IDIOPATHIC FACIAL PARALYSIS (BELL'S PALSY)

Bell's palsy is common; the incidence in the United States is 25 per 100,000 each year.[26] Although the condition was initially described over 150 years ago and has been thoroughly investigated, its pathology and pathogenesis are unknown and treatment is empirical.

Pathology and Pathogenesis

It is widely held that swelling of the facial nerve within the tight confines of the facial canal occurs in most cases; however, this notion stems from operative descriptions by enthusiastic surgeons and is not supported by morphologic data. Similarly, none of the postmortem studies used modern neuropathologic techniques. Most reports describe wallerian degeneration.[19,33] It is likely that mild cases with rapid recovery reflect conduction block from segmental demyelination. Axonal degeneration results in severe paralysis and prolonged or poor recovery. Hypotheses about the etiology of facial nerve dysfunction in Bell's palsy include injury from vasospasm, cold, viral or immunological inflammation, and venous thrombosis.[3,8,31] It is also claimed that Bell's palsy is part of a generalized cranial polyneuropathy (see later discussion) and is accompanied by subclinical dysfunction of other cranial nerves.[2,12,20]

Clinical Features

All age groups appear equally affected and both sides of the face are equally involved. Rarely the disorder is recurrent or familial.[26]

Unilateral facial paralysis usually develops rapidly within a few hours or evolves over one or two days and is often accompanied by pain behind the ear and excess tearing. Numbness of the face is a common complaint, but inevitably refers to the sensation that accompanies weakness. Rarely are hyperacusis and diminished taste significant to the patient. Global facial muscle weakness is the hallmark of this condition; it is partial in 40% of cases. Hyperacusis, diminished lacrimation, and abnormal taste sensation are present to variable degrees. Figure 19–1 delineates the putative levels of involvement of the facial nerve in the common Bell's palsy syndromes. Topographic locali-

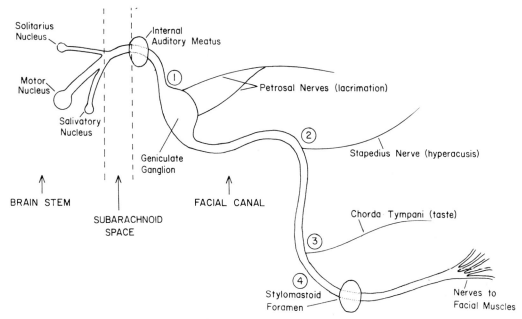

Figure 19–1. Diagram of four putative facial-canal lesion sites in the various Bell's palsy syndromes. *Site 1:* impaired lacrimation, hyperacusis, impaired taste, facial paralysis. *Site 2:* hyperacusis, impaired taste, facial paralysis. *Site 3:* impaired taste, facial paralysis. *Site 4:* facial paralysis.

zation is frequently not clinically helpful or accurate.

The prognosis for improvement is good; untreated, 80% to 85% of patients recover completely. A recent epidemiologic study has determined that risk factors for incomplete recovery include hypertension, old age, and complete facial paralysis.[26] Of patients without risk factors, 96% recover completely regardless of type of treatment. The most favorable prognosis is associated with lesions producing solely segmental demyelination and conduction block. In the small proportion in which all or most of the fibers undergo wallerian degeneration, recovery, which has to take place by axonal regeneration, is generally unsatisfactory and leads to persistent facial weakness and aberrant regeneration.

In occasional cases, motor recovery fails completely. Aberrant regeneration is frequently observed, leading to embarrassing synkinetic movements or excessive lacrimation, sometimes related to gustatory stimuli ("crocodile tears"). Patients destined to recover completely

usually begin to improve during the first three weeks, while those with permanent residual disability remain unchanged for approximately three months. Prognostic reliance is placed on careful observation and electrodiagnostic tests aimed at determining the extent of axonal degeneration (performed at least one week following onset). A good prognosis is portended by facial motor potential amplitude of at least 50% of normal with direct facial nerve stimulation at the stylomastoid foramen. In cases where no response is obtained, recovery is prolonged and rarely complete.

Treatment is with 1 mg/kg prednisone daily (in two divided doses) for 4 days, and tapered to 5 mg per day within 10 days.[1] It is claimed that prednisone should be instituted as soon as possible if it is to have an effect, that is, to prevent a mild lesion with only conduction block developing into axonal degeneration—and it should be continued an additional five days if the paralysis remains complete.[1,10] Prompt treatment may result in dramatic lessening of pain. Steroid administration is also alleged to decrease residual paralysis and synkinetic movements, although these notions are controversial.[47] Male patients frequently choose to grow a beard to lessen the cosmetic impact of facial paralysis. An eye patch may be necessary to avoid exposure keratitis. Hypoglossal-facial nerve anastomosis can restore facial tone. Gold implantation in the upper eyelid helps improve eye closure. Facial slings from the temporalis fascia to the angle of the mouth are generally unsatisfactory. Radical facial plastic surgery (face lift) may produce considerable improvement in facial symmetry and reduced lacrimation in persistent cases of bilateral facial palsy following AIDP. Even if the facial paralysis is recoverable, patients may develop exposure conjunctivitis or keratitis from lagophthalmos, and a lateral tarsorrhaphy may be necessary. There is no definitive treatment for synkinetic movements, although current trials of local botulinum are underway.

TRIGEMINAL SENSORY NEUROPATHY

Rare cases are encountered of slowly progressive bilateral sensory loss confined to the territory of one or both trigeminal nerves.[38] Sjögren's syndrome,[25] mixed connective tissue disease (CTD),[21] systemic sclerosis,[18] lupus erythematosus, and dermatomyositis[6] need to be excluded. Most patients experience steady progression of sensory impairment preceded by or associated with unpleasant sensations.[28] Pain, when present, differs from that of trigeminal neuralgia in that it is spontaneous and does not have a lancinating quality. Sensory loss usually commences unilaterally, soon becomes bilateral, and steadily progresses for months or years. Trigeminal motor impairment is rarely prominent. Facial hypalgesia may lead to tissue destruction, particularly around the nostrils, as a result of repeated picking and scratching. Clinical and electrophysio-

logical studies implicate the trigeminal ganglion as the site of the lesion. The pathogenesis of this isolated cranial neuropathy is unclear. A careful study of 22 patients with this disorder found systemic sclerosis or mixed connective tissue disease in 9 and either organ-specific or non-organ-specific autoantibodies in 9 of the others.[28] The explanation for the four remaining cases was obscure.

IDIOPATHIC CRANIAL POLYNEUROPATHY (CRANIAL POLYNEURITIS)

The idiopathic cranial polyneuropathy (ICP) syndrome depicts simultaneous dysfunction of several lower cranial nerves. The diagnosis is considered only when the following conditions are ruled out: congenital bone disease, infections, neoplasms, trauma, vascular compromise, diabetes, inflammatory disease, or the oligosymptomatic form of AIDP.[7]

The ICP syndrome is usually heralded by subacute onset of facial pain; it is aching and usually retro-orbital. Cranial nerve palsies develop soon after the onset of pain (occasionally they antecede it).[24] Nerves most commonly involved are the three oculomotor nerves, trigeminal nerve, facial nerve, and the lower three cranial nerves. Olfactory, optic, and auditory nerves are usually spared. A mild cerebrospinal fluid pleocytosis and elevated protein occurs in one half of the cases. Corticosteroid therapy lessens pain and appears to hasten recovery.[24] The prognosis is good; permanent deficits rare. Occasionally, the condition can recur.[22] The ICP syndrome is nonspecific and probably includes several entities; herpes virus infections, benign steroid-responsive granulomatous disease (Tolosa-Hunt syndrome), orbital pseudotumor, mycoplasma, and localized AIDP are among the possibilities.[23] It must be emphasized that *recognition of the syndrome of acute multiple cranial neuropathy allows for few diagnostic short-cuts, given the etiologically non-specific nature of its presentation.*[24]

PLEXUS NEUROPATHIES

Idiopathic Brachial Plexopathy

Idiopathic brachial plexopathy (IBP) is best defined as an idiopathic acute, painful, and usually monophasic illness characterized by brachial plexus dysfunction.[43] Synonyms for this disorder include neuralgic amyotrophy,[16,32] acute brachial radiculitis,[44] and acute shoulder-girdle neuritis.[32,37] The relationship of this condition to the syndromes of postvaccinial,[39] toxic,[11] and hereditary brachial plexopathies[36] is uncertain.

PATHOLOGY AND
PATHOGENESIS

There are no postmortem examinations and no animal model exists. Biopsy of cutaneous nerves has revealed nonspecific axonal degeneration. Plexus biopsies from two recurrent cases revealed swollen nerves histologically featured by microvasculitis, multifocal inflammation, and onion-bulb formation.[14]

The pathogenesis is unknown. Most cases have no common antecedent illness, immunization, or toxic exposure; some follow surgical procedures. The clinical profile is usually identical to the serum vaccine (presumably allergic) syndrome,[39] and a common immunologic basis has therefore been proposed, although not substantiated.[43]

CLINICAL FEATURES

IBP can occur at any age; in our experience it is especially common in males 18 to 40 years of age. A cardinal feature is severe shoulder-girdle-scapular pain, occasionally extending into the arm or hand, frequently so severe as to require narcotics. Pain typically persists for a few days to a week, and then subsides coincident with the appearance or appreciation of weakness, although it sometimes persists for several weeks or longer. It is often made worse by neck or arm movements. Proximal weakness, either partial or total paralysis, usually involves nerves originating from the upper plexus. Most commonly involved are the long thoracic, subscapular, and axillary nerves. The serratus anterior is the single most commonly affected muscle. Distal weakness less frequently occurs and, rarely, the entire arm and ipsilateral diaphragm are affected. Weakness may also appear in the other arm. Tendon reflexes are diminished in the involved extremity, but sensory loss is slight or negligible. Involvement is usually restricted to muscles innervated by the brachial plexus, although, as already stated, the diaphragm may become paralyzed, and accompanying laryngeal paralysis has been observed. Lower cranial nerves have rarely been involved. Some cases of isolated and unexplained unilateral or bilateral diaphragmatic paralysis may be from this cause.

Routine clinical laboratory tests are normal. The CSF is acellular and protein content is not elevated. Electrodiagnostic testing usually reveals evidence of denervation from axon loss affecting muscles subserved by the upper brachial plexus. Sensory potential amplitudes are variably diminished, the nerve most commonly affected is the musculocutaneous sensory nerve. Needle EMG reveals denervation potentials and reduced voluntary activity in affected muscles.

COURSE, PROGNOSIS,
AND TREATMENT

Persistent weakness and atrophy of involved muscles may develop in many cases. The prognosis is usually excellent; however, total recovery occurs in 90% of patients within two or three years.[43] Individuals with lower or pan-brachial plexus lesions have a worse prognosis. Treatment consists of physical therapy and orthotic devices to prevent joint damage. There is no evidence that corticosteroid therapy is of value except to relieve pain in the acute stage. Recurrences, sometimes multiple, occasionally occur.[14]

DIFFERENTIAL DIAGNOSIS

Brachial plexus neuropathies following serum or vaccine injections,[39] heroin abuse,[11] and of the hereditary type[36] are nearly identical clinical entities and can only be differentiated by careful history. Cervical spondylosis at the C_4–C_6 interspaces can cause severe shoulder pain and local weakness; myelography or magnetic resonance imaging (MRI) may be necessary to eliminate this possibility in an older patient. Thoracic outlet syndromes and invasive carcinoma of the lung and breast each may produce a painful brachial plexopathy, but these seldom appear suddenly and usually involve the lower plexus.[27] Anterior or posterior interosseous nerve entrapment neuropathies, in which the symptoms begin following unaccustomed exercise and are associated with pain, sometimes give rise to diagnostic difficulty. Radiation-induced injury to the brachial plexus, although associated with less pain, may be confused with IBP.[29]

Case History: Brachial Plexopathy

A 60-year-old man developed severe pain in the left shoulder that required opiates for relief. This persisted for two days, and because of some minor abnormalities on an electrocardiogram, he was admitted to a coronary care unit. Shortly following admission, the pain abated, and his left arm was noted to be weak. On examination, there was ⅖ weakness of the left supraspinatus, infraspinatus, deltoid, and biceps muscles. The triceps and wrist extensors were slightly weak. The remaining muscles of the arm were strong, and there was no weakness in the other limbs. All modalities of sensation were impaired in a band extending from the tip of the right shoulder down over the outer surface of the arm and fore-

arm. The biceps tendon reflex was absent. The remainder of the neurologic examination was normal. A myelogram revealed no significant evidence of cervical intervertebral-disc disease and the CSF was normal. He was discharged.

One month later, he was re-examined and there was striking atrophy and fasciculation of the left infraspinatus, supraspinatus, deltoid, and biceps muscles. The left biceps tendon reflex remained absent, but sensation was now normal over the left shoulder. An electromyogram disclosed fibrillation potentials in the atrophic muscles. Motor and sensory conduction studies of the ulnar and median nerves were normal.

Four months later, there was slight return of strength, and after one year the upper limb could be moved through a full range of motion against moderate resistance.

Comment. The signs and symptoms were largely confined to structures innervated by C_5–C_6 roots. Suspected cervical disc disease is a common diagnostic problem in older individuals with this condition. The degree of sensory loss was unusually great in this case. Typical for brachial plexus neuropathy were the predominance of motor over sensory signs, severe pain, and the satisfactory outcome.

Idiopathic Lumbar Plexopathy

This disorder is rare.[17,34] Except for location, it is similar to cryptogenic brachial plexopathy. Idiopathic lumbar plexopathy (ILP) may also accompany hereditary brachial plexopathy and can occur as a result of heroin abuse. It is suggested that some cases of progressive painful lumbar plexopathy associated with an elevated erythrocyte sedimentation rate constitute a distinct syndrome.[9] This variant is steroid-responsive and not associated with systemic vasculitis.

The clinical features of ILP include a painful prodome, unilateral or occasionally bilateral weakness in the distribution of several nerves, no underlying illness, and a gradual satisfactory recovery. As in brachial plexopathy, weakness is the predominant finding and sensory loss is trivial. The CSF is normal.

Differential diagnosis includes herniated lumbar intervertebral disc, proximal diabetic neuropathy, poliomyelitis, and cauda equina or pelvic tumors.

CRITICAL ILLNESS POLYNEUROPATHY

Critical illness polyneuropathy (CIP) is a poorly defined condition that appears in patients with sepsis and multiple organ failure. It is alleged that CIP, to some degree, occurs in 50% of patients who have been septic for longer than two weeks.[46] All cases have been admitted to intensive care units following intubation for cardiac or pulmonary disease; subsequently, they developed sepsis and multiple organ failure.[49] Limb weakness appears in about a month; CIP is first suspected when patients cannot be weaned from the respirator.

Weakness, loss of tendon reflexes, and electrophysiologic signs of motor and sensory axonal dysfunction are characteristic. Postmortem studies describe widespread axonal degeneration of motor and sensory fibers. Its pathogenesis is unclear; clearly, these are not instances of AIDP and there are no consistent toxic or metabolic features. It is suggested that the mechanism may be a defect somehow related to the accompanying failure of other organ systems.[49] About one half survive and gradually improve over the ensuing 6 months.

IDIOPATHIC SENSORY NEURONOPATHY SYNDROME

Idiopathic sensory neuronopathies (ISN) may be acute,[40,45] subacute, or chronic.[15] Sensory impairment develops over proximal and distal regions and, occasionally, over the trunk and face as well. The most dramatic and incapacitating feature is loss of kinesthesia with sensory ataxia. In some cases, the condition has an explosive onset with initial symptoms followed by inability to walk within one week and no further progression. In others diagnosed as having chronic idiopathic ataxic neuropathy, progression may continue for decades.[15] Large-fiber sensory modalities are most compromised, strength is spared, CSF is normal, no antineuronal antibodies are present, and signs of CNS involvement do not appear. Electrodiagnostic studies are consistent with widespread sensory axonal loss. There are no postmortem studies and the pathogenesis is unclear. Many cases follow a febrile illness treated with antibiotics. Penicillin was initially suggested as a toxic cause;[40] however, many subsequent cases have not been preceded by antibiotic therapy. The prognosis is highly variable.

ACUTE PANDYSAUTONOMIC NEUROPATHY

Clinical Features

Four types are known to exist. All are rare. Although referred to as acute, the disorder in all four types usually evolves over the course of a few weeks. In the first, acute pandysautonomia, the onset may follow a febrile illness and leads to widespread loss or impairment of autonomic function, both sympathetic and parasympathetic.[35,48] Sympathetic failure is manifested by orthostatic hypotension, failure of the pupils to dilate in the dark, ejaculatory failure, anhidrosis, and lack of piloerection. Parasympathetic disturbances consist of impaired lacrimation, salivation, and nasal secretion; loss of pupillary constriction to light; decreased gastrointestinal motility; bladder atonicity; and failure of penile erection. Somatic motor and sensory function is preserved. The second type, cholinergic dysautonomia, is clinically similar but sympathetic function is spared.[35,41,42] This form is more frequent in children. In the third, sympathetic and parasympathetic dysautonomia is combined with a sensory neuropathy.[13] The fourth type, which is also often associated with a sensory neuropathy, is paraneoplastic dysautonomia. Most often, the underlying tumor is bronchial carcinoma.[4]

Pathology and Pathogenesis

The pathology is poorly understood. Nerve biopsy has shown selective loss of small myelinated and unmyelinated axons and absence of the C fiber potential in in vitro recordings from the sural nerve.[30] During the recovery stage, there is an excess of very small unmyelinated axons, suggesting regeneration. The cause is obscure.[35] In those cases that follow a viral infection, an immunologic basis is possible.

Differential Diagnosis, Course, and Treatment

Exclusion of porphyria is important. For those cases with acute cholinergic dysautonomia, botulism, and intoxication with belladonna or anticholinergic drugs need to be considered. If abdominal symptoms are prominent, intestinal obstruction must be excluded and, in more protracted cases, Sjögren's syndrome. Case of acute pandysautonomia usually recovery over the course of weeks or months.[48] Corticosteroids have not been found to be beneficial. Reduced gastrointestinal mobility and bladder atony may require treatment with cholinergic drugs such as carbachol and orthostatic hypotension with fludrocortisone. Recovery may be imperfect for acute cholinergic dysautonomia.[5]

REFERENCES

1. Adour, KK: The diagnosis and management of facial paralysis. N Engl J Med 307:347, 1982.

2. Adour, KK: Cranial polyneuritis and Bell palsy. Arch Otolaryngol 102:262, 1976.

3. Adour, KK, Bell, DN, and Hilsinger, RL, Jr: Herpes simplex virus in idiopathic facial paralysis (Bell palsy). JAMA 233:527, 1975.

4. Anderson, NE, Rosenblum, MK, Graus, F, et al: Autoantibodies in paraneoplastic syndromes associated with small-cell lung cancer. Neurology 38:1391, 1988.

5. Anderson, O, Linberg, J, Modigh, K, et al: Subacute dysautonomia with incomplete recovery. Acta Neurol Scand 48:510, 1972.

6. Ashworth, B and Tait, GBW: Trigeminal neuropathy in connective tissue disease. Neurology (Minneapolis) 21:609, 1971.

7. Beal, MF: Multiple cranial nerve palsies—a diagnostic challenge. N Engl J Med 322:461, 1990.

8. Blunt, MJ: The possible role of vascular changes in the etiology of Bell's palsy. J Laryngol Otol 70:701, 1956.

9. Bradley, WG, et al: Painful lumbosacral plexopathy with elevated erythrocyte sedimentation rate: A treatable inflammatory syndrome. Ann Neurol 15:457, 1984.

10. Bustamante-Balcarcel, A (moderator): Panel discussion no. 9: Incidence and management of Bell's palsy according to geographic distribution. In Fisch, U (ed): Facial Nerve Surgery, ed 3. International Symposium on Facial Nerve Surgery, Zürich, 1976.

11. Challenor, YB, et al: Nontraumatic plexitis and heroin addiction. JAMA 225:958, 1973.

12. Charous, DI and Saxe, BI: The Landry-Guillain-Barré syndrome. Report of an unusual case, with a comment on Bell's palsy. N Engl J Med 267:1334, 1962.

13. Colan, RV, Snead, C, Oh, S, et al: Acute autonomic and sensory neuropathy. Ann Neurol 8:441, 1980.

14. Cusamino, MD, Bilbao, JM, and Cohen, SM: Hypertrophic brachial neuritis: A pathological study of two cases. Ann Neurol 24:615, 1988.

15. Dalakas, MC: Chronic idiopathic ataxic neuropathy. Ann Neurol 19:545, 1986.

16. England, JD and Sumner, AJ: Neuralgic amyotrophy: An increasingly diverse entity. Muscle Nerve 10:60, 1987.

17. Evans, BA, Stevens, JC, and Dyck, PJ: Lumbosacral plexus neuropathy. Neurology (Minneapolis) 31:1327, 1981.

18. Farrell, DA and Medsger, TA: Trigeminal neuropathy in progressive systemic sclerosis. Am J Med 73:57, 1982.

19. Fowler, EP: The pathologic findings in a case of facial paralysis. Acta Otolaryngol 68:1655, 1958.

20. Grose C, Feorina, PM, Dye, LA, and Rand J: Bell's palsy and infectious mononucleois. Lancet 2:231, 1973.

21. Hagen, NA, Stevens, JC, and Michet, CJ: Trigeminal sensory neuropathy associated with connective tissue diseases. Neurol 40:891, 1990.

22. Hokkanen, E, Haltia, T, and Myllyla, VV: Recurrent multiple cranial neuropathies. Eur Neurol 17:32, 1978.

23. Inzitari, D, Sitz, D, Marconi, GP, et al: The Tolosa-Hunt syndrome: Further clinical and pathogenic considerations based on the study of eight cases. J Neurol 224:221, 1981.

24. Juncas, JL and Beal, MF: Idiopathic cranial polyneuropathy. Brain 110: 197, 1987.

25. Kaltreider, HB and Talal, N: The neuropathy of Sjögren's syndrome: Trigeminal nerve involvement. Ann Intern Med 70:751, 1969.

26. Katusic, SK, Beard, M, Wiederholt, WC, et al: Incidence, clinical features and prognosis in Bell's palsy, Rochester, Minnesota, 1968–1982. Ann Neurol 20:622, 1986.

27. Kori, SH, Foley, KM, and Posner, JB: Brachial plexus lesions in patients with cancer: 100 cases. Neurology (Minneapolis) 31:45, 1981.

28. Leckey, BRF, Hughes, RAS, and Murray, NMF: Trigeminal sensory neuropathy. A study of 22 cases. Brain 110:1463, 1987.

29. Ledermann, RJ and Wilbourn, AJ: Brachial plexopathy: Recurrent cancer or radiation? Neurology 34:1331, 1984.

30. Low, PA, Dyck, PJ, Lambert, EH, et al: Acute panautonomic neuropathy. Ann Neurol 13:412, 1983.

31. Moldaver, J and Conley, J: The Facial Palsies. Charles C Thomas, Springfield, Ill, 1979.

32. Parsonage, JM and Turner, JWA: Neuralgic amyotrophy: The shoulder girdle syndrome. Lancet 1:973, 1948.

33. Reddy, JB, Liu, J, Balshi, S, et al: Histopathology of Bell's palsy. Eye, Ear, Nose, Throat Mon 45:62, 1966.

34. Sander, JE and Sharp, FR: Lumbosacral plexus neuritis. Neurology (Minneapolis) 31:470, 1981.

35. Serratrice, G: Acute pandysautonomia. New Iss Neurosci 1:311, 1988/89.

36. Smith, BH, Tamarkvisna, T, and Schlagenhauf, RE: Familial brachial neuropathy: Two case reports with discussion. Neurology (Minneapolis) 21:941, 1971.

37. Spillane, JD: Localized neuritis of the shoulder girdle: A report of 46 cases in the MEF. Lancet 2:532, 1943.

38. Spillane, JD and Wells, CDC: Isolated trigeminal neuropathy. A report of 16 cases. Brain 82:391, 1959.

39. Spillane, JD and Wells, CEC: The neurology of Jennerian vaccination. Brain 87:1, 1964.

40. Sterman, AB, Schaumburg, HH, and Asbury, AK: The acute sensory neuronopathy syndrome. A distinct clinical entity. Ann Neurol 7:354, 1980.

41. Takayama, H, Kazahaya, Y, Kashihara, N, et al: A case of postganglionic cholinergic dysautonomia. J Neurol Neurosurg Psychiatry 50:915, 1987.

42. Thomashefsky, AJ, Horwitz, SJ, and Feingold, MH: Acute autonomic neuropathy. Neurology 22:251, 1972.

43. Tsairis, P, Dyck, PJ, and Mulder, DW: Natural history of brachial plexus neuropathy: Report on 99 patients. Arch Neurol 27:109, 1972.

44. Turner, JWA: Acute brachial radiculitis. Br Med J 2:592, 1944.

45. Windebank, AJ, Blexrud, MD, Dyck, PJ,

et al: The syndrome of acute sensory neuropathy: Clinical features and electrophysiologic and pathologic changes. Neurology 40:584, 1990.

46. Witt, NJ, Boulton, CF, and Sibbald, WJ: The incidence and early features of the polyneurology of critical illness. Neurology (Suppl 1) 35:74, 1985.

47. Wolf, SM: Treatment of Bell's palsy with prednisone: A prospective, randomized study. Neurology (Minneapolis) 28:158, 1978.

48. Young, RR, Asbury, AK, Corbett, JL, et al: Pure pandysautonomia with recovery. Description and diagnostic criteria. Brain 98:613, 1975.

49. Zochodne, DW, et al: Critical illness polyneuropathy: A complication of sepsis and multiple organ failure. Brain 110:819, 1987.

Chapter 20

RARE OR POORLY VALIDATED NEUROPATHIES

PHARMACEUTICAL TOXINS
OCCUPATION-RELATED CONDITIONS
IATROGENIC CONDITIONS
INFECTIOUS DISEASE
RHEUMATOLOGIC-IMMUNOLOGIC
  DISEASE
NEOPLASTIC DISEASE
HEMATOLOGIC DISEASE
SYSTEMIC METABOLIC DISEASE
UNCLASSIFIED SYSTEMIC DISEASE
UNCLASSIFIED NEUROLOGIC
  CONDITIONS

This is a group of medical conditions whose association with peripheral neuropathy is rare and frequently supported solely by anecdotal case reports. Some, such as the neuropathies accompanying biliary cirrhosis or radiation therapy, display *truly* consistent clinical profiles; they appear genuine and are designated (TC). The others, with time, may prove to represent chance co-occurrences; their association with neuropathy is *not* certain and they are designated (NC). Within each group, conditions are listed in alphabetical order, given brief descriptions, and referenced.

PHARMACEUTICAL TOXINS

Amitriptyline (NC). Acute amitriptyline intoxication is occasionally associated with severe, distal symmetric axonal polyneuropathy of unclear pathogenesis;[25] chronic administration of amitriptyline is useful in treating painful peripheral nerve disease and does not consistently cause axonal dysfunction.

Cytosine Arabinoside (Ara C) (NC). A single report describes two patients with sensory neuropathy in concert with Ara C therapy;[40] Ara C inhibits Schwann-cell proliferation in newborn animals, suggesting possible fetotoxicity in humans.[1]

Lithium Neuropathy (NC). Acute lithium carbonate intoxication is occasionally associated with a diffuse polyneuropathy of unclear nature and pathogenesis; chronic intoxication causes tremor and ataxia, not polyneuropathy.[48]

Penicillamine (NC). Two instances of neuropathy, one motor and the other sensorimotor, have followed penicillamine therapy for rheumatoid arthritis.[34a,38]

Phenelzine (monoamine oxidase inhibitor) (NC). Phenelzine is a pyridoxine antagonist; a case report describes sensorimotor polyneuropathy and low serum pyridoxal phosphate levels in an individual who also had folic acid deficiency and renal cell carcinoma.[15]

Phenobarbital (NC). One study describes focal upper limb neuropathy, reflex sympathetic dystrophy, and shoulder-hand syndrome following prolonged barbiturate antiepileptic therapy in patients with brain tumors.[44]

Podophyllin Resin (TC). Podophyllin resin, used in the treatment of condylomata accuminata and as a chemotherapeutic agent, causes a severe distal axonopathy as well as central nervous system dysfunction; the active neurotoxic agent, podophyllotoxin, is a lipid-soluble mitotic spindle binder that crosses cell membranes with ease, accounting for its neurotoxicity with topical usage as well as ingestion.[12]

Tryptophan-induced Eosinophilia-Myalgia Syndrome (TC). Tryptophan abuse is associated with eosinophilia, myositis, fasciitis, and a multifocal neuropathy; possibly related to contaminants, this syndrome is assumed to be immunologically mediated and similar to the Spanish oil syndrome (Chapter 18).[43]

Vinarabine (Ara A) (TC). Ataxia and painful paresthesias are described in patients with renal or hepatic dysfunction receiving this agent or the related substance, adenine arabinoside monophosphate (Ara-AMP).[26]

OCCUPATION-RELATED CONDITIONS

Chlorpyriphos (Dursban) (TC). This organophosphate anticholinesterase with neuropathy target esterase (NTE) specificity causes severe sensorimotor neuropathy following massive suicidal ingestion;[27] the authors have observed sensory neuropathy in five individuals following environmental exposure to commercial Dursban, a preparation containing chlorpyriphos.

Dichlorophenoxyacetic Acid (2,4D) (NC). Isolated case reports describe a fulminant motor neuropathy following trivial skin contact with this widely deployed herbicide; these may represent coincidental instances of AIDP.[4]

2,3,7,8 Tetrachlorodibenzo-P-Dioxin (TCDD, dioxin) (NC). Industrial cutaneous and inhalation exposure to TCDD is associated with chronic skin infection (chloracne) and transient hepatic dysfunction; mild sensory polyneuropathy is alleged to occur, but is poorly documented.[42]

Instrumentalist's Neuropathy (NC). Several reports describe entrapment syndromes (e.g., thoracic outlet and ulnar neuropathy in violinists); none is a controlled study and the intuitive notion of occupational risk is unconfirmed.[23]

Organotin Intoxication (NC). Triethyltin is associated with widespread (PNS and CNS) intramyelinic edema and trimethyltin exclusively with a limbic-cerebellar syndrome; triphenyltin has been linked to occasional instances of polyneuropathy following massive overdose.[51]

Styrene (NC). Monomeric styrene (vinyl benzene) exposure is reported to cause a mild distal sensory neuropathy in several industrial surveys and in unconvincing case reports.[3]

IATROGENIC CONDITIONS

Coronary Artery Bypass Surgery (TC). Brachial radiculoplexopathy, attributed to jugular vein cannulation, trauma, or retraction stretch, and miscellaneous perioperative nerve compression syndromes accompany 13% of these surgical procedures; most recover and lasting disability is rare.[22]

Hyperalimentation with Hypophosphatemia (NC). Paresthesias and weakness accompany seizures and coma from hypophosphatemia during hyperalimentation; a single case report describes acute demyelinating neuropathy that appeared to respond to phosphate replacement.[41]

Postradiation Therapy (TC). Upper brachial plexopathy may occur long after (4 to 30 months) high-dose (6000 rads) radiotherapy for breast carcinoma; connective tissue proliferation is described in one autopsy report and may contribute to the axonal destruction and poor prognosis of most cases.[21] Neuropathy may also follow radiotherapy to the lumbosacral plexus.[45]

INFECTIOUS DISEASE

Hepatitis (TC). Acute inflammatory demyelinating polyradiculoneuropathy, brachial and lumbar plexus neuropathies, cranial and somatic mononeuropathies, and multiple mononeuropathy are all described in patients with viral hepatitis, presumably from immune dysfunction.[37]

HTLV-1 (TC). Human T-lymphotropic virus type 1 (HTLV-1) infection is occasionally associated with mild sensorimotor polyneuropathy of indeterminate type; its clinical features are generally overshadowed by those of myelopathy.[39]

Legionnaires' Disease (NC). Isolated reports describe both acute, reversible demyelinating neuropathy, indistinguishable from AIDP and distal axonopathy in patients with Legionnaires' disease; these are likely coincidental.[33]

RHEUMATOLOGIC-IMMUNOLOGIC DISEASE

Biliary Cirrhosis (TC). Primary biliary cirrhosis is sometimes accompanied by distal paresthesias and sensory loss; nodular and diffuse xanthomatous infiltrates of nerve are characteristic histopathologic features.[46] Not all neuropathies related to primary biliary cirrhosis are of this type.

Crohn's Disease (NC). Polyneuropathy in Crohn's disease is usually associated with vitamin B_{12} deficiency, metronidazole and sulfasalazine therapies; occasional cases of distal sensory neuropathy of uncertain causes accompany untreated Crohn's disease with normal B_{12} levels.[34]

Giant-Cell Arteritis (cranial arteritis) (NC). Polyneuropathy or a multiple mononeuropathy are described with giant-cell arteritis; the possibilities of co-existing systematic diseases and the peripheral nerve degeneration of aging have not received adequate consideration in these reports.[11]

Scleroderma (TC). Distal symmetric polyneuropathy, mononeuropathy multiplex, and trigeminal neuropathy (Chapter 19) occasionally accompany systemic sclerosis. The pathogenesis of these disorders is unclear; possibly they stem from endoneurial microvascular compromise.[24]

NEOPLASTIC DISEASE

Multiple Endocrine Neoplasia Type 2B (TC). In this autosomal dominant disorder facial dysmorphism and marfanoid habitus are accompanied by a generalized nonneoplastic proliferation of nerve fibers and ganglion cells (ganglioneuromatosis) that predominantly affects the alimentary tract and related viscera. Bilateral multimodular pheochromocytomas and medullary thyroid carcinomas may develop and a sensorimotor and autonomic neuropathy has been described.[5,6]

Multiple Symmetric Lipomatosis (TC). Distal axonopathy may be disabling in chronic symmetric lipomatosis (Madelung's disease); formerly held to be alcohol-related, distal axonopathy is now regarded as a primary facet of this disfiguring syndrome.[6]

Neurofibromatous Neuropathy (TC). Diffuse neurofibromatous infiltration of peripheral nerves may rarely accompany either type 1 or type 2 neurofibromatosis; such individuals display signs of a distal sensorimotor polyneuropathy.[47]

HEMATOLOGIC DISEASE

Castleman's Disease (TC). Castleman's disease, an abnormal lymphoid proliferation of unknown cause, is associated both with multifocal neuropathy and distal polyneuropathy.[9,19]

Chediak-Higashi Syndrome (TC). This rare autosomal recessive disorder is characterized by partial oculocutaneous albinism. The onset is in childhood with anemia, leukopenia, thrombocytopenia, and a tendency to lymphoreticular malignancy. Giant lysosomes are present in leukocytes. Mental retardation and an axonal sensorimotor neuropathy may be associated features.[36]

Hypereosinophilia Syndrome (TC). Both symmetric polyneuropathy and mononeuropathy occur in the idiopathic hypereosinophilia syndrome; endoneurial edema is present on nerve biopsy and may reflect alteration in the blood nerve barrier from eosinophil-derived products.[31]

Myelofibrosis (NC). A single report describes an elderly patient with paresthesias and gait difficulty with mild nerve conduction abnormalities.[30]

Neuroacanthocytosis (TC). Segmental weakness and atrophy in this disorder are attributed to motor neuron dysfunction or to a distal sensorimotor axonopathy; sural nerve biopsies display loss of large myelinated fibers.[14,49]

Polycythemia Vera (TC). A mild distal axonopathy, demonstrated by physiologic and morphologic studies, rarely accompanies long-standing polycythemia vera.[52]

SYSTEMIC METABOLIC DISEASE

Adrenomyeloneuropathy (TC) This hereditary disorder, a variant of adreno-leukodystrophy, is associated with a mild axonal neuropathy. The relationship of neuropathy to the underlying disorder of fatty acid metabolism is unclear.[45]

Gout Neuropathy (NC). An isolated report describes a generalized peripheral neuropathy that fluctuates with uric acid levels; more commonly, gouty tenosynovitis results in entrapment neuropathies; carpal tunnel syndrome is frequent.[8]

Hyperthyroidism (NC). Isolated case reports describe a tenuous relationship between hyperthyroidism and peripheral neuropathy.[13]

Hypoglycemia-Insulinoma (TC). Sensorimotor symptoms with distal limb distribution occur in hypoglycemia associated with pancreatic islet cell tumors and insulin shock therapy.[17]

Leigh's Disease (TC). Leigh's disease probably does not represent a single disorder. Otherwise known as subacute necrotizing encephalomyelopathy, it has a variety of clinical presentations, mostly with multifocal episodic brainstem disturbances developing during childhood. There may be an associated hypomyelinating neuropathy.[16]

Mitochondrial Encephalomyopathy (TC). Distal symmetric polyneuropathy accompanies myopathy and CNS manifestations in mitochondrial disease, which includes the syndromes of chronic progressive external ophthalmoplegia, Kearns-Sayre syndrome, mitochondrial encephalomyopathy, lactic acidosis and stroke-like episodes (MELAS), and myoclonic epilepsy with ragged red fibers (MERRF). The mechanism of the neuropathy is not known.[35]

Niemann-Pick Disease (TC). This disorder consists of a group of sphingomyelin lipidoses, all of autosomal recessive inheritance. A demyelinating neuropathy may occur in the acute neuronopathic form (type A) in association with diffuse CNS disease.[20]

Primary Hyperoxaluria (TC). A neuropathy occurs in which oxalate crystals in nerve are described; the main features of hyperoxaluria are renal calculi and renal failure.[32]

UNCLASSIFIED SYSTEMIC DISEASE

Cockayne's Syndrome (TC). This autosomal recessive disorder is associated with growth and mental retardation, progeria, pigmentary retinopathy, deafness, and ataxia. A hypomyelinating neuropathy is a frequent accompaniment.[50]

Cerebrotendinous Xanthomatosis (TC). In this rare autosomal recessive disorder, otherwise known as cholestanolosis, xanthomatous deposits containing cholestanol and cholesterol develop on tendons and are accompanied both by a progressive CNS disorder and an axonal sensory neuropathy. The metabolic defect is an impairment in bile acid synthesis; the disorder is partially treatable with chenodeoxycholic acid.[10]

Xeroderma Pigmentosum (TC). A slowly progressive sensory and motor neuronopathy and CNS disorder are common in type A xeroderma pigmentosum, an autosomal recessive cutaneous disorder.[18]

UNCLASSIFIED NEUROLOGIC CONDITIONS

Chronic Idiopathic Anhidrosis (TC). This rare disorder is characterized by adult onset of heat intolerance and inability to sweat; it does not progress to generalized autonomic dysfunction.[28]

Migrant Sensory Neuropathy of Wartenberg (NC). This benign syndrome of pain and sensory loss in the distribution of cutaneous nerves is usually provoked by banal movements, e.g., key turning, kneeling; stretch injury of tethered nerves is suggested to be the cause.[29] This may be a heterogenous disorder.

Myotonic Dystrophy (TC). Axonal polyneuropathy, frequently mild or asymptomatic, may accompany myopathy in myotonic dystrophy; its severity appears unrelated to the myopathy and the pathogenesis is unclear.[7]

Sensory Perineuritis (NC). An isolated report describes patients with distal, painful, partially remitting lower limb cutaneous neuropathy with Tinel's sign; pathologic features include inflammatory scarring of the perineurium with random compression of nerve fibers.[2]

REFERENCES

1. Aguayo, AJ, Romine, JS, and Bray, GM: Experimental necrosis and arrest of proliferation of Schwann cells by cytosine arabinoside. J Neurocytol 4:663, 1975.
2. Asbury, AK, Picard, EH, and Baringer, JR: Sensory perineuritis. Arch Neurol 26:302, 1972.
3. Behari, M, Choudhary, C, Roy, S, et al: Styrene-induced peripheral neuropathy. Eur Neurol 25:424, 1986.
4. Berkley, MC and Magee, KR: Neuropathy following exposure to a dimethylamine salt of 2,4D. Arch Intern Med 111:133, 1963.
5. Carney, JA: Multiple endocrine neoplasia, type 2B. In Dyck, PJ, Thomas, PK, Lambert, EH, and Bunge, R (eds): Peripheral Neuropathy, ed 2, Vol 2. WB Saunders, Philadelphia, 1984, p 1642.
6. Chalk, CH, Mills, KR, Jacobs, JM, et al: Familial multiple symmetric lipomatosis with peripheral neuropathy. Neurology 40:1246, 1990.
7. Cros, D, Harnden, P, Pouget, J, et al: Peripheral neuropathy in myotonic dystrophy: A nerve biopsy study. Ann Neurol 23:470, 1988.
8. Delaney, P: Gouty neuropathy. Arch Neurol 40:823, 1983.
9. Donaghy, M, Hall, P, Gawler, J, et al: Peripheral neuropathy in Castleman's disease. J Neurol Sci 89:253, 1989.
10. Donaghy, M, King, RH, McKeran, RO, et al: Cerebrotendinous xanthomatosis: Clinical, electrophysiological and nerve biopsy findings, and response to treatment with chenodeoxycholic acid. J Neurol 237:216, 1990.
11. Feigal, DW, Robbins, DL, and Leek, JC: Giant cell arteritis associated with mononeuritis multiplex and complement-activating 19S IgM rheumatoid factor. Am J Med 79:495, 1985.
12. Filley, CM, Graff-Radford, NR, Lacy, JR, et al: Neurologic manifestations of podophyllin toxicity. Neurology (NY) 32:308, 1982.

13. Fisher, M, Mateer, JE, Ullrich, I, and Gutrecht, J: Pyramidal tract deficits and polyneuropathy in hyperthyroidism. Am J Med 78:1041, 1985.

14. Hardie, RJ, et al: Neuroacanthocytosis. A clinical, haematological and pathological study of 19 cases. Brain 114:13, 1991.

15. Heller, CA and Friedman, PA: Pyridoxine deficiency and peripheral neuropathy associated with long-term phenelzine therapy. Am J Med 75:887, 1983.

16. Jacobs, JM, Harding, BN, Lake, DB, et al: Peripheral neuropathy in Leigh's disease. Brain 113:447, 1990.

17. Jakobsen, J and Sidenius, P: Hypoglycemic neuropathy. In Dyck, PJ, Thomas, PK, Lambert, EH, and Bunge, R (eds): Peripheral Neuropathy, ed 2, Vol 2. WB Saunders, Philadelphia, 1984, p 94.

18. Kanda, T, Oda, M, Yonzawa, M, et al: Peripheral neuropathy in xeroderma pigmentosum. Brain 113:1025, 1990.

19. Landis, DMD: Case records of the Massachusetts General Hospital. N Engl J Med 311:388, 1984.

20. Landrieu, P and Said, G: Peripheral neuropathy in type A Niemann-Pick disease. A morphological study. Acta Neuropathol (Berl) 63:66, 1984.

21. Lederman, RL and Wilbourn, AJ: Brachial plexopathy: Recurrent cancer or radiation? Neurology 34:1331, 1984.

22. Lederman, R, Breuer, A, Hanson, M, et al: Peripheral nervous system complications of coronary artery bypass graft surgery. Ann Neurol 12:297, 1982.

23. Lederman, RJ: Peripheral nerve disorders in instrumentalists. Ann Neurol 26:640, 1989.

24. Lee, P, Bruni, J, and Sukenik, S: Neurological manifestations in systemic sclerosis (scleroderma). J Rheumatol 11:480, 1984.

25. LeWitt, PA and Forno, LS: Peripheral neuropathy following amitriptyline overdose. Muscle Nerve 8:723, 1985.

26. Lok, ASF, Novick, DM, Karayiannis, P, et al: A randomized study of the effects of adenine arabinoside 5'-monophosphate (short or long courses) and lymphoblastoid infection on hepatitis B virus replication. Hepatology 5:1132, 1985.

27. Lotti, M, Moretto, A, Zoppellari, R, et al: Inhibition of lymphocyte neuropathy target esterase predicts the development of organophosphate-induced delayed polyneuropathy. Arch Toxicol 59:176, 1986.

28. Low, PA, Fealey, RD, Sheps, SG, et al: Chronic idiopathic anhydrosis. Ann Neurol 18:344, 1985.

29. Matthews, WB and Esiri, M: The migrant sensory neuritis of Wartenberg. J Neurol Neurosurg Psychiatry 46:1, 1983.

30. McLeod, JG and Walsh, JC: Peripheral neuropathy associated with lymphomas and other reticuloses. In Dyck, PJ, Thomas, PK, Lambert, EH, and Bunge, R (eds): Peripheral Neuropathy, ed 2, Vol 2. WB Saunders, Philadelphia, 1984, p 2201.

31. Monaco, S, Lucci, B, Laperchia, N, et al: Polyneuropathy in hypereosinophilic syndrome. Neurology 38:494, 1988.

32. Moorhead, PJ, Cooper, DJ, and Timperley, WR: Progressive peripheral neuropathy in a patient with primary hyperoxaluria. Br Med J 2:312, 1975.

33. Morgan, DJR and Gauler, J: Severe peripheral neuropathy complicating legionnaire's disease. Br Med J 283:1577, 1981.

34. Nemni, R, Fazio, R, Corbo, M, et al: Peripheral neuropathy associated with Crohn's disease. Neurology 37:1414, 1987.

34a. Pedersen, PB and Hogenhaven, H: Penicillamine-induced neuropathy in rheumatoid arthritis. Acta Neurol Scand 81:188, 1990.

35. Pezeshkpour, G, Krarup, C, Buchthal, F, et al: Peripheral neuropathy in mitochondrial disease. J Neurol Sci 77:285, 1987.

36. Pezeshkpour, G, Kurent, JS, Krarup, C, et al: Peripheral neuropathy in

Chediak-Higashi syndrome. J Neuropathol Exp Neurol 45:353, 1986.

37. Plough, JC and Ayerle, RS: The Guillain-Barré syndrome associated with acute hepatitis. N Engl J Med 249:61, 1953.

38. Pool, KD, Feit, H, and Kirkpatrick, J: Penicillamine-induced neuropathy in rheumatoid arthritis. Ann Intern Med 95:457, 1981.

39. Roman, GC, Osame, M, and Igata, H: HTLV-1—associated myelopathy (HAM) and tropical spastic paraparesis (TSP). In Roman, GC, Vernant, J-C, and Osame, M (eds.): HTLV-1 and the Nervous System. Alan R Liss, New York, 1989, p 93.

40. Russell, JA and Powles, RL: Neuropathy due to cytosine arabinoside. Br Med J 14 December 1974.

41. Silvis, SE and Paragas, PD: Paresthesias, weakness, seizures, and hypophosphatemia in patients receiving hyperalimentation. Gastroenterology 62:513, 1972.

42. Singer, R, Moses, R, Valcuikas, J, et al: Nerve conduction velocity studies of workers employed in the manufacture of phenoxy herbicides. Environ Res 29:297, 1982.

43. Smith, BE and Dyck, PJ: Peripheral neuropathy in the eosinophilia-myalgia syndrome associated with L-tryptophan ingestion. Neurology 40:1035, 1990.

44. Taylor, LP and Posner, JB: Phenobarbital rheumatism in patients with brain tumor. Ann Neurol 25:92, 1989.

44a. Thomas, PK: Other inherited neuropathies. In Dyck, PJ, Thomas, PK, Lambert, EH, et al (eds): Peripheral Neuropathy, ed 2, Vol 2. WB Saunders, Philadelphia, 1984, p 1457.

45. Thomas, PK and Holdorff, B: Neuropathy due to physical agents. In Dyck, PJ, Thomas, PK, Lambert, EH, and Bunge, R (eds): Peripheral Neuropathy, ed 2, Vol 2. WB Saunders, Philadelphia, 1984, p 1479.

46 Thomas, PK and Walker, JG: Xanthomatous neuropathy in primary biliary cirrhosis. Brain 88:1079, 1965.

47. Thomas, PK, King, RHM, Chiang, TR, et al: Neurofibromatous neuropathy. Muscle Nerve 13:93, 1990.

48. Vanhooren, G, Dehaene, I, Van Zandycke, M, et al: Polyneuropathy in lithium intoxication. Muscle Nerve 13:204, 1990.

49. Vita, G, Serra, S, Dattola, R, et al: Peripheral neuropathy in amyotrophic choreoacanthocytosis. Ann Neurol 28:538, 1989.

50. Vos, A, Gabreels-Festen, A, Joosten, A, et al: The neuropathy of Cockayne syndrome. Acta Neuropathol (Berl) 61:153, 1983.

51. Wu, R-M, Chang, Y-C, and Chiu, H-C: Acute triphenyltin intoxication: A case report. J Neurol Neurosurg Psychiatry 53:356, 1990.

52. Yiannikas, C, McLeod, JG, and Walsh, JC: Peripheral neuropathy associated with polycythemia vera. Neurology 33:139, 1983.

Part VIII
REHABILITATION

Chapter 21

REHABILITATION IN CHRONIC POLYNEUROPATHIES

RANGE-OF-MOTION (ROM) EXERCISES
NEUROMUSCULAR RE-EDUCATION
SENSORY RE-EDUCATION
ORTHOSES
GAIT TRAINING
OCCUPATIONAL THERAPY

This chapter addresses the major rehabilitative facets in the care of chronic polyneuropathies.[1] Discussion of the care of acute polyneuropathies is found in Chapter 5 and that of specific nerve injuries in Chapter 16.

Individuals with peripheral neuropathies requiring neurologic rehabilitation generally fall into one of two groups: acute and chronic. Demyelinating radiculopathies account for most of the acute cases, although occasional toxic axonopathies and toxic or infectious neuronopathies may have sudden onset and rapid progression. Acute neuropathies are characterized by diffuse weakness or sensory loss; the patients are bedbound and may require treatment for ventilatory and autonomic failure. They are additionally subject to complications from immobility. The care of acute polyneuropathies is found in the AIDP section of Chapter 5. The chronic group are patients with mild demyelinating neuropathies, axonopathies that are expected to improve (e.g., toxic neuropathies), or steadily progressing neuropathies with a slow deterioration of function and poor prognosis (e.g., metabolic or hereditary neuropathies). Such patients frequently remain ambulatory and functional. Rehabilitative efforts in this group center around maximizing residual function by customizing the home or work environment, minimizing traumatic insults to insensitive limbs, maintaining adequate joint mobility, treating the consequences of autonomic dysfunction, and relieving pain.

Rehabilitative goals must be realistic, sufficiently detailed, individually tailored to the degree of residual capacity, and mindful of the stage of emotional adjustment. Rehabilitative efforts in patients with recovering peripheral neuropathies clearly affect the degree of functional recovery. Even in chronically progressive neuropathies, proper bracing and environmental adaptations can make the difference between a productive life and a dependent existence.

Certain commonsense guidelines are applicable to all generalized neuropathies. Excessive weight gain should be avoided as it places stress on an already compromised motor system. The home environment should be customized, if possible, to minimize the chance of tripping. Plush carpeting, throw rugs, doorway molding extending across the entrance way, and small objects lying around the floor should all be avoided. Chairs and toilet seats should be of sufficient height and have handrails to allow patients with proximal muscle weakness to rise easily. Feet should routinely be checked for small cuts or bruises that may become infected; minor abrasions often go unnoticed because of sensory loss. A nightlight should be employed to aid the sensory-impaired patient in nocturnal ambulation.

RANGE-OF-MOTION (ROM) EXERCISES

Full ROM must be maintained in all affected joints. This is simple if effective passive ROM exercises have been employed in earlier stages. During a period of receding neurologic deficit, the patient with satisfactory motor coordination is encouraged to carry the limb through the ROM with assistance by the therapist. Fatigue should be avoided, since this decreases the accuracy of motor coordination.

NEUROMUSCULAR RE-EDUCATION

Joint movement reflects the combined actions of a number of related muscles. Usually, however, one muscle (the prime mover) has the predominant responsibility for effecting the desired movement. Effective joint function requires the central nervous system to integrate the orderly inhibition of antagonist muscles and proper sequencing of synergistic and stabilizing muscles.[6] When a muscle normally employed in moving a joint is inactive, either from paralysis or spasm, other muscles, not normally active, come into play. Without motor re-education, attempts at voluntary activity may result in unwanted spread of motor activity. In patients with severe demyelinating or axonal neuropathies resulting in significant residual weakness or incoordination, neuromuscular re-education is often valuable in re-establishing coordinated and efficient joint movement by teaching selective contraction of individual muscles or groups of muscles again. Re-education proceeds stepwise from initial passive ROM exercises to active resistive strengthening exercises.

Many with generalized neuropathies do not require specialized motor re-education. Since most chronic progressive neuropathies do not produce periods of total muscle inactivity, there is little need to re-establish acceptable motor patterns. It is unclear if such patients benefit from intensive strengthening exercises directed at intact residual motor units. It is usually sufficient to keep the patient as active as possible

through normal daily activities, maintain adequate joint mobility, and augment impaired function by orthotics and functional aids.

The proper approach for motor re-education is unclear. Some believe that the most efficient way to regain previous motor patterns is to practice the desired movement exactly as it will normally be performed. In this manner, central sequencing for prime mover and synergist activation and antagonist inhibition become well established by the time active assistance is possible. There is some suggestion, however, that normal movements occur over diagonal or spiral patterns. That is to say, muscle contraction is facilitated when the movement extends over diagonal pathways such as moving the hand from one's side to one's mouth. Various diagonal patterns have been developed that encourage activation of previously inactive motor neurons. Muscles may also work in bulk fashion; the contraction of one muscle is facilitated by the activity of other limb muscles. Maximal contraction of the wrist extensors is, for example, facilitated by simultaneous extension of the arm and fingers.

Depression often follows loss of motor function. Patients are remarkably reassured when residual function or, even better, a return of movement is demonstrated. Muscle re-education should begin as soon as a body part can be moved through a small ROM without pain. In general, the earlier motor re-education is begun, the better the results.

Resistive exercises designed to develop strength may be initiated as soon as coordinated movements are performed. Flaccid muscle loses about 3% to 7% of its contractile force daily. Resistive exercise, if done properly, results in muscle fiber hypertrophy and an increase in the cross-sectional diameter of the muscle capillary bed. Increased contractile strength develops from the constant demand for greater tensile force needed to overcome progressive resistance. Strengthening occurs whenever 35% or more of the maximal force is applied in a resistive manner. Strength may be increased by performing frequent repetitions with heavy resistance, whereas endurance is developed by multiple repetitions using a light weight. The intensity of exercise should be kept low to avoid fatiguing already damaged motor units. Exercising denervated motor units to fatigue may be harmful. Muscle fatigue and aching which lasts into the next day indicates that the exercises are too strenuous. To avoid motor unit fatigue, endurance training using high-repetition–low-weight resistive exercises should not commence until the patient is able to tolerate moderate exercise without reverting to uncoordinated motor patterns.

At this point the program should be reduced to about two to three times per week using lower weights to maintain strength.

Once sufficient strength and endurance have been established, fluidity and coordination become the focus of rehabilitation. This requires accurate repetition of the components movements involved in the desired movement. Repetition develops improved central processing of motor patterns, resulting in increased speed and efficiency of movement. Coordination exercises consist initially of repeating the desired movement with the patient paying close attention to the correct sequencing of muscles.

SENSORY RE-EDUCATION

Rehabilitative efforts have predominantly centered around motor re-education. The restoration of adequate sensory appreciation has been considered less important, for it was believed that denervated sensory receptors could not be functionally reinnervated. It is now clear that the regenerative capacity of the sensory system is extraordinary; sensory receptors cannot only be effectively reinnervated after long periods of time but some may, if previously degenerated, regrow and have their modality specificity determined by the stress pat-

terns to which they are currently exposed.

Receptors may respond to more than one type of stimulus by having a low threshold to one stimulus and high thresholds to others. Sensory appreciation is dependent not only on the specific nature of the sensory receptor but also on the spatial and temporal patterns of discharges in various groups of nerve fibers. Re-education is possible because the detection of sensation is a dynamic process; sensory appreciation depends on cortical interpretations of specific nerve impulse patterns and receptor specificity varies with demand. Reinnervated sensory receptors may transmit sensory information in a pattern not recognized by sensory cortices accustomed to previous sensory codes. It is likely that functional sensory appreciation is not limited by the degree of nerve regeneration but by the failure of the central nervous system to recognize new patterns of sensory coding. Retraining involves teaching the central nervous system to recognize new sensory patterns.

The goal of sensory re-education is restoration of adequate hand function. Unlike the many patients with focal nerve injuries that require sensory re-education, only a limited number of patients with generalized neuropathies will be candidates. Re-education can begin once the patient is able to appreciate moving touch or vibration at 30 Hz.[7] Hypersensitivity must be decreased before re-education can proceed. The patient should be able to tolerate the stroking of the limb, both proximal and distally, with cotton wool initially and then with felt and other coarser materials.[7]

In hand injuries, the blindfolded patient initially attempts to determine the relative weights of different-sized wood blocks. In future sessions, objects of different texture and shapes are introduced. In each case, the patient is eventually allowed to view the object in order to correlate new sensations with visual ones. Visual assistance is eventually eliminated. The recovery of adequate sensory appreciation is often remarkable. Pinprick has been reported to be felt first, followed by 30-Hz vibration, moving touch, constant touch, and eventually 256-Hz vibration.[2] Training sessions should be kept short, since sensory re-education is both tiring and frustrating. About two to four 10-minute sessions per day are adequate. The time required to achieve each correct answer is charted. Training cannot be hurried, but, by careful attention to re-education, good sensory appreciation may be restored in relatively short periods of time.

ORTHOSES

Splinting may be required during the recovery stage to prevent contractures and muscle overstretch or to support a limb in a position of maximum function. Patients with long-standing intrinsic hand weakness may require a combination of splints designed both to prevent hyperextension of the metacarpophalangeal (MP) joint (ulnar nerve dysfunction) and to keep the thumb in palmar abduction, thereby preventing thumb web contracture (median nerve dysfunction).[4] Wrist drop is corrected by a cock-up splint supporting the wrist in mild extension, thereby allowing a more functional position of the hand. The addition of an outrigger to create a dynamic splint will provide MP joint extension and yet allow finger flexion.

Lower-extremity orthoses may be needed to compensate for foot-drop.[5] Isolated dorsiflexion weakness frequently requires only a lightweight plastic splint that fits into the shoe and provides just enough foot support to allow the toes to clear the floor while walking (Fig. 21–1). If mediolateral instability is present, a double-upright brace (Klenzac) may be needed with or without a T-strap (Fig. 21–2). The T-strap pulls one side of the shoe toward the opposite upright, providing stability. Klenzac braces are hinged to permit dorsi and plantar flexion. Spring and

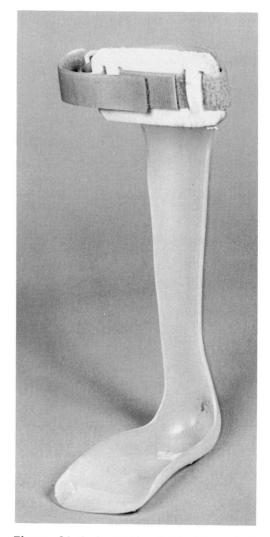

Figure 21–1. A simple, lightweight, plastic splint for prevention of foot drop.

pin stops may be inserted as needed. Klenzac braces are heavy and expensive; they are generally used in cases of severe and prolonged disability.

The degree of dorsiflexion support needed is determined by the required amount of toe clearance and the resultant effect on the knee joint. Support at the ankle always affects the knee. Prevention of plantar flexion during the stance phase of walking causes the patient to rock over the posterior portion of the heel to achieve the flat-footed position. This results in bending at the

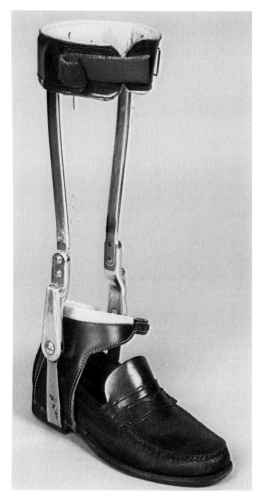

Figure 21–2. A double-action, hinged Klenzac brace with T-strap for ankle stability.

to the center of the knee, thereby reducing the amount of knee bending. The addition of a dorsiflexion stop often stabilizes patients in whom accompanying plantar flexion weakness predisposes to forward instability. Patients with impaired pushoff may benefit from a rocker bar, which, when attached to the sole of the shoe, facilitates the shift of weight forward during initiation of the swing phase. The proper brace is one that provides the least amount of plantar flexion restriction needed for the toes to clear the floor during walking, making walking more efficient and reducing energy requirements.

Isolated quadriceps weakness is usually not a problem for patients recovering from polyneuropathies, since the weakness is predominantly distal. Occasionally, quadriceps weakness may be so severe as to require knee bracing to avoid buckling during the stand phase of gait. Strong quadriceps muscles, in concert with weak hip extensors, results in mild hip flexor contractures. Knee-stabilizing braces, fastened to the shoe, have a knee joint that can be unlocked when sitting. Posteriorly, the brace extends up the thigh to near the ischial tuberosity. Patients with weak quadriceps but strong gluteal muscles usually walk well, without bracing, except for stairs or inclines.

knee, which must be overcome by the knee extensors before the swing phase can be initiated. The greatest degree of knee bending, and therefore the greatest stress on the knee extensors, results when rigid pin stops are used in a Klenzac brace. Use of a spring stop instead of a rigid pin stop allows gradual lengthening of the dorsiflexors, thereby taking the strain off the knee extensors. Positioning the foot in 5° of plantar flexion significantly reduces knee bending, as does inserting a cushion wedge or cutoff heel. This results in the foot contacting the ground in front of the heel bringing the line of force closer

GAIT TRAINING

A cardinal goal of re-educating lower-extremity muscles is satisfactory gait. In recovering neuropathies, the important muscles in walking are (in order of importance): trunk muscles, upper-extremity muscles (needed for crutch walking with braces on legs), hip extensors (to stabilize the thighs on the pelvis), and ankle muscles for stabilization.[3]

The quadriceps muscles are relatively unimportant when walking on a level surface but are needed for inclines.

Normal gait sequencing involves bending of hip flexors and knees, bringing the leg forward, extending the knee, dorsiflexing the foot, and striking the surface with the heel, toes off the floor and knee extended. The weight of the body is then brought over the extended leg. After the center of gravity passes the center of the foot, the patient rises on the toes, flexes the knee and hip, and starts the process over again. Normal foot clearance of the floor is only about 6 cm, making even minimal dorsiflexion weakness a hazard for tripping.[3] Effective gait training aims at providing just enough toe elevation to clear the floor. Too high a step results in a marked increase in energy expenditure required for balance or support. Parallel bars are initially used to teach shifting and pivoting. Patients regain balance by initially practicing with one hand support and then without any. Once balance is obtained, patients simulate walking by practicing the component steps in place. When proper sequencing is mastered and balance is adequate, the patient attempts single steps outside the parallel bars. Eventually, uninterrupted walking is attempted. Gait training is tedious; there is no shortcut. Frequent repetition, similar to other techniques for motor re-education, is needed to develop effective central programming.

OCCUPATIONAL THERAPY

After sufficient ROM and strength have developed in the arms, occupational therapy commences to develop coordination and endurance through repetition of basic stereotyped motions that comprise the movements needed to perform writing, feeding, and grooming. Once individual component motions are mastered they are combined to form the desired activity. At this stage of rehabilitation, patients are evaluated concerning their ability to perform activities of daily living.

Patients with impaired manual dexterity, ROM, or power may need special techniques and equipment to develop partial or complete functional independence. The occupational therapist can select, modify, and apply adaptive equipment as needed and train the patients in their use. Frequent shifts from one activity to another before proficiency is obtained is frustrating.

Hand crafts are often used to develop strength, endurance, and coordination and are more interesting than stereotypical repetitive movements. Although endurance is often increased by repetition, there is less improvement in ROM, as movements tend to be confined to the available range.

REFERENCES

1. Berger, AR and Schaumburg, HH: Rehabilitation of peripheral neuropathies. J Neurol Rehab 2:25–26, 1988.
2. Dellon, AL, Curtis, RM, and Edgerton, M: Re-education of sensation in the hand after nerve injury and repair. Plast Reconstr Surg 53:297, 1974.
3. Knapp, ME: Exercise for lower motor neuron lesions. In Basmajian, JV (ed): Therapeutic Exercise. Williams & Wilkins, Baltimore, 1981.
4. Long, D and Schutt, A: Upper limb orthotics. In Redford, JB (ed): Orthotics Etcetera, ed 3, Williams & Wilkins, Baltimore, 1986, p 198.
5. Lehmann, JF: Lower limb orthotics. In Redford, JB (ed): Orthotic Etcetera, ed 3. Williams & Wilkins, Baltimore, 1986, p 278.
6. Loyd, D: Facilitation and inhibition of spinal motorneurons. J Neurophysiol 9:421, 1946.
7. Weeks, PM and Wray, CR: Management of Acute Hand Injuries. CV Mosby, St Louis, 1978.

Appendix A

CUTANEOUS FIELDS OF PERIPHERAL NERVES (Figures A–1, A–2, and A–3*)

Please see next page.

*(From Haymaker, W: Bing's Local Diagnosis in Neurological Disease, ed 15. CV Mosby, St Louis, 1969, with permission.)

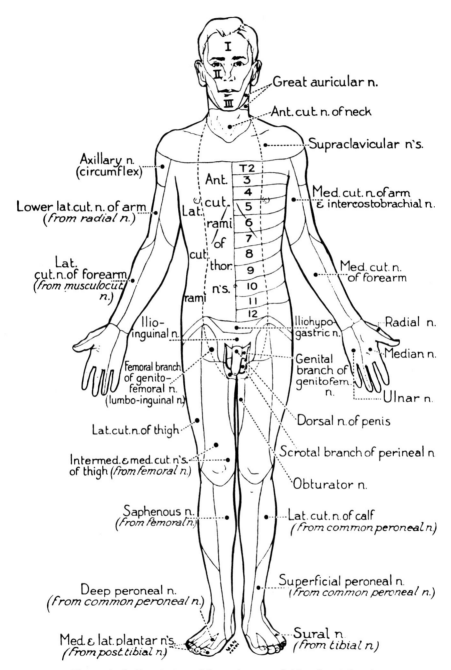

Figure A–1. Front view of the cutaneous fields of peripheral nerves.

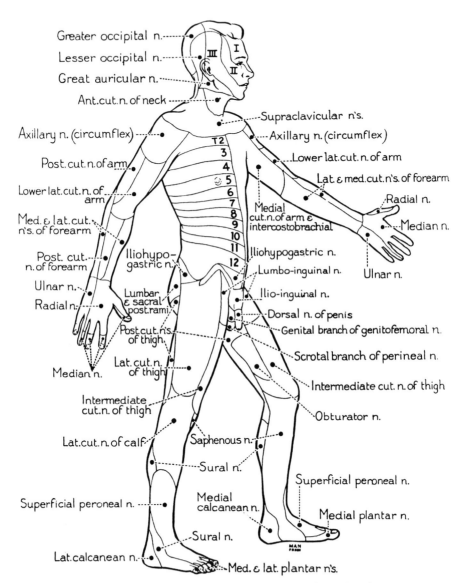

Figure A–2. Side view of the cutaneous fields of peripheral nerves.

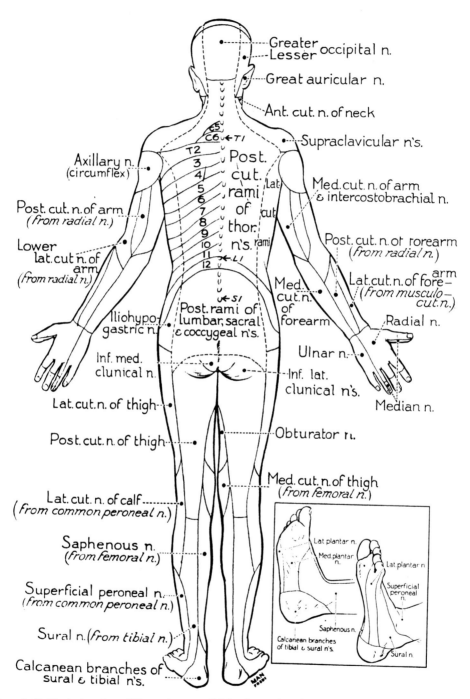

Figure A–3. Posterior view of the cutaneous fields of peripheral nerves.

Appendix B

SEGMENTAL INNERVATION OF SOMATIC MUSCLES (Figures B–1, B–2, and B–3*)

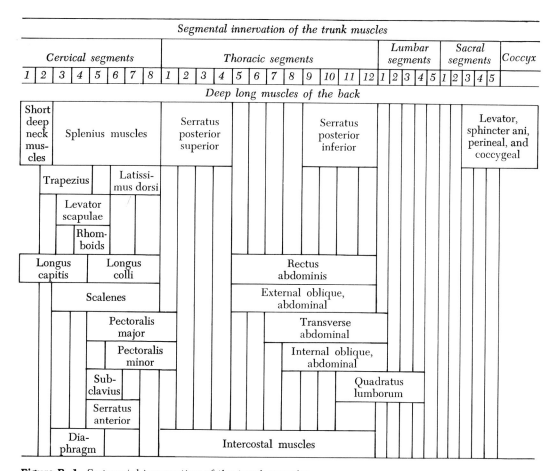

Figure B–1. Segmental innervation of the trunk muscles.

*(From Haymaker, W: Bing's Local Diagnosis in Neurological Disease, ed 15. CV Mosby, St Louis, 1969, with permission.)

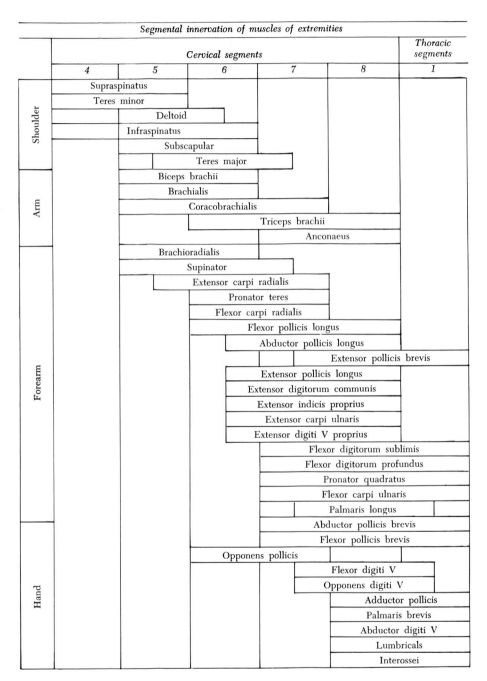

Figure B–2. Segmental innervation of upper-limb muscles.

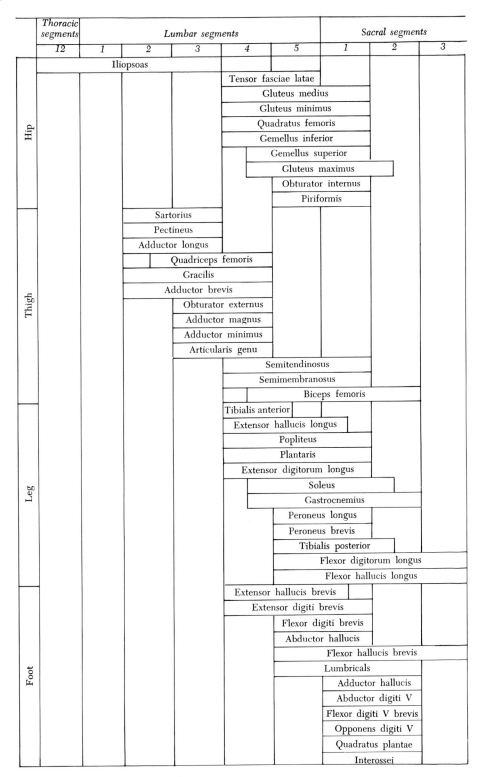

Figure B–3. Segmental innervation of the lower limbs.

Index

An *italic* number indicates a figure. A "t" indicates a table.